ARMY, MARINE CORPS, NA

MW01320967

POTENTIAL MILITARY CHEMICAL/BIOLOGICAL AGENTS AND COMPOUNDS

FM 3-11.9
MCRP 3-37.1B
NTRP 3-11.32
AFTTP(I) 3-2.55

JANUARY 2005

MULTISERVICE TACTICS, TECHNIQUES, AND PROCEDURES

Published by

Eximdyne
2208 Autumn Trace Parkway
Wentzeville, Missouri 63385
United States

http://www.eximdyne.com

Printed in the United States

ISBN (Paperback): 0–9677264—0—9

Editor's note: This is a print version of Field Manual FM 3-11.9 *Potential Military Chemical/Biological Agents and Compounds* (2005). The original FM 3-11.9 is an eBook. This version has been enhanced to improve print quality. Appendi A, B and C have been replaced by the editor to be more comprehensive and of higher print quality.

FOREWORD

This publication has been prepared under our direction for use by our respective commands and other commands as appropriate.

STANLEY H. LILLIE
Brigadier General, USA
Commandant
US Army Chemical School

EDWARD HANLON, JR.
Lieutenant General, USMC
Deputy Commandant
for Combat Development

JOHN M. KELLY
Rear Admiral, USN
Commander
Navy Warfare Development Command

BENTLEY B. RAYBURN
Major General, USAF
Commander
Headquarters Air Force Doctrine Center

PREFACE

1. Scope

This document provides commanders and staffs with general information and technical data concerning chemical/biological (CB) agents and other compounds of military interest such as toxic industrial chemicals (TIC). It explains the use; classification; and physical, chemical, and physiological properties of these agents and compounds. Users of this manual are nuclear, biological, and chemical (NBC)/chemical, biological, and radiological (CBR) staff officers, NBC noncommissioned officers (NCOs), staff weather officers (SWOs), NBC medical defense officers, medical readiness officers, medical intelligence officers, field medical treatment officers, and others involved in planning battlefield operations in an NBC environment.

2. Purpose

This publication provides a technical reference for CB agents and related compounds. The technical information furnished provides data that can be used to support operational assessments based on intelligence preparation of the battlespace (IPB).

3. Application

The audience for this publication is NBC/CBR staff personnel and commanders tasked with planning, preparing for, and conducting military operations.

4. Implementation Plan

Participating service command offices of primary responsibility (OPRs) will review this publication, validate the information, and reference and incorporate it in service and command manuals, regulations, and curricula as follows:

Army. The United States Army (USA) will incorporate this publication in USA training and doctrinal publications as directed by the Commander, United States Army Training and Doctrine Command (TRADOC). Distribution is in accordance with Department of the Army (DA) Form 12-99-R (Initial Distribution Requirements for Publications).

Marine Corps. The United States Marine Corps (USMC) will incorporate the procedures in this publication in USMC training and doctrinal publications as directed by the Commanding General (CG), United States Marine Corps Combat

Development Command (MCCDC). Distribution is in accordance with Marine Corps Publication Distribution System (MCPDS).

Navy. The United States Navy (USN) will incorporate these procedures in USN training and doctrinal publications as directed by the Commander, Navy Warfare Development Command (NWDC). Distribution is according to the military standard requisitioning and issue procedures (MILSTRIP).

Air Force. The United States Air Force (USAF) will validate and incorporate appropriate procedures according to applicable governing directives.

5. User Information

a. The United States Army Chemical School (USACMLS) developed this publication with the joint participation of the approving service commands.

b. We encourage recommended changes for improving this publication. Please reference the specific page and paragraph, and provide a rationale for each recommendation. Send comments and recommendations directly to—

Army

Commandant
US Army Chemical School
ATTN: ATSN-CM-DD
401 MANSCEN Loop, Suite 1029
Fort Leonard Wood, MO 65473-8926
COMM (573) 596-0131, extension 3-7364

Marine Corps

Commanding General
US Marine Corps Combat Development Command
ATTN: C42 (Director)
3300 Russell Road
Quantico, VA 22134-5001
DSN 278-6234; COMM (703) 784-6234

Navy

Commander
Navy Warfare Development Command
ATTN: N5
686 Cushing Road
Newport, RI 02841-1207
DSN 948-4201; COMM (401) 841-4201

Air Force

HQ Air Force Doctrine Center
ATTN: DJ
155 North Twining Street
Maxwell AFB, AL 36112-6112
DSN 493-7224; COMM (334) 953-7224

*FM 3-11.9
MCRP 3-37.1B
NTRP 3-11.32
AFTTP(I) 3-2.55

FM 3-11.9	US Army Training and Doctrine Command Fort Monroe, Virginia
MCRP 3-37.1B	Marine Corps Combat Development Command Quantico, Virginia
NTRP 3-11.32	Naval Warfare Development Command Newport, Rhode Island
AFTTP(I) 3-2.55	Headquarters Air Force Doctrine Center Maxwell Air Force Base, Alabama

10 January 2005

POTENTIAL MILITARY CHEMICAL/BIOLOGICAL AGENTS AND COMPOUNDS

TABLE OF CONTENTS

Page

DISTRIBUTION RESTRICTION: Approved for public release; distribution is unlimited.

*This publication supersedes FM 3-9, 12 December 1990.

EXECUTIVE SUMMARY

Potential Military Chemical/Biological Agents and Compounds

Chapter I
Introduction

Chapter I briefly addresses the threat and the significant military aspects of chemical and biological agents and TIC.

Chapter II
Chemical Warfare Agents and Their Properties

Chapter II provides information on chemical agents that might be encountered in the field. It discusses chemical agent physical characteristics and toxicity data of choking, nerve, blood, blister, and incapacitating agents.

Chapter III
Military Chemical Compounds and Their Properties

Chapter III discusses military chemical compounds such as the riot control agents (RCAs). It provides physical and chemical characteristics and toxicity data for military chemical compounds.

Chapter IV
Biological Agents and Their Properties

Chapter IV addresses general characteristics of biological agents (including toxins) and provides a summary of selected antipersonnel agents that may be employed in weapons systems.

Chapter V
Toxic Industrial Chemicals and Their Properties

Chapter V addresses TIC and a summary of available TIC information sources.

PROGRAM PARTICIPANTS

The following commands and agencies participated in the development of this publication:

Joint

Joint Requirements Office, 401 MANSCEN Loop, Suite 1309, Fort Leonard Wood, MO 65473

Army

United States Army Chemical School, 401 MANSCEN Loop, Suite 1029, Fort Leonard Wood, MO 65473

United States Army Edgewood Chemical and Biological Center, Aberdeen Proving Ground, MD 21040

Marine Corps

United States Marine Corps Combat Development Command, 3300 Russell Road, Suite 318A, Quantico, VA 22134-5021

Navy

United States Navy Warfare Development Command, 686 Cushing Road, Sims Hall, Newport, RI 02841

United States Navy Surface Warfare Development Group, 2200 Amphibious Drive, Norfolk, VA 23521

Air Force

HQ Air Force Doctrine Center, ATTN: DJ, 155 North Twining Street, Maxwell AFB, AL 36112-6112

HQ Air Force Civil Engineer Support Agency, 139 Barnes Drive, Suite 1, Tyndall AFB, FL 32403-5319

Chapter I

INTRODUCTION

1. Background

The threat or use of CB weapons is a possible condition of future warfare and could occur in the early stages of war to disrupt United States (US) operations and logistics. In many of the regions where the US is likely to deploy forces, potential adversaries may use CB weapons. Potential adversaries may seek to counter US conventional military superiority using less expensive and more attainable, asymmetrical means. To meet this challenge, US forces must be properly trained and equipped to operate effectively and decisively in the face of NBC attacks.[1] Additionally, US forces could be confronted in an environment where TIC present a hazard to US forces.[2]

a. Use of CB Weapons.[3] Adversaries may employ CB agents and other toxic materials to achieve specific effects. In addition to the physical effects, there exist psychological effects, both in the immediate target area and in other vulnerable areas that may be potential targets.

(1) Chemical agents have effects that can be immediate or delayed, can be persistent or nonpersistent, and can have significant physiological effects. While relatively large quantities of an agent are required to ensure an area remains contaminated over time, small-scale selective use that exploits surprise can cause significant disruption and may have lethal effects.

(2) Biological agents can produce lethal or incapacitating effects over an extensive area and can reproduce. The delayed onset of symptoms and detection, identification, and verification difficulties for biological agents can also confer important advantages to adversaries who decide to use biological agents.

(3) The means available to adversaries for delivery of CB weapons range from specially designed, sophisticated weapon systems developed by nations to relatively inefficient improvised devices employed by terrorists and other disaffected individuals and groups.

b. US Policy.[3] This paragraph contains brief descriptions of treaty, legal, and policy strictures on chemical and biological warfare (CBW).

(1) The *Protocol for the Prohibition of the Use in War of Asphyxiating, Poisonous or Other Gases, and of Bacteriological Methods of Warfare,*" also known as the *Geneva Protocol of 1925*, prohibits chemical and bacteriological methods of warfare. Most parties interpret the protocol as a prohibition only of the first use of these agents in war. It did not ban the development, production, or stockpiling of these weapons. In 1974, the US Senate gave advice and consent to ratification of this protocol, subject to the reservation that the US would not be bound by the provisions with respect to an enemy state or its allies who fail to respect the prohibitions of the protocol. On 22 January 1975, the US ratified the protocol subject to this reservation. The protocol entered into force for the US on 10 April 1975. The relevance of the *Geneva Protocol* is largely superseded by the more

restrictive *Convention on the Prohibition of the Development, Production, Stockpiling, and Use of Chemical Weapons and on their Destruction* (also known as the *Chemical Weapons Convention [CWC]* and by the *Convention on the Prohibition of Bacteriological and Toxic Weapons* (also known as the *Biological Weapons Convention [BWC])* summarized below.

(2) The *Presidential Statement on Chemical and Biological Weapons*, 25 November 1969, renounced the US use of lethal biological agents and weapons and confined biological research to defensive measures such as immunization and safety. Under the terms of the BWC, parties undertake not to develop, produce, stockpile, or acquire biological agents or toxins "of types and in quantities that have no justification for prophylactic, protective and other peaceful purposes," as well as weapons and means of delivery. The BWC does not establish a specific verification regime. The US ratified the BWC on 29 March 1975.

(3) Executive Order No. 11850, 8 April 1975, *Renunciation of Certain Uses in War of Chemical Herbicides and Riot Control Agents*, renounced first use of herbicides in war (except for specified defensive uses) and first use of RCAs in war except for defensive military modes to save lives.

(4) The CWC, which entered into force on 29 April 1997, bans the development, production, acquisition, stockpiling, transfer, or use of chemical weapons. It provides for the destruction of all chemical weapons stocks within 10 years after entry into force. It contains a vigorous challenge regime to ensure compliance. The US ratified the CWC on 25 April 1997.

2. Threat

a. Changes.[4] Countries with chemical weapons programs are adding agents and more sophisticated delivery systems. Similarly, the sophistication of CBW capabilities is increasing. Proliferation of weapons technology, precision navigation technology, and CBW technology in developing nations presents the US with a complicated national security challenge. Intelligence efforts include collection and analysis of nations' dual-use, CB industrial capabilities, and development of the indications and warning of adversarial use of dual-use capabilities.

b. Challenges. The US faces a number of regional proliferation challenges. Many of these are detailed in the January 2001 report published by the Office of the Secretary of Defense (OSD), *Proliferation: Threat and Response*. At least 25 countries now possess—or are in the process of acquiring and developing—capabilities to inflict mass casualties and destruction: NBC weapons or the means to deliver them.[5]

c. Proliferation.[4] Proliferation of CBW technology also raises several important issues. Various nations could export a wide array of chemical products, including Australian group-controlled items to numerous countries of proliferation concern. The controlled items include specific chemical agent precursors, pathogens with biological warfare (BW) applications, and dual-use equipment that can be used in both CBW programs.

d. Increases in Proliferation.[4] In the next several years, the threat from the proliferation of CBW may increase. This could result from the development of CB agents that are more difficult to detect and from the adoption of more capable delivery systems. States with existing programs may master the production processes for complete weapons development and will be less dependent on outside suppliers.

(1) Any nation with the political will and a minimal industrial base could produce CBW agents suitable for use in warfare. Efficient weaponization of these agents, however, does require design and production skills usually found in countries that possess a munitions development infrastructure or access to such skills from cooperative sources.

(2) On the other hand, almost any nation or group could fabricate crude agent dispersal devices. Such weapons might be capable of inflicting only limited numbers of casualties; nevertheless, they could have significant operational repercussions due to the psychological impact created by fears of CBW agent exposure.[4]

(3) Genetic engineering gives BW developers the tools to pursue agents that could defeat the protective and treatment protocols of the prospective adversary. Genetically engineered microorganisms also raise the technological hurdle that must be overcome to provide for effective detection, identification, and early warning of BW attacks.

(4) Numerous characteristics need to be controlled for a highly effective BW agent. Historically, the accentuation of one characteristic often resulted in the attenuation of one or more other characteristics, possibly even rendering the modified agent ineffective as a weapon. Advances in biotechnology, genetic engineering, and related scientific fields provide ever-increasing potential to control more of these factors, possibly leading to an enhanced ability to use BW agents as battlefield weapons.

e. Novel BW Agents.[1] Advances in biotechnology and genetic engineering may facilitate the development of potentially new and more deadly BW agents. The ability to modify microbial agents at a molecular level has existed since the 1960s, when new genetic engineering techniques were introduced, but the enterprise tended to be slow and unpredictable. With today's techniques, infectious organisms can be modified to bring about disease in different ways. The current level of sophistication for many biological agents is low, but there is enormous potential—based on advances in modern molecular biology, fermentation, and drug delivery technology—for making more sophisticated weapons. The BW agents may emerge in two likely categories: man-made manipulations of classic BW agents and newly discovered or emerging infectious diseases. An example of a recent new pathogen (though not necessarily ideal BW agents) includes streptococcus pneumonia S23F, a naturally occurring strain of pneumonia resistant to at least six of the more commonly used antibiotics.

(1) The potential types of novel biological agents that could be produced through genetic engineering methodologies are listed below. Each of these techniques seeks to capitalize on the extreme lethality, virulence, or infectivity of BW agents and exploit this potential by developing methods to deliver more efficiently and to control these agents on the battlefield.

(a) Benign microorganisms genetically altered to produce a toxin, venom, or bioregulator.

(b) Microorganisms resistant to antibiotics, standard vaccines, and therapeutics.

(c) Microorganisms with enhanced aerosol and environmental stability.

(d) Immunologically altered microorganisms able to defeat standard identification, detection, and diagnostic methods.

(e) Combinations of the above four types with improved delivery systems.

(2) The future likelihood of infectious agents being created for BW purposes will be influenced by technological trends such as—

(a) Genetically engineered vectors in the form of modified infectious organisms may become increasingly available as medical tools and techniques become more widely available.

(b) Strides will be made in the understanding of infectious disease mechanisms and in microbial genetics that are responsible for disease processes.

(c) An increased understanding of the human immune system function and disease mechanisms will shed light on the circumstances that cause individual susceptibility to infectious disease.

(d) Vaccines and antidotes will be improved over the long term, perhaps to the point where classic BW agents will offer less utility as a means of causing casualties.

(e) Many bioengineering companies (both US and foreign) now sell all-in-one kits to enable researchers to perform recombinant deoxyribonucleic acid (DNA) experiments. The availability of free online gene sequence databases and analytic software over the Internet further amplifies and disseminates this capability. It is now possible to transform relatively benign organisms to cause harmful effects.

3. Militarily Significant Aspects of Toxic Chemical Agents

a. Classification. A toxic chemical agent is any chemical which, through its chemical action on life processes, can cause death, temporary incapacitation, or permanent harm to humans or animals.[6] For the purpose of this manual, chemical agents are further divided into chemical warfare (CW) agents and military chemical compounds. The terms "persistent" and "nonpersistent" describe the time chemical agents remain in an area and do not classify the agents technically.

(1) CW Agents. The CW agents are toxic chemicals and their precursors prohibited under the CWC. These agents include choking, nerve, blood, blister, and incapacitating agents. Their physiological actions are as follows:

(a) Choking Agents.[7] Choking agents cause damage to the lungs, irritation to the eyes and the respiratory tract, and pulmonary edema ("dry-land drowning").

(b) Nerve Agents. Nerve agents inhibit cholinesterase (ChE) enzymes. This inhibition permits acetylcholine (ACh), which transmits many nerve impulses, to collect at its various sites of action.[7] The body's muscles and glands become overstimulated due to excessive amounts of ACh. At sufficient doses, this can lead to an inability of the body to sustain breathing.

(c) Blood Agents.[7] The blood transports these agents to all body tissues. Hydrogen cyanide (AC) and cyanogen chloride (CK) are cellular poisons, and they disrupt the oxidative processes used by the cells.[7] Arsine (SA) is different. It causes hemolysis of the red blood cells.[8] The central nervous system (CNS) is especially vulnerable to lack of oxygen regardless of the etiology, and respiratory and cardiovascular collapse resulting from AC and CK poisoning. In the case of SA poisoning, the proximal cause of death is myocardial failure.

(d) Blister Agents (Vesicants). Blister agents are noted for producing reddening and blistering of the skin, but the eyes and respiratory tract are more sensitive than the skin. Eye exposure results in reddening of the eyes and temporary blindness or permanent effects. Inhaled mustard damages mucous membranes and the respiratory tract.[7]

(e) Incapacitating Agents.[9] Used in a military context, incapacitation is understood to mean inability to perform one's military mission. Since missions vary, for the purpose of this manual, incapacitation means the inability to perform any military task effectively. An incapacitating agent is an agent that produces temporary physiological or mental effects, or both, which will render individuals incapable of concerted effort in the performance of their assigned duties. Medical treatment is not essential but can facilitate a more rapid recovery.[7]

(2) Military Chemical Compounds. Military chemical compounds are less toxic and include materials such as respiratory irritant agents, RCAs, smoke and obscurants, and incendiary materials. The term excludes CW agents. Their physiological actions are as follows:

(a) RCAs (Lacrimators). The RCAs are chemicals that rapidly produce in humans sensory irritation or disabling physical effects which disappear within a short time following termination of exposure.[6] They are local irritants that, in very low concentrations, act primarily on the eyes, causing intense pain and tearing. At high concentrations they irritate the respiratory tract and the skin. They sometimes cause nausea and vomiting.

(b) Respiratory Irritant Agents. These agents were previously called vomiting agents. Their primary action is irritation of the respiratory tract.[10] In addition, these agents cause lacrimation (tearing), irritation of the eyes, uncontrollable coughing, sneezing, nausea, and a general feeling of bodily discomfort. Usually symptoms disappear in 20 minutes to 2 hours, leaving no residual injury.[7]

b. Duration of Effectiveness. Several factors determine the time a chemical agent remains effective. These include, but are not limited to, the method of dissemination, weather and terrain conditions, and the physical and chemical properties of the agent.

(1) Method of Dissemination.[11] Chemical agents are usually disseminated in the field in the form of vapors (gases), aerosols, or liquids. When a chemical agent is disseminated as a vapor from a bursting munition, initially the cloud expands, grows cooler and heavier, and tends to retain its form. Aerosols are finely divided liquid and/or solid substances suspended in the atmosphere and behave in much the same manner as vaporized agents. Liquid agents can be absorbed (soaked into) and adsorbed (adhered to) by surfaces. They can then be evaporated or desorbed (off-gas) from surfaces, causing a vapor hazard.

(2) Weather and Terrain Conditions.[11] Many weather factors and terrain conditions influence the duration of effectiveness of chemical agents. Most important weather factors include temperature, temperature gradient, wind speed, relative humidity, and precipitation. Important terrain conditions include vegetation, soil, and terrain contours.

(3) Physical Properties. Some of the important physical properties are vapor density, vapor pressure (VP), volatility, freezing point (FP), and melting point (MP). Vapor

density determines whether the agent is lighter or heavier than air, thus determining whether the agent will settle to low areas or float away and dissipate in the atmosphere. Vapor pressure is used to determine the volatility of an agent. The volatility has an effect upon the vapor concentration. It also affects the duration of an agent hazard after dissemination. The boiling and freezing points of chemical agents influence their operational use and the means of disseminating them. See Appendix A for information on table of equivalents and commonly used prefixes, Appendix B for information on temperature conversions, and Appendix C for the periodic table of elements. See Chapter II for definitions of selected physical properties.

(4) Chemical Properties. The chemical properties of an agent include its stability and reactivity with water and other substances. See Chapter II for definitions of selected chemical properties.

c. Potency and Physiological Actions. Factors that contribute to the adverse human health effects of chemical agents include toxicity, route of exposure (ROE), dosage, exposure duration, minute volume (MV), temperature, endpoint, physiological stressors, rate of detoxification (ROD), and rate of action (ROA). Note that not all factors are applicable to all exposure scenarios. For example, MV is not applicable to a percutaneous liquid exposure. Dosages are given for a 70-kilogram (kg) male with an MV of 15 liters per minute (L/min). Additional toxicological data are required to determine if the toxicity estimates can be applied to women. Emphasis is placed on acute toxic effects. Acute toxic effects are those occurring within moments to a few days of the toxic exposure. The toxicity estimates provided are not applicable to the general population.[10, 12, 13]

NOTE: Occupational health guidelines for the evaluation and control of occupational exposure to nerve and blister agents is promulgated separately by the US Army Surgeon General in DA Pamphlets 40-8 and 40-173. These references define the medical surveillance program for personnel who support chemical demilitarization operations.

d. CWC Chemicals. Appendix D contains the list of toxic chemicals, groups of chemicals, and precursors subject to the CWC. The examples given in Appendix D are not all-inclusive. There are, by conservative estimates, 25,000 or more chemicals subject to the CWC regulation—listing each chemical by name is not practical.[14] Chemicals covered under the CWC are divided into three categories as follows:

(1) Schedule 1 chemicals (See Table D-1, page D-1) have little or no use in industrial and agricultural industries. They pose a high risk to the object and purpose of the CWC by virtue of their high potential for use in activities prohibited under the CWC.[14]

(2) Schedule 2 chemicals (See Table D-2, page D-2) may be useful in the production of chemical weapons; however, they also have legitimate uses in other industrial areas. They pose a significant risk to the object and purpose of the CWC.[14]

(3) Schedule 3 chemicals (See Table D-3, page D-3) have legitimate uses in industrial areas and pose a risk to the object and purpose of the CWC.[14]

e. Dual-Use Precursors. Precursors for CW agents also have civil uses in industrial and agricultural industries (see Appendix E).

f. CW Agents and Other Military Chemical Compounds. See Appendix F for the symbols of CW agents and other military chemical compounds. See Appendix G for a consolidation of the information given in the chemical agent tables and toxicity tables in Chapters II and III.

NOTE: See information in Chapters II, III, and Appendix H for detailed information on toxicity of CW agents and military chemical compounds.

g. Agent Mixtures. Mixing chemical agents with each other or with other materials can alter the characteristics and effectiveness of the agents. Mixtures may lower the freezing point, increasing agent effectiveness over a wider temperature range. The addition of thickeners or thinners to agents will increase or decrease persistency: for example, soman (GD) mixed with thickeners will increase persistency; RCAs mixed with thinners will decrease persistency. In addition to changing the physical properties, mixing agents together will create special problems through their physiological effects. These problems can produce difficulty in identification, immediate and delayed effects, or contact and vapor hazards occurring simultaneously. Some mixtures would make it difficult to maintain the seal of the protective mask. Mixing some agents can also increase the toxic effects, either by a synergistic effect or by an improved absorption through the skin.

4. Militarily Significant Aspects of Biological Agents

a. Classification. A biological agent is a microorganism that causes disease in personnel, plants, or animals or causes the deterioration of material.[6] Biological agents can be classified as pathogens, toxins, bioregulators, or prions.

(1) Pathogens. Pathogens are disease-producing microorganisms,[6] such as bacteria, rickettsiae, or viruses. Pathogens are either naturally occurring or altered by random mutation or recombinant DNA techniques.

(2) Toxins. Toxins are poisons formed as a specific secreting product in the metabolism of a vegetable or animal organism, as distinguished from inorganic poisons. Such poisons can also be manufactured by synthetic processes.[6] Toxins are produced by a variety of organisms, including microbes, snakes, insects, spiders, sea creatures, and plants.[15]

(3) Bioregulators. Bioregulators include biochemical compounds that regulate cell processes and physiologically active compounds such as catalysts and enzymes. Although they can be found in the human body in small quantities, introduction of large quantities can cause severe adverse effects or death.[15]

(4) Prions. Prions are proteins that can cause neurodegenerative diseases in humans and animals.[16] Proteins have a unique, genetically defined amino acid sequence that determines their specific shapes and functions. Normal cell proteins have the same amino acid building blocks but they fold differently than prions. When prions enter brain cells, they apparently convert normal proteins into prions. Ultimately, the infected brain cells die and release prions into the tissue. These prions enter, infect, and destroy other brain cells. Prions entered the public's consciousness during the mad cow epidemic that hit England in 1996.[17] Transmission of the prions from cows to man is suspected to cause human illness. There are no known therapies effective against prions.[18]

b. Uses. Biological agents can be disseminated and used against personnel, animals, plants, or material. Food and industrial products can be rendered unsafe or unfit for use by contamination or by the effects resulting from contamination with biological agents. The US military forces are deployed throughout the world. Associated with the movement of troops are risks of introduction of exotic agricultural pests and animal disease agents through soil contamination and transportation of regulated items such as fruits, vegetables, meat, and dairy products, other food items, and animal products (e.g., trophies). The United States Department of Agriculture (USDA), Animal and Plant Health Inspection Service (APHIS) oversees the entry of cargo, personnel, equipment, personal property, mail, and their means of conveyance into the US.[19] (See Appendix I for selected properties of some biological agents.)

(1) Antipersonnel. Biological antipersonnel agents are those that are effective directly against humans. The threat would select these agents on the basis of the agents' ability to cause death or disability. Potential biological antipersonnel agents include toxins, bacteria, rickettsiae, viruses, and toxins.

(2) Antianimal. Biological antianimal agents are those that could be employed against animals to incapacitate or destroy them through disease. The purposeful spreading of infectious agents that attack cattle or other domestic animals can lead to serious consequences for a country's food supply or export of animal products (hides, wool, fats, and biological medicinal products such as adrenalin, insulin, pituitary extracts, cortisone, vaccines, and antisera).[20] See Appendix J.

(3) Antiplant. Biological antiplant agents are organisms that cause disease or damage to plants. These agents may be used intentionally by an enemy to attack food or economically valuable crops, thereby reducing a nation's ability to resist aggression.[20] See Appendix K.

(4) Antimaterial. Antimaterial agents are organisms that degrade or break down some item of material. For example, fungi may damage fabrics, rubber products, leather goods, or foodstuffs. Some bacteria produce highly acidic compounds that cause pitting in metals; these agents could create potential problems with stockpiled material. Some bacteria can use petroleum products as an energy source and cause residues that might clog fuel or oil lines.[20]

c. Duration of Effectiveness. The duration of effectiveness of a biological agent refers to the persistency of the agent in the environment. It depends on the characteristics of the agent and environmental factors.[3]

(1) Biological agent characteristics such as encapsulation (natural, such as bacterial spores, or manmade protective coverings), addition of dyes to the spray fluid, or possibly genetic engineering (of pathogens) may protect some agents from sunlight and other destructive natural forces.[3] Bacteria that are resistant to environmental extremes frequently produce spores to allow survival during adverse conditions. Spore formation is not a method of reproduction inasmuch as each vegetative cell forms only a single spore and each spore germinates to form a single vegetative cell. The bacterium (vegetative cell) makes a copy of its DNA. The DNA becomes surrounded by a series of membranes that accumulate calcium, dipicolinic acid (heat-resistant factor), and protein layers. The resistant spore might remain dormant for years without requiring nutrients or water and might survive under extreme ranges of temperature. When conditions become favorable, the spore develops into an actively growing vegetative cell.[21]

(2) Ultraviolet (UV) radiation, relative humidity, wind speed, and temperature gradient are important weather factors in determining duration of effectiveness.

d. Methods of Dissemination.[22] Biological agents may be disseminated as aerosols, liquid droplets (toxins only), or dry powders. See Appendix L for additional information on dissemination of biological agents.

(1) Biological agents may be delivered in either wet or dry form. Dry powders composed of very small particles tend to have better dissemination characteristics and have advantages in storage. Dried agents require an increased level of technological sophistication to produce, although freeze-drying and spray-drying technologies have been available in the industry for a number of years.

(2) The BW agents might be released against our forces or against civilian populations by means of sprays, explosive devices, and contamination of food and water. Most commonly, delivery methods use aerosolized agents.

(a) A BW agent can be released as a line source. A line source would be released perpendicular to the direction of the wind, upwind of the intended target area.

(b) A second type of aerosol source is a point source, which is a stationary device for aerosolization of the agent, such as a stationary sprayer. A modified point source would be a group of spray devices, such as specially designed bomblets dispersed in a pattern on the ground or a missile or artillery shell designed to release such bomblets.

e. Physiological Aspects. Employment considerations for BW agents include the following:

(1) ROE. The important portals of entry are the respiratory tract, the exposed mucosal surfaces (moist surfaces of nose, mouth, and eyes), and the digestive tract.[23] In a biological attack the respiratory route would be the primary route of entry.[3] The respiratory system is much more susceptible to penetration. The body is more resistant to invasion by microorganisms through the skin; however, penetration across the skin can occur. This is particularly true of abraded (broken) surfaces and some toxins such as mycotoxins.[23] Toxins absorbed through the respiratory tract can produce signs and symptoms different from those acquired through natural occurrence.[24] For example, staphylococcal enterotoxin B when ingested in food causes acute gastrointestinal (GI) illness; however, when delivered via aerosol to the respiratory tract, it produces respiratory disease.[23] Personnel can encounter biological agents by natural routes, such as in water and food or by vectors.

(2) Dosage. The BW agents are inherently more toxic than CW nerve agents on a weight-for-weight basis and can potentially provide broader coverage per pound of payload than CW agents.[15]

(a) Infective Dose.[25] The infectivity of an agent reflects the relative ease with which microorganisms establish themselves in a host species. Pathogens with high infectivity cause disease with relatively few organisms.

(b) Lethal Dose. Some pathogens produce toxins that can result in disease (for example, anthrax, botulinum, cholera, diphtheria, and typhus). The extreme toxicity of many toxins causes the lethal dose to be much smaller than that of chemical agents. Hence, units of micrograms (µg) or even nanograms (ng) may be used instead of milligrams (mg) in expressing toxicity. Human toxicity estimates are based on animal data, and the ROE for the animals is not always what would be expected on the battlefield.

Some human toxicity data are based on accidental contact, ingestion, or inhalation of these natural poisons.

(3) ROA. The rate of reaction to toxins varies widely. Rapid-acting toxins generally incapacitate within minutes. Delayed-acting agents may take several hours to days to incapacitate. The time for maximum effects for pathogens is normally more than 24 hours (unless the pathogen produces a toxin). However, the incubation periods of microorganisms used in BW may be far shorter than those expected by examining the natural disease.

f. Requirements for a Weaponized BW Agent.[22] The key factors that make a biological agent suitable for an attack include availability or ease of production in sufficient quantity; the ability to cause either lethal or incapacitating effects in humans at doses that are achievable and deliverable; appropriate particle size in aerosol; ease of dissemination; stability (while maintaining virulence) after production in storage, weapons, and the environment; and susceptibility of intended victims with nonsusceptibility of friendly forces.

(1) Availability or Ease of Production. Many replicating agents (bacteria and viruses) can be produced in large quantities with modern fermentation and viral production technologies. Some toxins, like ricin, are widely available because their source in nature is ubiquitous and the process necessary to harvest the toxin is technically straightforward. On the other hand, some replicating agents are very difficult to grow in quantity, and many toxins are produced in nature in such low quantities that harvesting them is impractical (shellfish toxins are a good example).

(2) Incapacitation and Lethality. BW agents are likely to be selected for their ability to either incapacitate or kill the human targets of the attack. A BW agent does not necessarily have to be lethal to be useful as a military weapon. An agent such as Venezuelan equine encephalitis (VEE) virus could cause incapacitation among large numbers of unit personnel. If lethality is desired, agents such as anthrax have high case fatality rates once infection is established in unimmunized hosts.

(3) Appropriate Particle Size in Aerosol. An effective weaponized BW agent is of a particle size that would allow it to be carried for long distances by prevailing winds and inhaled deeply into the lungs of the unsuspecting victims. The size range of particles that meets both of these conditions is 1 to 5 microns in diameter. Particles larger than this would either settle out into the ground or more likely be filtered out in the upper respiratory tract of those who inhale them. Particles in this size range are invisible to the human eye; thus, a cloud of such particles would not generally be detected by those attacked, even if such a cloud were to be carried through their position. It is worth noting, however, that particles outside this size range are still dangerous and able to cause deadly illnesses, even though their transmission efficiency is less.

(4) Ease of Dissemination. An effective weaponized BW agent is easily disseminated in the open air by using off-the-shelf devices such as industrial sprayers or other types of aerosol-producing devices. These could be mounted on an airplane, boat, car, or other moving vehicle, or even placed in a stationary position. An alternative method would be to disseminate the agent in an enclosed space (e.g., a building) where it could more efficiently infect or intoxicate humans living or working in the area.

(5) Stability after Production. Once an adversary produces a BW agent in quantity, it must be fairly stable—either in bulk storage or once put into a weapon or delivery system. It must, therefore, retain its viability and virulence or toxicity during production, storage, transportation, and delivery.

(6) Susceptibility and Nonsusceptibility. An effective BW agent is one to which the target force is known to be susceptible (i.e., not immunized against), but to which the adversary possesses high levels of immunity, usually via vaccination.

5. Militarily Significant Aspects of Toxic Industrial Chemicals

a. Classification. The TIC are chemicals that are toxic to plants, animals, or humans.

b. Uses. The TIC are found in abundance in all countries, and are used in chemical manufacturing processes, agriculture (pesticides), water treatment (chlorination), and many other areas. Each year, more than 70,000 different chemicals amounting to billions of tons of material are produced, processed, or consumed by the global chemical industry. A large portion of these chemicals may exhibit characteristics or be sufficiently hazardous to be a threat in a military situation.[2]

c. Characteristics of TIC. The TIC of military concern may exist as solids, liquids, or gases. For many cases, release of a TIC may involve a change of the state of the chemical, therefore making protection difficult. Like CW agents, TIC include many lethal compounds.

(1) Toxicity. Many TIC, due to their toxicity, can cause incapacitation or death.

(2) Corrosiveness. Many TIC are highly corrosive. Special equipment containers and procedures are necessary to ensure safe handling.

(3) Flammability. Many TIC are highly flammable and present a major fire hazard.

(4) Explosiveness. Unlike CW agents, TIC can be highly explosive and present a serious threat when handled.

(5) Reactivity. Many TIC react violently with water or other materials, and thus present dangers upon contact with other materials, including air.

(6) Byproducts. When burned, mixed, or exploded, many TIC produce additional highly toxic byproducts.

(7) Quantities available. The sheer volume and widespread availability of TIC present a serious danger in the event of a release.

d. Duration of Effectiveness. A number of factors determine the amount of time a TIC would present a danger after release. Factors include the physical properties of the TIC as well as weather, terrain, and conditions at the release site. These factors affect TIC in the same manner as that for chemical agents.[3]

e. Physiological Aspects. Exposure to TIC affects the body in a variety of ways. Generally, they disrupt bodily functions. The effects are dependent on the routes of entry, toxicity of the chemical, and the concentration to which exposed.

(1) ROE. The TIC can enter the body through inhalation, ingestion, dermal absorption, or a combination of these methods. The primary concern for exposure is that of the inhalation of a TIC as a gas.[2]

(2) Exposure Concentration and Levels of Concern. The type and seriousness of effects from exposure to TIC, like any chemical is dependent upon the concentration and length of time one is exposed. This concentration and time relationship is unique to every chemical. The dosages of TIC are expressed in parts per million (ppm). In general, TIC tend to be at least one order of magnitude less potent than nerve agents and tend not be rapidly lethal in small quantities. Standards have been developed for industry for different exposure scenarios.

(a) Immediately Dangerous to Life and Health (IDLH):[24] The definition of IDLH that was derived during the Standards Completion Program (SCP) was based on the Mine Safety and Health Administration (MSHA) definition stipulated in 30 CFR 11.3(t). The purpose for establishing an IDLH value in the SCP was to ensure that a worker could escape without injury or irreversible health effects from an IDLH exposure in the event of the failure of respiratory protection equipment. The highly reliable breathing apparatus providing maximum worker protection was permitted. In determining IDLH values, the ability of a worker to escape without loss of life or irreversible health effects was considered along with severe eye or respiratory irritation and other deleterious effects (e.g., disorientation or lack of coordination) that could prevent escape. As a safety margin, the SCP IDLH values were based on the effects that might occur as a consequence of a 30-minute exposure. However, the 30-minute period was not meant to imply that workers should stay in the work environment any longer than necessary. In fact, every effort should be made to exit immediately.

(b) Refer to the United States Army Center for Health Promotion and Preventive Medicine (USACHPPM) Technical Guide 230, *Chemical Exposure Guidelines for Deployed Military Personnel*, for obtaining the military exposure guidelines for assessing exposure concentrations for TIC.

f. TIC Hazard Assessment. As part of the IPB process, a planner must assess the likelihood of a release or exposure as well as the actual TIC material. Some example considerations are[3]—

(1) Accidents in civilian operations significantly increase when technically trained personnel flee an area, such as a combat zone (CZ). Civilian personnel remaining may be pressured to operate equipment beyond their training/technical expertise in a area of combat.

(2) Pipelines can offer a very attractive target for terrorists because actions can be planned well in advance of execution and pipelines do not rely on shipping or transportation scheduled.

(3) Storage yards, ports, airfields and rail yards often contain significant amounts of transiting TIC. This not only presents opportunities for improvised use against US forces, but also presents increased possibility of accidents and targets for those who want to destroy the TIC (such as ammunition precursor chemicals).

g. Pesticides. Large stockpiles of obsolete pesticides have been accumulated in virtually all developing countries over periods sometimes exceeding four decades.[28] The term "pesticides," as used by US forces include insecticides, rodenticides, fungicides, and herbicides. The health effects of pesticides depend on the type of pesticide. Some, such as the organophosphates and carbamates, affect the nervous system. Others may irritate the skin or eyes. Some pesticides may be carcinogens. Others may affect the hormone or endocrine system in the body.[29] The US Environmental Protection Agency (EPA) has recognized the dangers of many pesticides and publishes lists of those pesticides that are either banned or severely restricted in their use. Applicable service personnel (e.g., Army preventive medicine (PVNTMED), Air Force civil engineering, public health) can provide information on specific pesticides that could be used in specific areas of operation (AOs).

NOTES

[1]Office of the Secretary of Defense, *Proliferation: Threat and Response,* ISBN: 0-16-042727-4, US Government Printing Office, November 1997.

[2]A.K. Steumpfle et al., *Final Report of International Task Force-25: Hazard From Toxic Industrial Chemicals,* March 18, 1996.

[3]Joint Publication 3-11, *Joint Doctrine of Operations in NBC Environment,* 11 July 2000.

[4]*DOD Chemical and Biological Defense Program Annual Report to Congress,* Vol. I, April 2002.

[5]Office of the Secretary of Defense, *Proliferation: Threat and Response,* US Government Printing Office, January 2001.

[6]Joint Publication 1-02, *Department of Defense Dictionary of Military and Associated Terms,* as amended through 05 June 2003.

[7]FM 8-285/Navy Medical (NAVMED) P-5041/Air Force Joint Manual (AFJMAN) 44-149/Fleet Marine Force Manual (FMFM) 11-11, *Treatment of Chemical Agent Casualties and Conventional Military Chemical Injuries,* 22 December 1995.

[8]L. Fishbein and S. Czerczak, *Concise International Chemical Assessment Document 47: Human Health Aspects,* WHO, 2002.

[9]Brigadier General (BG) Russ Zajtchuck, et al. (eds.), *Textbook of Military Medicine: Medical Aspects of Chemical and Biological Warfare,* Office of the Surgeon General, 1997, Chapter 11, "Incapacitating Agents."

[10]Sharon Reutter et al., *Review and Recommendations for Human Toxicity Estimates for FM 3-11.9,* ECBC-TR0349, September 2003.

[11]FM 3-6/FMFM 7-11-H/Air Force Manual (AFM) 105-7, *Field Behavior of NBC Agents (Including Smoke and Incendiaries),* 3 November 1986.

[12]Jeffrey H. Grotte and Lynn I Yang, *Report of the Workshop on Chemical Agent Toxicity for Acute Effects: Institute for Defense Analyses, May 11-12, 1998,* IDA Document D-2176, June 2001.

[13]Anna Johnson-Winegar, PhD, Assistant to the Secretary of Defense, *Memorandum, Subject: Interim Certification of Chemical and Biological Data,* December 27, 2001.

[14]Federal Register, Department of Commerce, Bureau of Export Administration, "15 CFR Part 710 et al., Chemical Weapons Convention Regulations; Final Rule," December 30, 1999.

[15]Office of the US President, *The Biological and Chemical Warfare Threat,* 1999.

[16]Centers for Disease Control and Prevention (CDC), Office of Health and Safety (OHS), "BMBL Section VII: Agent Summary Statements, Section VII-D: Prions," 17 June 1999, http://www.cdc.gov/OD/OHS/BIOSFTY/bmb14/bmbl4s7d.htm, 19 August 2003.

[17]Ruth Levy Guyer, "Research in the News: Prions: Puzzling Infectious Proteins," http://science-education.nih.gov/nihHTML/ose/snapshots/multimedia/ritn/prions/prions1.html, 8 August 2003.

[18]AFMAN 10-2602, *Nuclear, Biological, Chemical, and Conventional (NBCC) Defense Operations and Standards (Operations),* 1 December 2002.

[19]USDA, APHIS, "Protocol for Military Clearance," 18 June 2001.

[20]BG Russ Zajtchuk, et al. (eds.), *Textbook of Military Medicine: Medical Aspects of Chemical and Biological Warfare,* Office of the Surgeon General, 1997, Chapter 21, "The Biological Warfare Threat."

[21]TM 3-216/AFM 355-6, *Technical Aspects of Biological Defense,* 12 January 1971.

[22]BG Russ Zajtchuk, et al. (eds.), *Textbook of Military Medicine: Medical Aspects of Chemical and Biological Warfare,* Office of the Surgeon General, 1997, Chapter 20, "Use of Biological Weapons."

[23]FM 8-284/NAVMED P-5042/AFMAN (I) 44-156/Marine Corp Reference Publication (MCRP) 4-11.1C, *Treatment of Biological Warfare Agent Casualties,* 17 July 2000.

[24]BG Russ Zajtchuk, et al. (eds.), *Textbook of Military Medicine: Medical Aspects of Chemical and Biological Warfare,* Office of the Surgeon General, 1997, Chapter 30, "Defense Against Toxin Weapons."

[25]FM 8-9/NAVMED P-5059/AFJMAN 44-151, *NATO Handbook on the Medical Aspects of NBC Defense Operations AMEDD-6(B),* 1 February 1996.

[26]CDC, NOISH, *Documentation for Immediately Dangerous to Life or Health Concentrations,* National Technical Information Service (NTIS) Publication No. PB-94-195047, May 1994.

[27]Mark Davis, *Baseline Study on the Problem of Obsolete Pesticide Stocks,* Food and Agriculture Organization of the United Nations (FAO) Pesticide Disposal Series N.9, 2001.

[28]US EPA, "Pesticides: Health and Safety: Human Health Issues," 19 May 2003, http://www.epa.gov/pesticides/health/human.htm, 19 August 2003.

Chapter II

CHEMICAL WARFARE AGENTS AND THEIR PROPERTIES

1. Background

The CW agents can be classified according to their physiological effects or their military use; they include choking, nerve, blood, blister, and incapacitating agents.[1] This chapter contains definitions for selected physical and chemical properties; definitions for selected toxicity terms; and the physical, chemical, and physiological properties of selected CW agents and precursors listed in Table II-1.

Table II-1. List of Selected CW Agents and Precursors

2. Definitions of Selected Physical and Chemical Properties

The definitions for the physical and chemical properties are given in the same order listed in the chemical agent tables.

a. Molecular Weight (MW). MW is the value represented by the sum of the atomic weights of all the atoms in a molecule.[2] See Table C-1 (page C-1) for the Periodic Table of Elements and Table C-2 (page C-2) for the list of elements and their symbols. For example, the MW of ethyldichloroarsine (ED), $C_2H_5AsCl_2$, is computed as follows:

C (atomic weight = 12.011) x 2 = 24.02

H (atomic weight = 1.0079) x 5 = 5.04

As (atomic weight = 74.9216) x 1 = 74.92

Cl (atomic weight = 35.453) x 2 = 70.91

MW = 174.89

b. Physical State. Chemical agents may exist as solids, liquids, or gases.[2] To a certain extent the state in which an agent normally exists determines its use, duration of effectiveness, and physiological action. It also determines the type of munitions used for its dissemination.

c. Odor. Odor is the emanation from any substance that stimulates the olfactory cells in the organ of smell.[3]

d. Boiling Point. The boiling point is the temperature at which the vapor pressure of a liquid equals the pressure of the gas above it. The normal boiling point is the temperature at which the vapor pressure of a liquid equals one atmosphere (atm). At high altitudes where the atmospheric pressure is less than one atm, water boils below 100 degrees Celsius (C) or 212 degrees Fahrenheit (F).[2]

e. FP/MP. The FP is the temperature at which the solid and liquid phases of a given substance are in equilibrium and is generally equivalent to the MP.[2]

NOTE: Some liquids can be cooled well below their freezing temperatures and still remain in a liquid state. This extended form of the liquid physical state is called "supercooling." Supercooled liquids are unstable and can crystallize

spontaneously. Constant agitation and/or the use of seed crystals can sometimes prevent or reduce the amount of supercooling that occurs. However, many chemical agents experience some degree of supercooling, especially the G agents. Due to the potential for supercooling, freezing point values should be used with caution. Whenever possible, MP valves should be used because they are more thermodynamically reproducible than the FP.[4]

f. Density (Liquid/Solid). The density of a chemical agent is the mass per unit volume of the substance. Because volume varies with temperature, a specified temperature should be given. The density of a liquid or solid is usually given as grams per milliliter (g/ml) or grams per cubic centimeter (g/cm^3).[2]

NOTE: Solid density can be further specified as bulk (or apparent) density or crystalline (or true) density. Both properties describe the mass per unit volume. Bulk density includes the volume of the voids, pores, or empty spaces between particles, whereas crystalline density includes only the volume occupied by the material itself.[5]

g. Vapor Density. Vapor density is the ratio of the weight of a given volume of a gaseous substance and that of the same volume of another gas measured under the same conditions of pressure and temperature.[6] For the purpose of this manual, the other gas is air. To calculate the vapor density, divide the MW of the compound of interest by 29 (the average MW of air). If the vapor density is less than 1, the gas will generally rise in the air. If the vapor density is greater than 1, the gas will generally settle on the ground.

h. VP. The VP is the pressure exerted by a vapor when a state of equilibrium exists between the vapor and its liquid (or solid) state. It is the pressure in a closed space above a substance when no other gas is present. The VP varies with temperature, so the temperature should be stated. The VP increases as temperature increases.[2]

i. Volatility. Volatility is the tendency of a solid or liquid material to pass into the vapor state at a given temperature.[7] The volatility depends on vapor pressure and varies with temperature. Volatility is expressed as milligrams of vapor per cubic meter (mg/m^3). It is calculated numerically by an equation derived from the perfect gas law:

$$V = \frac{16020 \times MW \times VP}{T}$$

Where

V = Volatility (mg/m^3)

MW = molecular weight

VP = vapor pressure (in torr at a specified temperature)

T = Kelvin temperature (degrees C + 273.15)

j. Latent Heat of Vaporization. The latent heat of vaporization is the quantity of energy absorbed or given off as a substance undergoes a change in state with no change in temperature.[7] It is calculated using the following equation:[8]

$$\Delta H_v = \frac{\ln 10\ RBT^2}{(C + t)^2}$$

Where

ΔH_v = Enthalpy of vaporization (latent heat of vaporization)

T = Kelvin temperature = (degrees C + 273.15)

R = Ideal gas law constant ($1.987\ cal\ K^{-1}\ mol^{-1}$)

B,C = Vapor pressure constants (from Antoine or Clausius Clapeyron fit)

t = Temperature in degrees C

k. Viscosity. Viscosity is resistance that a gaseous or liquid system offers to flow when it is subjected to a shear stress.[9] The more complex the molecules in a liquid and the stronger the intermolecular forces between them, the more difficult it is for the molecules to move past each other and the greater the viscosity of the liquid. A fluid with a large viscosity resists motion. Also, as temperatures increase, the viscosity of the liquid decreases.[2] Units for viscosity are given in centipoises (cP).

l. Surface Tension. Surface tension is the force that causes the surface of a liquid to contract, reducing its surface area to a minimum. The molecules within a liquid are attracted equally in all directions by the cohesive forces within the liquid. However, the molecules on the surface of a liquid are attracted only into the liquid and to either side. This unbalanced molecular attraction tends to pull the surface molecules back into the liquid such that the minimum number of molecules possible are on the surface.[2] Units for surface tension are dynes per centimeter (dynes/cm).

m. Flash Point. The flash point is the temperature at which a liquid or volatile solid gives off sufficient vapor to form an ignitable mixture near the surface of the liquid.[7]

n. Decomposition Temperature. The decomposition temperature is the temperature at which a chemical breaks down into two or more substances.[4] Because reaction rates vary; decomposition temperature is a function of both temperature and time (some reactions are slower than others).

o. Solubility. The solubility of a solute is the quantity that will dissolve in a given amount of solvent to produce a saturated solution.[2]

p. Hydrolysis. Hydrolysis is the reaction of a compound with water whereby decomposition of the substance occurs.[2] New substances (hydrolysis products) form when a compound reacts with water.

q. Half-Life of a Reaction ($t_{1/2}$). This is the time required for half of the original concentration of the limiting reactant to be consumed.[4]

r. Stability in Storage. Stability in storage determines the practical usefulness of a compound. If a compound decomposes in storage, it will have little military operational value. The addition of stabilizers will typically slow down decomposition and polymerization in storage.

s. Action on Metals, Plastics, Fabrics, and Paint. This item describes the action between a given compound and different materials. Depending on their activity, some chemicals can react with and degrade materials they contact. Chemical agent-resistant coating (CARC) can minimize this effect.

t. Specific Heat. The specific heat is the quantity of heat required to raise the temperature of 1 gram of a substance 1 degree C[2] (given in lieu of latent heat of vaporization for military chemical compounds only).

3. Definitions of Toxicity-Related Terms

a. ROE. Chemical agents enter the body through the respiratory tract, skin, eyes, and by ingestion. Any part of the respiratory tract, from the nose to the lungs, may absorb inhaled gases and aerosols. For some agents, effects are more severe in normally sweaty areas. The skin can also absorb vapors. The surface of the skin, eyes, and mucous membranes can absorb droplets of liquids and solid particles. Wounds or abrasions are probably more susceptible to absorption than the intact skin. Chemical agents can contaminate food or drink, and therefore, the body can absorb them through the gastrointestinal tract. Nerve agents exert their toxic effects through the skin, eyes, and lungs. Inhalation is the usual route of exposure for blood agents. Blister agents damage skin and other tissues that they contact, to include eyes and lungs. The choking agents exert their effects only if inhaled. The onset and severity of signs may vary, depending upon the ROE and dosage.[1] The ROEs have been limited to those that are likely to be encountered in the field: vapor inhalation with eye exposure, skin exposure to vapor, and skin exposure to liquid. The toxicity estimates given for percutaneous vapor exposure for the nerve agents are based on bare skin exposures. The toxicity estimates for mustard agent percutaneous vapor exposure are based on clothed skin exposures. Effective dosages of liquid on the skin are based upon data for bare skin.[9]

b. Dosage. Dosage is the amount of substance administered (or received) per body weight.[11] In this manual, dosage is usually expressed as milligrams per kilogram (mg/kg) of body weight for liquid agents and as milligrams-minute per meter cubed for (mg-min/m^3) for vapor exposure. Dosages are given for a 70-kg man.[9]

(1) Median Lethal Dosage (LD_{50}) of Liquid Agent. The LD_{50} is the amount of liquid agent expected to kill 50 percent of a group of exposed, unprotected individuals.

(2) Median Effective Dosage (ED_{50}) of Liquid Agent. The ED_{50} is the amount of liquid agent expected to cause some defined effect (e.g., severe, such as prostration,

collapse, convulsions; mild, such as erythema) in 50 percent of a group of exposed, unprotected individuals.

(3) Median Lethal Dosage (LCt_{50}) of a Vapor or Aerosol. The LCt_{50} of a chemical agent in vapor form is the dosage that is lethal to 50 percent of exposed, unprotected personnel for a defined MV and exposure duration.

(4) Median Effective Dosage (ECt_{50}) of a Vapor or Aerosol. The ECt_{50} is the effective dosage of a chemical agent vapor that is sufficient to cause some defined effect in 50 percent of exposed, unprotected personnel for a defined MV and exposure duration.

NOTES:

1. Effective dosages can be calculated for more or less than the median dosage (e.g., LCt_{25}, ED_{84}). Such calculations require knowledge of the probit (Bliss) slope and use the probit equation.[10]

2. Selected toxicity estimates are assigned provisional values. A toxicity estimate is considered provisional based on only having limited data, but the data are within the range of available animal data or when the agent is presumed to be comparable in toxicity to a related agent.[10]

3. Physiological Stressors. Physiological stressors include anxiety, heat, and humidity and are likely to reduce toxicity estimates. In other words, the agent is potentially more effective.[10]

c. Modifying Factors. After exposure to a chemical agent vapor, a person may show signs and symptoms that are less or more severe than expected. The severity of the effects may depend upon some of the following potential variables:

(1) How long the person held his or her breath during short exposure.

(2) Speed with which he or she donned the mask.

(3) Proper fit of the mask.

(4) Whether the body absorbed the agent through the skin.

(5) Whether the agent increased the MV.

(6) MV of the person at the time of exposure.

(7) Physical exertion of the person at the time of exposure.

(8) Rate of detoxification, especially if exposure was long.

(9) Previous exposure to chemical agents and type of agent.

d. MV. The MV is the volume of air exchanged in one minute. In general, as the MV increases, the apparent dosage decreases because more agent is inhaled into the lungs. It is important to note that increasing respiratory rate does not necessarily increase MV and may actually decrease it. The relationship of MV to dosage is approximately linear over ranges of MV from 10 to 50 liters. For example, if the LCt_{50} is given as 35 mg-min/m^3, for an MV of 15 liters, it would be approximately 50 mg-min/m^3 for an MV of 10 liters and approximately 15 mg-min/m^3 for an MV of 30 liters.[3] Where available, MV profile tables are provided in Appendix H.

e. Exposure Duration. The official interim standards for inhalation/ocular exposures are for a 2-minute duration; those for percutaneous vapor exposure (masked personnel) are for a 30-minute duration. More data have become available since the interim standards were defined, and they are also given for longer exposures, as data are available.[10]

f. Temperature. Selected agents have toxicity estimates for hot and moderate temperatures. Hot temperatures are defined as greater than 85 degrees F. Moderate temperatures are defined as 65 to 85 degrees F. In general, as the temperature increases, the effective dosage decreases. The dosages for hot temperatures are about half of those for moderate temperatures.[10]

g. Endpoint (Physiological Effects). Toxicity estimates are provided for different endpoints to include lethality, severe effects, threshold effects, and mild effects.

(1) For nerve agents, severe effects include prostration, collapse, and/or convulsions. Some deaths will occur. Threshold effects for percutaneous vapor exposure include slight, not necessarily significant, ChE inhibition and/or localized sweating. Mild effects following inhalation/ocular exposure include miosis, rhinorrhea, and tight chest. These effects can occur in the absence of measurable ChE inhibition in the blood.[10]

(2) For percutaneous exposure to blister agents, severe effects include vesication and mild effects include erythema, edema, pain, and itching—depending upon the agent. Severe effects following ocular exposure include pain, conjunctivitis, blepharospasm, and temporary blindness; mild effects consist of erythema and minimal conjunctivitis.[10]

h. ROD. The human body can detoxify some toxic materials. The ROD is an important factor in determining the hazards of repeated exposure to CW agents. Many CW agents are essentially cumulative in their effects.

i. ROA. The ROA of a chemical agent is the rate at which the body reacts to or is affected by that agent. The rate varies widely, even between those of similar tactical or physiological classification.

j. Concentration-Time (Ct) Profile. Dosage is a function of Ct; however, the equation k=Ct does not describe all cases of injury from chemical agent exposure. As seen in some Ct profile tables, a given effective dosage does not always produce a specific effect for all duration exposures. In order to describe the effective dosages better mathematically, the equation $C^n t=k$ is used. The exponent "n" is called the toxic load exponent (TLE). See Figure II-1 and II-2 (page II-8)

(a) When the TLE value is greater than 1, effective dosages increase as the exposure duration increases (and the exposure concentration decreases).

(b) When the TLE value is less than 1, effective dosages decrease as the exposure duration increases (the agent is more potent following long exposures to low concentrations than short exposures to high concentrations).

(c) When the TLE value equals 1 (*Haber's Law*), a given dosage (Ct) produces a given effect—independent of concentration or exposure duration. When the Ct profile is unknown, a default exponent value of one is used (TLE is assumed to be 1).

The effect the TLE has on the effective dosage is illustrated in the following Ct profile tables. In the equation $k = C^n t$, there are three possible cases for the value of the TLE(n).

Case 1: TLE>1

Ct Profile (15L MV)		
Exposure Duration (min)	LCt_{50} (mg-min/m^3)	Concentration (mg/m^3)
2	35	17.5
60	70	1.2
120	80	0.7

NOTES:

- **LCt_{50}/ECt_{50} increases as exposure duration increases.**
- **Concentration decreases; exposure to low levels for an extended period of time can cause effects.**
- **There is no single LCt/ECt for a specific type of exposure and endpoint.**

Case 2: TLE<1

Ct Profile (15L MV)		
Exposure Duration (min)	LCt_{50} (mg-min/m^3)	Concentration (mg/m^3)
2	35	17.5
60	20	0.3
120	15	0.125

NOTES:

- **LCt_{50}/ECt_{50} decreases as exposure duration increases.**
- **The agent is more potent following long exposures to low concentrations than short exposures to high concentrations.**
- **There is no single LCt/ECt for a specific type of exposure and endpoint.**

Case 3: TLE =1 (*Haber's Law)*

Ct Profile (15L MV)		
Exposure Duration (min)	LCt_{50} (mg-min/m^3)	Concentration (mg/m^3)
2	35	17.5
60	35	0.5
120	35	0.3

NOTES:

- **LCt_{50}/ECt_{50} remains constant as exposure duration increases.**
- **Concentration decreases; exposure to low levels for an extended period of time can cause effects.**
- **There is a single LCt/ECt for a specific type of exposure and endpoint.**

Figure II-1. The TLE Effect on the Ct Profile

Dosage vs. Exposure Duration

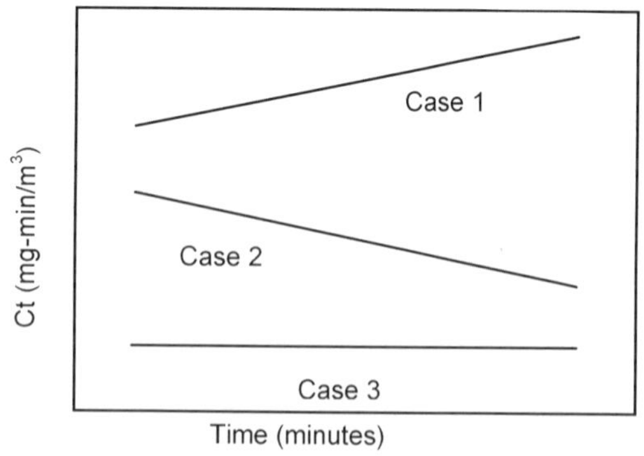

The graph on the left is a representation of a Ct profile for dosage versus exposure duration. The graph is constructed by plotting the exposure duration (t) on the horizontal axis and the dosage (k) on the vertical axis. In the quotation $k=C^nt$ there are three possible cases for the value of the TLE (n). The graph shows the three cases.

Case 1: TLE > 1; LCt50/ECt50 increases as exposure duration increases.
Case 2: TLE < 1; LCt50/ECt50 decreases as exposure duration increases.
Case 3: TLE = 1; LCt50/ECt50 remains constant as exposure duration increases.

CW Agent Vapor: Dosage versus Exposure Duration

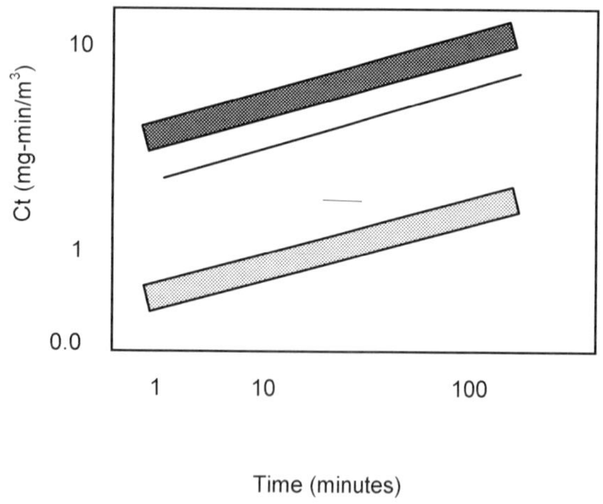

This is an example of the Ct profile graphs for dosage versus exposure duration given in Appendix H. this is for an inhalation/ocular vapor exposure to a CW agent.

Given:

Upper Band: LCt_{16} and LCt_{84} region
Line: ECt_{16} (severe) [roughly LCt_{01}]
Lower Band: ECt_{16} and ECt_{84} region for mild effects

Upper Band: The endpoint is lethality. The region shows the effective dosages that would be lethal to 16-84% of 70-kg males exposed.

Line: The endpoint is severe effects (specific symptoms depend on the type of CW agent exposure). The line gives the effective dosages where 16 percent of 70-kg males would experience severe effects. This is roughly equivalent to the lethal dosage for 1 percent. Severe effects can include some deaths.

Lower Band: The endpoint is mild effects (specific symptoms depend on the type of CW agent exposure). The region shows the effective dosages that would cause mild effects in 16-84% of 70-kg males exposed.

Notice as exposure duration (time) increases, the effective dosage (LCt/ECt) increases. There is no single LCt or ECt for an exposure; they are time-dependent.

Figure II-2. Ct Profile for Dosage versus Exposure Duration

k. Probit Slope. Dose response curves can be used to assess from a graphical depiction the results of an exposure to a CW agent. The dose response curve is impacted by the concentration of the CW agent and duration of the exposure (i.e., a graphical x-axis plot) and response (i.e., a graphical y-axis plot). The probit slope is derived from dose response curve data and corresponds to the variability in the response of the population to the chemical agent.

(a) A high probit slope value means that a small change in the dose will make a large change in the number of individuals responding to the chemical agent. With very potent agents, a small change in the dose can also make a big change in the level and severity of the effects produced.

(b) A low probit slope value means that a relatively large change in the dose will make for a relatively minor change in the number of individuals responding to the given dose.

l. Degree of Confidence (DOC).[10] The DOC is an indication of the level of confidence in each toxicity estimate and is provided to indicate uncertainty. It is a subjective evaluation based on the quality and quantity of the underlying data and the method(s) by which the estimate was derived. The following definitions are provided:

(1) Low. There are no primary data, and/or the data are extremely limited.

(2) Moderate. There are primary data for both humans and animals, and there are sufficient data for mathematical modeling.

(3) High. There are ample primary data for both humans and animals, and there is good statistical confidence in the value.

4. Choking Agents

Choking agents are CW agents that attack lung tissue, primarily causing pulmonary edema. They cause irritation to the bronchi, trachea, larynx, pharynx, and nose. Initial symptoms may include tears, dry throat, coughing, choking, tightness of chest, nausea, vomiting, and headache.[1] In extreme cases, membranes swell, lungs become filled with liquid, and death results from lack of oxygen; thus, these agents "choke" an unprotected person. Fatalities of this type are called "dry-land drownings." Of the choking agents, phosgene (CG) is the only one considered likely to be used in the future.[12] The protective mask gives protection against choking agents.[1]

a. CG (see Table II-2 [page II-10]). CG is a colorless gas with an odor similar to musty hay or rotting fruit.[13] Vapors can linger for some time in trenches and low-lying areas under calm or light winds.[12] The severity of poisoning cannot be estimated from the immediate symptoms, and the full effect can be delayed up to 72 hours after exposure.[14] Any activity or stress after exposure is likely to exacerbate the effects and turn a sublethal exposure into a lethal exposure.[10] Lung damaging concentrations may not be detected by smell.[10]

Table II-2. CG

Alternate Designations: Collongite (French); Zusatz (German); Green Cross (German); D–gas (German); Fosgeen (Dutch); Fosgen (Polish); Fosgene (Italian); Phosgen (German); NCI-C60219	
Chemical Name: Carbonyl chloride	
Synonyms: Carbon oxychloride; Carbon dichloride oxide; Carbone (oxychlorure de) (French); Carbonic chloride; Carbonio (ossiclorurodi) (Italian); Carbonylchlorid (German); Carbonyl dichloride; Chloroformy chloride; Koolstofoxychloride (Dutch)	
CAS Registry Number: 75-44-5	
RTECS Number: SY5600000	
Physical and Chemical Properties	
Structural Formula: 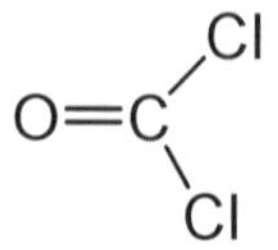 Molecular Formula: $COCl_2$ Molecular Weight: 98.92	
Physical State	Colorless gas that is readily liquefied [1]
Odor	Musty hay or rotting fruit [2]
Boiling Point	7.8°C [3,4]
FP/MP	-128°C (MP) [5]
Liquid Density (g/mL)	Liquefied phosgene 1.360 @ 25°C; 1.402 @ 7.8°C; 1.420 @ 0°C [6]
Vapor Density (relative to air)	3.4 (calculated)
Vapor Pressure (torr)	1.40 x 10^3 @ 25°C; 7.60 x 10^2 @ 7.8°C; 5.60 x 10^2 @ 0°C [3,4]
Volatility (mg/m^3)	7.46 x 10^6 @ 25°C; 4.29 x 10^6 @ 7.8°C; 3.53 x 10^6 @ 0°C (calculated from vapor pressure) [3,4]
Latent Heat of Vaporization (kcal/mol)	5.92 @ 25°C; 5.95 @ 7.8°C; 5.96 @ 0°C (calculated from vapor pressure) [3,4]
Viscosity (cP)	Data not available
Viscosity of Vapor (cP)	Data not available
Surface Tension (dynes/cm)	Data not available
Flash Point	Nonflammable [1]
Decomposition Temperature	Complete @ 800°C [7]
Solubility	Limited in water; [8] miscible with common organic solvents, petroleum, and lubricating oil [9,10]
Rate of Hydrolysis	$t_{1/2}$ = 0.25 sec. @ 13°C; does not react quickly with water vapor, but it immediately reacts with liquid water to yield carbon dioxide and hydrochloric acid [8,11]
Hydrolysis Products	Hydrochloric acid and carbon dioxide [9]
Stability in Storage	Stable in steel containers @ ambient temperatures for at least one year if CG is dry; stability decreases at elevated temperatures [12]
Action on Metals or Other Materials	None when CG is dry; acidic and corrosive when moist [13]
Other Data	
Eye toxicity	Initial effects resemble those of tear gas. [14]
Inhalation toxicity	Causes pulmonary edema [15]
Rate of action	Immediate to 3 hours, depending on concentration [16]
Means of detection in field	M18A2 CADK, MM1 [17]
Protection required	Protective mask [15]
Decontamination	Not required in the field except in very cold climates [18]
Use	Delayed action casualty agent [19]

Table II-2. CG (Continued)

NOTES
[1]*Matheson Gas Data Book*, 4th ed., p. 411, The Matheson Company, Inc., East Rutherford, NJ, 1966.
[2]Franke, S., *Manual of Military Chemistry Volume I-Chemistry of Chemical Warfare Agents*, ACSI-J-3890, Chemie der Kampfstoffe, East Berlin, April 1968, UNCLASSIFIED Technical Manual (AD849866).
[3]Abercrombie, P., ECBC Notebook # NB 98-0079, p. 7 (U).
[4]Germann, A.F.O. and Taylor, Q.W., "The Critical Constants and Vapor Tension of Phosgene," *J. Amer. Chem. Soc.,* Vol. 48*(5)*, pp. 1154 –1159, 1928.
[5]Giauque, W. F., and Jones, W.M., "Carbonyl Chloride. Entropy. Heat Capacity. Vapor Pressure. Heats of Fusion and Vaporization. Comments on Solid Sulfur Dioxide Structures," *J. Amer. Chem. Soc.,* Vol. 70, p. 120, 1948.
[6]Davies, C. N., "The Density and Thermal Expansion of Liquid Phosgene," *J. Chem. Phys.*, Vol. 14, p. 48, 1946.
[7]Bodenstein, M., and Durant, G., "Die Dissociation des Kohlenoxychlorids," *Z. Physik. Chem.*, Vol. 61, p. 437, 1908.
[8]Hall, R.W., *An Investigation on the Solubility and Rate of Hydrolysis of Phosgene in Water*, Porton Report 2663, Chemical Defence Experimental Establishment, Porton, England, 19 December 1944, UNCLASSIFIED Report.
[9]Atkinson, R.H., et al., "The Preparation and Physical Properties of Carbonyl Chloride," *J. Chem. Soc.*, Vol. 117, pp. 1410-1426, 1920.
[10]Baskerville, C., and Cohen, P.W., "Solvents for Phosgene," *J. Ind. Eng. Chem.*, Vol. 13, p. 333, 1921.
[11]*Properties of War Gases Volume III: Vomiting & Choking Gases & Lacrimators (U)*, ETF 100-41/Vol-3, Chemical Corps Board, Army Chemical Center, MD, December 1944, CONFIDENTIAL Report (AD108458).
[12]Henley, F.M., *Surveillance Tests on 75 MM. Steel Gas Shell Extending Over a Period of One Year*, EACD 11, Chemical Warfare Service, Edgewood Arsenal, MD, June 1920, UNCLASSIFIED Report (ADB959731).
[13]Patten, H.E., and Bouder, N.M., *Chemical Properties of Phosgene*, EACD 124, Chemical Warfare Service, Edgewood Arsenal, MD, March 1923, UNCLASSIFIED Report (ADB955133).
[14]BG Russ Zajtchuk, et al. (eds.), *Textbook of Military Medicine: Medical Aspects of Chemical and Biological Warfare*, Office of the Surgeon General, 1997, Chapter 4, The Chemical Warfare Threat and the Military Healthcare Provider."
[15]FM 8-285/NAVMED P-5041/AFJMAN 44-149/FMFM 11-11, *Treatment of Chemical Agent Casualties and Conventional Military Chemical Injuries*, 22 December 1995.
[16]*NIOSH-DOD-OSHA Sponsored Chemical and Biological Respiratory Protection Workshop Report*, February 2000.
[17]AFMAN 10-2602 *Nuclear, Biological, Chemical, and Conventional (NBCC) Defense Operations and Standards (Operations)*, 29 May 2003.
[18]FM 8-9/NAVMED P-5059/AFJMAN 44-151, *NATO Handbook on the Medical Aspects of NBC Defense Operations AMEDP-6(B)*, 1 February 1996.
[19]W.R. Kirner, *Summary Technical Report of Division 9, NDRC Volume 1, Chemical Warfare Agents, and Related Chemical Problems Part I-II*, Office of Scientific Research and Development, Washington, DC, 1946, UNCLASSIFIED Report (AD234270).

b. CG Toxicity Estimates (see Table II-3). Although the probit slope for CG is unknown, it is known to be so steep that "incapacitating" or "severe effects" dosages could include some lethality.[10]

Table II-3. CG Toxicity Estimates[10]

Endpoint	Toxicity (mg-min/m³)	MV (L)	Exposure Duration	ROE	Probit Slope	TLE	ROD	DOC
Lethality	LCt_{50}: 1500 [a]	15	2-60 min	Inhalation/Ocular	Unknown	1 [b]	Probably Insignificant	Moderate
Odor Detection	EC_{50}: 6 mg/m³ [c]	N/A	Few Seconds	Inhalation	N/A	N/A	Probably Insignificant	Low

NOTES

[a]Based on ECBC modeling of 10 mammalian species.
[b]See Appendix H for supporting toxicity profile estimates.
[c]Based on secondary human data and TM 3-215 (1952).

c. Diphosgene (DP) (see Table II-4 [page II-12]). DP is a colorless liquid with an odor similar to that of musty hay.[13] DP is not a polymer of CG, but does produce similar physiological effects. DP is described as a respiratory irritant and a lachrymator. It is more easily detected than CG because of its lacrimatory effects.[9] See Table II-5 (page II-13) for DP toxicity estimates.

Table II-4. DP

Alternate Designations: Difosgene; Superpalite (British); Perstoff (German); Surpalite(French); Green Cross (German)	
Chemical Name: Trichloromethyl chloroformate	
Synonyms: Trichloromethyl chlorocarbonic acid ester; Chloroformic acid trichloromethyl ester; Trichloromethyl chlorocarbonate; Trichloromethyl carbonochloridate; Formic acid, chloro-, trichloromethyl ester; Carboncohloridic acid trichloromethyl ester	
CAS Registry Number: 503-38-8	
RTECS Number: LQ7350000	
Physical and Chemical Properties	
Structural Formula: 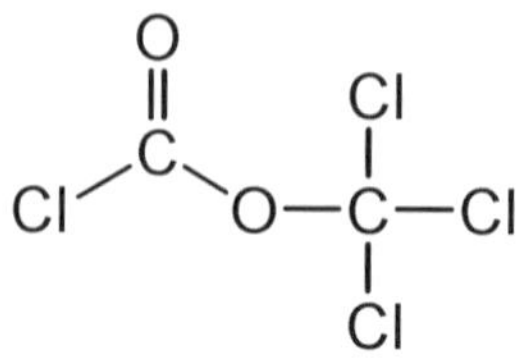 Molecular Formula: $C_2Cl_4O_2$ Molecular Weight: 197.83	
Physical State	Colorless oily liquid [1]
Odor	Musty hay [2]
Boiling Point	127°C [3,4]
FP/MP	-57°C (MP) [2]
Liquid Density (g/mL)	Munitions grade: 1.656 @ 20°C; 1.687 @ 0°C [2]
Vapor Density (relative to air)	6.8 (calculated)
Vapor Pressure (torr)	4.41 @ 20°C; 9.14×10^{-1} @ 0°C [3,4]
Volatility (mg/mL)	4.77×10^{4} @ 20°C; 1.06×10^{4} @ 0°C (calculated from vapor pressure) [3,4]
Latent Heat of Vaporization (kcal/mol)	12.2 @ 20°C; 12.8 @ 0°C (calculated from vapor pressure) [3,4]
Viscosity (cP)	Data not available
Viscosity of Vapor (cP)	Data not available
Surface Tension (dynes/cm)	Data not available
Flash Point	None [5]
Decomposition Temperature	300°C to 350°C (yields two molecules of CG) [1]
Solubility	Solubility in water is 44.6 g DP/L solution @ 20°C; [6] readily soluble in common organic solvents [2]
Rate of Hydrolysis	Slow @ ambient temperature and fairly rapid @ 100°C [1]
Hydrolysis Products	Hydrogen chloride (HCl) and carbon dioxide [1]
Stability in Storage	Unstable; converts to CG. [1]
Action on Metals or Other Materials	Metals act as catalyzers in conversion to CG. [1,2] Also attacks rubber, cork, [2,4] and cement. [2]
Other Data	
Eye toxicity	Lachrymator [7]
Inhalation toxicity	Causes pulmonary edema [8]
Rate of action	Immediate to 3 hours depending on concentration [9]
Means of detection	MM1
Protection required	Protective mask [8]
Decontamination	Not required in the field except in very cold climates. [10]
Use	Delayed or immediate action casualty agent, depending on dosage rate

Table II-4. DP (Continued)

NOTES
[1]Hood, H.P., and Murdock, H.R., "Superpalite," *J. Phys. Chem.,* Vol. 23, p. 498, 1919.
[2]Potts, A.M.,*The Physical and Chemical Properties of Phosgene and Diphosgene*, OEMCMR-114, 1945, UNCLASSIFIED Report.
[3]Abercrombie, P., ECBC Notebook # NB 98-0079, p. 34 (U).
[4]Herbst, V.H., "Uber die Fluchtigkeit und Vernebelung einer Reihe organischer Stoffe," *Kolloid Beihefte,* Vol. 23, p. 330, 1927.
[5]TM 3-215/AFM 355-7, *Military Chemistry and Chemical Agents*, December 1963, UNCLASSIFIED Technical Manual.
[6]Carter, R.H., and Knight, H.C., *Fundamental Study of Toxicity Solubilities of Certain Toxics in Water and in Olive Oil*, EACD 445, Chemical Warfare Service, Edgewood Arsenal, Edgewood, MD, May 1928, UNCLASSIFIED Report (ADB955216).
[7]Sharon Reutter, et al., *Review and Recommendations for Human Toxicity Estimates for FM 3-11.9*, ECBC-TR-349, September 2003.
[8]FM 8-285/NAVMED P-5041/AFJMAN 44-149/FMFM 11-11, *Treatment of Chemical Agent Casualties and Conventional Military Chemical Injuries*, 22 December 1995.
[9]*NIOSH-DOD-OSHA Sponsored Chemical and Biological Respiratory Protection Workshop Report,* February 2000.
[10]FM 8-9/NAVMED P-5059/AFJMAN 44-151, *NATO Handbook on the Medical Aspects of NBC Defense Operations AMEDP-6(B),* 1 February 1996.

Table II-5. DP Toxicity Estimates[10]

Endpoint	Toxicity (mg-min/m^3)	MV (L)	Exposure Duration	ROE	Probit Slope	TLE	OD	DOC
Lethality	LCt_{50}: 1500[a] (Provisional)	15	10-60 min	Inhalation/ Ocular	Unknown	1 [b,c]	Unknown, Probably Insignificant	Low
Odor Detection	EC_{50}: 4 mg/m^3 [d]	N/A	Few Seconds	Inhalation	N/A	N/A	N/A	Low

NOTES

[a]Based on recommendations for CG.
[b]The TLE value is assumed to be 1 because the Ct profile is unknown.
[c]See Appendix H for supporting toxicity profile estimates.
[d]Based on secondary human data.

5. Nerve Agents

Nerve agents are more toxic than other CW agents. They may cause effects within seconds and death within minutes.[15] The nerve agents are all liquids, not nerve gas per se. They can be absorbed through any body surface and can penetrate ordinary clothing rapidly.[1] They are divided into the G agents and V agents. The V agents have high boiling points, low volatility, and resultant high persistency.[12] Even though the V agents are considered primarily a contact hazard[12]; they are at least twice as potent as GB, and even a minute amount of airborne material is extremely hazardous.[10] Nerve agents are cumulative poisons. Repeated exposure to low concentrations may produce symptoms.[1] Level 4 mission-oriented protective posture (MOPP4) is required for protection.[1]

a. Physiological Effect. Both the G and V agents have the same physiological action on humans. Normally, the enzyme acetylcholinesterase (AChE) binds and hydrolyzes the neurotransmitter ACh, which terminates the activity of ACh at the receptor sites. Upon exposure, the nerve agents bind to AChE, making it unable to bind with ACh. As a result, ACh is not hydrolyzed. The accumulation of ACh causes hyperactivity of the body organs stimulated by cholineraic neruons.[15] Individuals poisoned by nerve agents may experience symptoms in the following order:

- Miosis, runny nose, and chest tightness.
- Dim vision and headache.
- Nausea, vomiting, and cramps.

- Drooling, excessive sweating, drowsiness, and confusion.
- Difficulty breathing, twitching, jerking, and staggering.
- Convulsions and coma.

b. Miosis. When airborne vapor comes in contact with the eyes, miosis occurs as a result of a direct local effect of the nerve agent on the eyes and can occur prior to any inhibition of ChE in the blood. This type of exposure is frequently accompanied by tightness of the chest and/or rhinorrhea, and any or all other symptoms can occur. In cases of nerve agent exposure not involving vapor contact with the eyes, miosis is one of the last effects to occur before death.[10]

c. Treatment. Treatment of nerve agent poisoning includes use of atropine, 2-PAM chloride, convulsant antidote for nerve agents (CANA), and pyridostigmine bromide (PB).

(1) Atropine binds to receptor sites blocking the excess acetylcholine caused by nerve agent poisoning.

(2) 2-PAM Cl acts by reactivating ChE inhibited by a nerve agent.[15] Prompt treatment is essential because after the agent binds to AChE, a second reaction occurs in which the agent loses one alkyl or alkoxy group. This phosphorylated AChE is called an "aged" enzyme and is completely resistant to both spontaneous and oxime-medicated (2-PAM Cl) reactivation. The "aging" period varies from minutes to hours depending on the type of agent.[16] For GD, the "aging" half-time is within 2 minutes.[17]

(3) The CANA prevents and treats convulsions caused by exposure to nerve agents in moderate to severe cases.[1]

(4) PB is a pretreatment for exposure to GD. PB and AChE bind and form what is called a carbamoylated AChE. Although PB is also an AChE inhibitor, it is unlike nerve agents in that the interaction between PB and AChE is freely and spontaneously reversible and it does not undergo the aging process. The carbamoylated AChE is fully protected from attack by nerve agents. Atropine is still needed to counteract the excess ACh and 2-PAM Cl is still needed to reactivate AChE active sites that were protected by PB.[17] PB is available to US forces in active theaters of operation (TOs).[1]

d. Tabun (GA) (see Table II-6). GA was the first of the nerve agents developed by the Germans.[15] GA is primarily an inhalation hazard. See Table II-7 (page II-17) for GA toxicity estimates.

Table II-6. GA

Alternate Designations: EA 1205; Le-100 (German), T-83 (German); MCE; FM-511; T-2104 (British); TL-1578 (UCTL); Trilon 83 (German); Gelan I (German); Taboon A
Chemical Name: Ethyl N, N-dimethylphosphoroamidocyanidate
Synonyms: Ethyl dimethylamidocyanophosphate; Dimethylaminoethoxyphosphoryl cyanide; Dimethylaminocyanophosphoric acid ethyl ester; Cyanodimethylaminoethoxyphosphine; Dmethylaminecyanoethoxyphosphine oxide; Ethyl dimethylaminocyanophosphonate Phosphoramidocyanidic acid, dimethyl, ethyl ester; Dimethylamidoethoxyphosphoryl cyanide; Dimethylaminocyanphosphorsaeureaethylester (German); Dimethylphosphoramidocyanidic acid, ethyl ester; Ethyl dimethylphosphoramidocyanidate; Ethylester-dimethylamid kyseliny kyanfosfonove (Czech)
CAS Registry Number: 77-81-6
RTECS Number: TB4550000

Table II-6. GA (Continued)

Physical and Chemical Properties

Structural Formula:

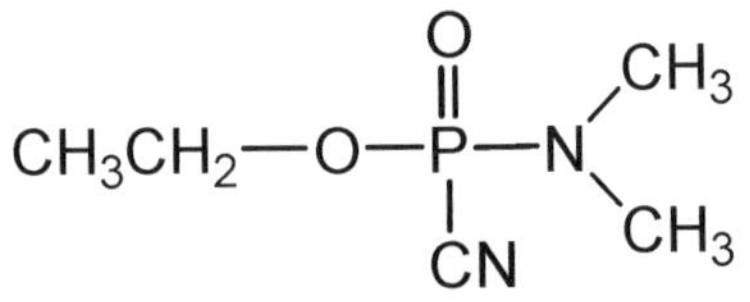

Molecular Formula: $C_5H_{11}N_2O_2P$

Molecular Weight: 162.13

Physical State	Colorless to brown liquid [1]
Odor	Faintly fruity; none when pure [2]
Boiling Point	248°C (extrapolated) [3-7]
FP/MP	-50°C (FP) [5]
Liquid Density (g/mL)	1.0756 @ 25°C; 1.0999 @ 0°C [3]
Vapor Density (relative to air)	5.6 (calculated)
Vapor Pressure (torr)	5.70×10^{-2} @ 25°C; 4.75×10^{-3} @ 0°C (extrapolated) [3-7]
Volatility (mg/m^3)	4.97×10^{2} @ 25°C; 4.52×10^{1} @ 0°C (calculated from vapor pressure) [3-7]
Latent Heat of Vaporization (kcal/mol)	15.5 @ 25°C; 16.7 @ 0°C (calculated from vapor pressure) [3-7]
Viscosity (cP)	2.277@25.0° C, 4.320@ 0°C[3]
Viscosity of Vapor (cP)	6.20×10^{-3} @ 25.0°C, 5.60×10^{-3} @ 0°C[3]
Surface Tension (dynes/cm)	32.5 @ 25.0°C, 35.0 @ 0°C[3]
Flash Point	78°C (closed cup) [8]
Decomposition Temperature	Decomposes completely @ 150°C after about 3 to 3 1/4 hrs [9]
Solubility	Solubility in water is approximately 7.2g GA/100g solution @ 20°C and 9.8g GA/100g @ 0°C; [3] readily soluble in common organic solvents [2]
Rate of Hydrolysis	$t_{1/2}$ = 8.5 hrs @ 20°C and pH 7;[10] slow in water but fairly rapid with strong acids and alkalis with self-buffering @ pH 4 to 5; [1] autocatalytic below pH 4 [11]
Hydrolysis Products	AC, dimethylaminocyanophosphonic acid, and other products [10]
Stability in Storage	When stabilized with 5% chlorobenzene, GA can be stored in steel containers for several years @ ambient temperatures. The degree of stability decreases @ elevated temperatures with decomposition occurring within 6 months @ 50°C and 3 months @ 65°C. [12]
Action on Metals or Other Materials	Corrosion rate of steel on crude GA with 5 to 20% chlorobenzene is 0.000034 inch/month @ 65°C. [12]
Other Data	
Skin and eye toxicity	Eyes: very high; much greater through eyes than through skin. Skin: highly toxic, decontamination of smallest drop of liquid agent is essential; liquid penetrates skin. [13]
Inhalation Toxicity	Primarily inhalation hazard [14]
Rate of action	Rapid [15]
Means of detection	M8 paper, M9 paper, M256A1 CADK, M8A1 ACAA, M90 AMAD, M21 ACAA, M22 ACADA, CAM/ICAM, M272 Water Testing Kit, M18A3 CADK[16], M18A2 CADK, MM1, CAPDS, IPDS, AN/KAS-1 CWDD[16]
Protection required	MOPP4; liquid nerve agents penetrate ordinary clothing rapidly [13]
Decontamination	Flush eyes with water immediately. Use the M291 SDK to remove any liquid nerve agent on skin or clothing. Use the M295 IEDK for individual equipment.[13] STB is effective on equipment. Water, steam, and absorbents (earth, sawdust, ashes, and rags) are effective for physical removal. [17] **NOTE: GA may react to form CK in bleach slurry.**[18]
Use	Quick-acting casualty agent

Table II-6. GA (Continued)

NOTES
[1]Witten, Benjamin, *The Hydrolysis of MCE*, Technical Division Memorandum Report 1121, USA Chemical Research and Development Laboratories, Army Chemical Center, MD, August 1945, UNCLASSIFIED Report (ADB964102). [2]Welchman, R.M.A., *Preliminary Report on the Potential Value of Nerve Gases as C.W. Agents*, Porton Report No. 2747 (PR 2747), Chemical Defence Experimental Establishment, Porton, England, January 1947, UNCLASSIFIED Report. [3]Samuel, J.B., et al., *Physical Properties of Standard Agents, Candidate Agents, and Related Compounds at Several Temperatures (U)*, ARCSL-SP-83015, USA Armament Research and Development Command, Aberdeen Proving Ground, MD, June 1983, UNCLASSIFIED Report (ADC033491). [4]Abercrombie, P., ECBC Notebook # NB 98-0079, p. 36 (U). [5]Harris, B.L., *Physical Constants of MCE*, Technical Division Memorandum Report 1094, USA Chemical Research and Development Laboratories, Army Chemical Center, MD, July 1945, UNCLASSIFIED Report (ADB 964103). [6]Balson, E.W., *Determination of the Vapor Pressure of T.2104,* A.3804/3, Military Intelligence Division, Chemical Defence Experimental Establishment, Porton, England, April 1945, UNCLASSIFIED Report. [7]Belkin, F., and Brown, H.A., *Vapor Pressure Measurements of Some Chemical Agents Using Differential Thermal Analysis, Part III*, ECTR-75032, Edgewood Arsenal, Aberdeen Proving Ground, MD, June 1975, UNCLASSIFIED Report (ADA010666). [8]Walpole, J.L., *Determination of the Flash Points of GA and GB*, Porton Technical Paper No. 45 (PTP 45), Chemical Defence Experimental Establishment, Porton, England, March 1948, UNCLASSIFIED Report. [9]Miller, C.E., *Thermal Studies on MCE*, Technical Division Memorandum Report 1132, USA Chemical Research and Development Laboratories, Army Chemical Center, MD, September 1945, UNCLASSIFIED Report (ADB964104). [10]Marsh, D.J. et al., *Kinetics of the Hydrolysis of Ethyl Dimethylamino Cyanophosphonate (and certain other related compounds) in Water*, Proton Technical Paper No. 85 (PTP-85), Chemical Defense Experimental Establishment, Porton, England, December 1948, UNCLASSIFIED Report. [11]Clark, D.N, *Review of Reactions of Chemical Agents in Water, Final Report to USA Biomedical Research and Development Laboratory*, Battelle, Columbus, OH, January 1989, UNCLASSIFIED Report (ADA213287). [12]Harris, B.L. and Macy, R., *Storage Stability of German GA in Uncoated and Lacquered 75mm Shell at 50°C and 65°C. Corrosion Rate of Steel by GA at 65°C*, Technical Division Memorandum Report 1299, USA Chemical Research and Development Laboratories, Army Chemical Center, MD, December 1946, UNCLASSIFIED Report (ADB964902). [13]FM 8-285/NAVMED P-5041/AFJMAN 44-149/FMFM 11-11, *Treatment of Chemical Agent Casualties and Conventional Military Chemical Injuries*, 22 December 1995. [14]Sharon Reutter, et al., *Review and Recommendations for Human Toxicity Estimates for FM 3-11.9*, ECBC-TR-349, September 2003. [15]*NIOSH-DOD-OSHA Sponsored Chemical and Biological Respiratory Protection Workshop Report,* February 2000. [16]DOD Chemical And Biological Defense Program Annual Report to Congress, Volume I, April 2003. [17]FM 3-5/MCWP 3-37.3, *NBC Decontamination,* 28 July 2000. [18]W.R. Kirner, *Summary Technical Report of Division 9, NDRC Volume 1, Chemical Warfare Agents, and Related Chemical Problems Part I-II,* Office of Scientific Research and Development, Washington, DC, 1946, UNCLASSIFIED Report (AD234270).

Table II-7. GA Toxicity Estimates[10]

Endpoint	Toxicity (mg-min/m^3)	MV (L)	Exposure Duration	ROE	Probit Slope	TLE	ROD	DOC
Lethality	LD_{50}: 1500 mg [a]	N/A	N/A; 70-kg man	Percutaneous Liquid [b]	5	N/A	Unknown	Low
	LCt_{50}: 70 [a]	15	2 min	Inhalation/Ocular	12	1.5 [c]	Some	Moderate
	LCt_{50}: 15,000 [a,d]	N/A	30-360 min	Percutaneous Vapor [e]	5	1 [c,f]	Unknown	Low
	LCt_{50}: 7500 [g,h] (Provisional)	N/A	30-360 min	Percutaneous Vapor [e]	5	1 [c,f]	Unknown	Low
Severe effects, includes deaths	ED_{50}: 900 mg[a]	N/A	N/A; 70-kg man	Percutaneous Liquid [b]	5	N/A	Unknown	Low
	ECt_{50}: 50 [a]	15	2 min	Inhalation/Ocular	10	1.5 [c]	Some	Moderate
	ECt_{50}: 12,000[a,d]	N/A	30-360 min	Percutaneous Vapor [e]	5	1 [c,f]	Unknown	Low
	ECt_{50}: 6000 [g,h] (Provisional)	N/A	30-360 min	Percutaneous Vapor [e]	5	1 [c,f]	Unknown	Low
Threshold effects (slight ChE inhibition)	ECt_{50}: 2000 [a,d]	N/A	30-360 min	Percutaneous Vapor [e]	5	1 [c,f]	Unknown	Moderate
	ECt_{50}: 1000 [g,h] (Provisional)	N/A	30-360 min	Percutaneous Vapor [e]	5	1 [c,f]	Unknown	Moderate
Mild effects (miosis, rhinorrhea)	ECt_{50}: 0.4 [h]	N/A	2 min	Inhalation/Ocular	10	1.5 [c]	Some	Low

NOTES

[a]Based on Grotte and Yang (2001).
[b]Bare skin.
[c]See Appendix H for supporting toxicity profile estimates.
[d]Moderate temperatures (65-85°F).
[e]Assumes personnel are masked with eye protection and bare skin.
[f]The TLE value is assumed 1 because the Ct profile is unknown.
[g]Hot temperatures (greater than 85°F).
[h]Based on recommendations for GB.

e. GA Toxicity Estimates. Note that for an inhalation/ocular expouse the TLE is greater that 1. This means that the effective dosage increases with longer exposure durations and the concentration of the agent decreases.

f. Sarin (GB) (see Table II-8 [page II-18]). The Germans developed GB after they developed GA,[15] hence the designation GB. Pure GB is odorless and colorless.[13] It is a volatile liquid at room temperature. Unlike many other agents, for which clothing affords some protection against a liquid agent, clothing may enhance the potency of GB liquid on the skin. It is hypothesized that clothing retards evaporation, thereby increasing the effective dose.[10] See Table II-9 (page II-20) for toxicity estimates.

Table II-8. GB

Alternate Designations: EA 1208; T-144 (German); Trilon 144 (German); Trilon 46 (German); T 46 German); TL-1618 (UCTL); T-2106 (British); MFI; IMPF; Sarin II

Chemical Name: Isopropyl methylphosphonofluoridate

Synonyms: Fluorisoproopoxymethylphosphine oxide; Isopropyl methylfluorophosphate; Isopropyl methanefluorophosphonate; Isopropoxymethylphosphoryl fluoride; Propoxyl-^{2}-methylphosphoryl fluoride; Phosphonofluridic acid, methyl-, isopropyl ester; Isopropylester kyseliny methylfluorfosfonove (Czech); O-Isopropyl methylphosphonofluoridate; Isopropyl-methylphosphoryl fluoride; Methylphosphonofluoridic acid isopropyl ester; Methylphosphonofluoridic acid 1-methylethyl ester; Phosphine oxide, fluoroisopropoxymethyl-; Phosphoric acid, methylfluoro-, isopropyl ester; Methylfluorphosphorsaeureisopropylester (German)

CAS Registry Number: 107-44-8

RTECS Number: TA8400000

Physical and Chemical Properties

Structural Formula:

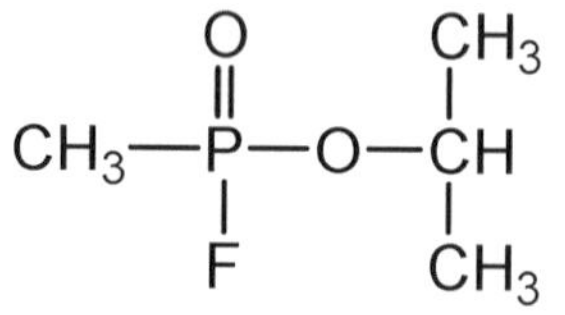

Molecular Formula: $C_4H_{10}FO_2P$

Molecular Weight: 140.09

Physical State	Colorless liquid [1]
Odor	None when pure [2]
Boiling Point	150°C (extrapolated) [3]
FP/MP	-56°C (FP) [3-5]
Liquid Density (g/mL)	Pure: 1.0887 @ 25°C; 1.1182 @ 0°C (extrapolated) [6] Munitions grade: 1.0964 @ 25°C; 1.1255 @ 0°C (extrapolated) [6]
Vapor Density (relative to air)	4.8 (calculated)
Vapor Pressure (torr)	2.48×10^{0} @ 25°C; 4.10×10^{-1} @ 0°C [3]
Volatility (mg/m^3)	1.87×10^{4} @ 25°C; 3.37×10^{3} @ 0°C (calculated from vapor pressure) [3]
Latent Heat of Vaporization (kcal/mol)	11.6 @ 25°C; 11.7 @ 0°C (calculated from vapor pressure) [3]
Viscosity (cP)	1.397 @ 25.0°C, 2.583 @ 0°C (extrapolated)[7]
Viscosity of Vapor (cP)	7.19×10^{-3} @ 25.0°C, 5.51×10^{-3} @ 0°C[7]
Surface Tension (dynes/cm)	25.9 @ 25.0°C, 28.8 @ 0°C (extrapolated)[7]
Flash Point	Nonflammable [8]
Decomposition Temperature	Complete decomposition occurs within 2 1/2 hr @ 150°C [9]
Solubility	Completely miscible with water and common organic solvents [1,2]
Rate of Hydrolysis	Varies with pH and temperature; at 20°C, $t_{1/2}$ = 27 min. @ pH 1; $t_{1/2}$ = 3 1/2 hr @ pH 2; $t_{1/2}$ = 80 hr @ pH 7; $t_{1/2}$ = 5.4 min @ pH 10; and $t_{1/2}$ = 0.6 min @ pH 11. [10]
Hydrolysis Products	Under acidic conditions, hydrogen fluoride (HF) and isopropyl methylphosphonic acid (IMPA) are formed which further hydrolyze to produce methylphosphonic acid (MPA) and isopropanol. Under alkaline conditions, methylfluorophosphonic acid (MFPA) and isopropyl alcohol are initially formed which further hydrolyze to produce MPA and HF. [11]
Stability in Storage	GB stabilized with tributylamine can be stored in steel containers for at least 5 to 10 years @ ambient temperature. At elevated temperatures up to 71°C, storage life decreases slightly. [12]
Action on Metals or Other Materials	At 71°C, slightly corrosive on steel, copper, brass, inconel, K-monel, and lead as well as slight to severe amounts of corrosion on aluminum, depending on the type. [13]

Table II-8. GB (Continued)

Other Data	
Skin and eye toxicity	Eyes: very high; much greater through eyes than through skin; Skin: highly toxic; decontamination of smallest drop of liquid agent is essential; liquid penetrates skin [14]
Inhalation toxicity	Most toxic route of exposure [15]
Rate of action	Rapid [16]
Means of detection	M8 paper, M9 paper, M256A1 CADK, M8A1 ACAA, M90 AMAD, M21 ACAA, M22 ACADA, CAM/ICAM, M272 Water Testing Kit, , CAPDS, IPDS, AN/KAS-1 CWDD, M18A3 CADK[16], M18A2 CADK, MM1
Protection required	MOPP4; liquid nerve agent penetrates ordinary clothing rapidly, clothing may enhance the potency of GB liquid on the skin [15]
Decontamination	Flush eyes with water immediately. Use the M291 SDK to remove any liquid nerve agent on skin or clothing. Use the M295 IEDK for individual equipment.[14] STB is effective on equipment. Water, steam, and absorbents (earth, sawdust, ashes, and rags) are effective for physical removal. [18]
Use	Quick-acting casualty agent

NOTES

[1]Franke, S., *Manual of Military Chemistry Volume I- Chemistry of Chemical Warfare Agents*, ACSI-J-3890, Chemie der Kampfstoffe, East Berlin, April 1968, UNCLASSIFIED Report (AD849866).
[2]Welchman, R.M.A., *Preliminary Report on the Potential Value of Nerve Gases as C.W. Agents*, Porton Report No. 2747 (PR 2747), Chemical Defence Experimental Establishment, Porton, England, January 1947, UNCLASSIFIED Report.
[3]Penski, Elwin C., *The Properties of 2-Propyl Methylfluorophosphonate (GB) I. Vapor Pressure Data Review and Analysis*, ERDEC-TR-166, USA Chemical and Biological Defense Command, Aberdeen Proving Ground, MD, June 1994, UNCLASSIED Report (ADB187225).
[4]Zeffert, B.M., et al., *Slow Fractional Crystallization of GB*, CRLR 2, USA Chemical and Radiological Laboratories, Army Chemical Center, MD, April 1951, UNCLASSIFIED Report (AD498968).
[5]Tannenbaum, H., and Zeffert, B.M., *Crystallization of GB*, TCIR-513, USA Chemical and Radiological Laboratories, Army Chemical Center, MD, November 1949, UNCLASSIFIED Report (ADE471275).
[6]Wardrop, A.W.H., and Bryant, P.J.R., *Physico-Chemical Properties of Phosphorus Esters Part II: Some Constants of Isopropyl methylfluorophosphinate (GB)*, Porton Technical Paper No. 278 (PTP-278), Chemical Warfare Laboratories, Army Chemical Center, MD, March 1952, UNCLASSIFIED Report (ADE481544).
[7]Samuel, J.B., et al., *Physical Properties of Standard Agents and Related Compounds at Several Temperatures (U)*, ARCSL-SP-83015, USA Armament Research and Development Command, Aberdeen Proving Ground, MD, June 1983, UNCLASSIFIED Report (ADC033491).
[8]Walpole, J.L., *Determination of the Flash Points of GA and GB*, Porton Technical Paper No. 45 (PTP 45), Chemical Defence Experimental Establishment, Porton, England, March 1948, UNCLASSIFIED Report (ADE481350).
[9]Perry, B.J., et al., *The Chemistry of the Alkylfluorophosphonites and Related Compounds*, Porton Technical Paper No. 258, Chemical Defense Experimental Establishment, Porton, England, 31 August 1951. UNCLASSIFIED Report (ADE481528).
[10]Epstein, J., *Studies on Hydrolysis of GB I. Effect of pH and Temperature on Hydrolysis Rates. II. Observations on Hydrolysis of GB in Sodium Bicarbonate Buffered Waters*, MDR 132, Chemical Warfare Laboratories, Army Chemical Center, MD, February 1948, UNCLASSIFIED Report.
[11]Clark, D.N, *Review of Reactions of Chemical Agents in Water, Final Report to USA Biomedical Research and Development Laboratory*, Battelle, Columbus, OH, January 1989, UNCLASSIFIED Report (ADA213287).
[12]*Comparison of GA and GB as Chemical Warfare Agents (U)*, CWL-SP-1, USA Chemical Warfare Laboratories, Army Chemical Center, MD, November 1957, UNCLASSIFIED Report.
[13]Hutchcraft, A.S. Jr., et al., *Special Report: Corrosion Resistance of Metals Toward Isopropyl Methylphosphonofluoridate (GB)*, CRLR 510, USA Chemical and Radiological Laboratories, Army Chemical Center, MD, May 1955, UNCLASSIFIED Report (AD474404).
[14]FM 8-285/NAVMED P-5041/AFJMAN 44-149/FMFM 11-11, *Treatment of Chemical Agent Casualties and Conventional Military Chemical Injuries*, 22 December 1995.
[15]Sharon Reutter, et al., *Review and Recommendations for Human Toxicity Estimates for FM 3-11.9*, ECBC-TR-349, September 2003.
[16]*NIOSH-DOD-OSHA Sponsored Chemical and Biological Respiratory Protection Workshop Report,* February 2000.
[17]DOD Chemical And Biological Defense Program Annual Report to Congress, Volume I, April 2003.
[18]FM 3-5/MCWP 3-37.3, *NBC Decontamination,* 28 July 2000.

g. GB Toxicity Estimates (Table II-9). Note that for an inhalation/ocular expouse the TLE is greater than 1. This means that the effective dosage increases with longer exposure durations and the concentration of the agent decreases.

Table II-9. GB Toxicity Estimates[10]

Endpoint	Toxicity (mg-min/m^3)	MV (L)	Exposure Duration	ROE	Probit Slope	TLE	ROD	DOC
Lethality	LD_{50}: 1700 mg [a]	N/A	N/A; 70-kg man	Percutaneous Liquid [b]	5	N/A	Unknown	Low [c]
	LCt_{50}: 35 [a]	15	2 min	Inhalation/ Ocular	12	1.5 [d]	Some	Moderate
	LCt_{50}: 12,000 [a,e]	N/A	30-360 min	Percutaneous Vapor [f]	5	1 [d,g]	Unknown	Low
	LCt_{50}: 6000 [h,i] (provisional)	N/A	30-360 min	Percutaneous Vapor [f]	5	1 [d,g]	Unknown	Low
Severe effects, includes some deaths	ED_{50}: 1000 mg [a]	N/A	N/A; 70-kg man	Percutaneous Liquid [b]	5	N/A	Unknown	Low
	ECt_{50}: 25 [a]	15	2 min	Inhalation/ Ocular	12	1.5 [d]	Some	Moderate
	ECt_{50}: 8000 [a,e]	N/A	30-360 min	Percutaneous Vapor [f]	5	1 [d,g]	Unknown	Low
	ECt_{50}: 4000 [h,i] (provisional)	N/A	30-360 min	Percutaneous Vapor [f]	5	1 [d,g]	Unknown	Low
Threshold effects (Slight ChE inhibition)	ECt_{50}: 1200 [a,e]	N/A	30-360 min	Percutaneous Vapor [f]	5	1 [d,g]	Unknown	Moderate
	ECt_{50}: 600 [h, i] (provisional)	N/A	30-360 min	Percutaneous Vapor [f]	5	1 [d,g]	Unknown	Low
Mild effects (miosis, rhinorrhea)	ECt_{50}: 0.4 [j]	N/A	2 min	Inhalation/ Ocular	10	1.5 [d, j]	Some	Moderate

NOTES

[a]Based on Grotte and Yang (2001).
[b]Bare skin.
[c]LD_{50} could be less with clothing.
[d]See Appendix H for supporting toxicity profile estimates.
[e]Moderate temperatures (65-85°F)
[f]Assumes personnel are masked with eye protection and bare skin.
[g]The TLE value is assumed 1 because the concentration-time profile is unknown.
[h]Hot temperatures (greater than 85°F)
[i]Based on Grotte and Yang (2001) and Letter (25 Mar 03).
[j]Based on human data and recent rat data of Mioduszewski et al. (2002).

h. GD (see Table II-10). GD is a colorless liquid when pure.[13] 2 PAM Cl is not as effective for GD poisoning as it is for other nerve agents[1] because the "aging" process half-time is within 2 minutes.[15] See Table II-11 (page II-23) for toxicity estimates.

Table II-10. GD

Alternate Designations: EA 1210 (US); Zoman (USSR); T-2107 (British, UK); Trilon (German); PMFP

Chemical Name: Pinacolyl methyl phosphonofluoridate

Synonyms: 3,3-Dimethyl-n-but-2-yl methylphosphonofluoridate; 3,3-Dimethyl-2-butyl methylphosphonofluridate; 2-Butanol, 3,3-dimethyl-, methylphosphonofluoridate; Methylphosphonofluoridic acid, 3,3-dimethyl-2-butyl ester; 1,2,2-Trimethylpropyl methyphosphonofluoridate; 1,2,2-Trimethylpropylester kyseliny methylfluorfosfonove (Czech); Methylphosphonofluoridic acid 1,2,2-trimethylpropyl ester; Phosphonofluoridic acid, methyl-, 1,2,2-trimethylpropyl ester; Phosphine oxide, fluoromethyl (1,2,2-trimethylpropoxy)-; Methyl pinacolyl phosphonofluoridate; Pinacolyl methylfluorophosphonate; Fluoromethylpinacolyloxyphosphine oxide; Methyl pinacolyloxyfluorophosphine oxide; Pinacolyl methane fluorophosphonate; Pinacoloxymethylphosphoryl fluoride; Methylfluoropinacolylphosphonite; Methylfluorphosphorsaeurepinakolylester (German); Methyl pinacolyloxy phosphorylfluoride; Methyl pinacolyl phosphonofluoridate; Pinacoloxymethylphosphoryl fluoride; Pinacolyl methylphosphonofluoride; Pinacolyloxy methylphosphoryl fluoride; Pynacolyl methylfluorophosphonate

CAS Registry Number: 96-64-0

RTECS Number: TA8750000

Physical and Chemical Properties

Structural Formula:

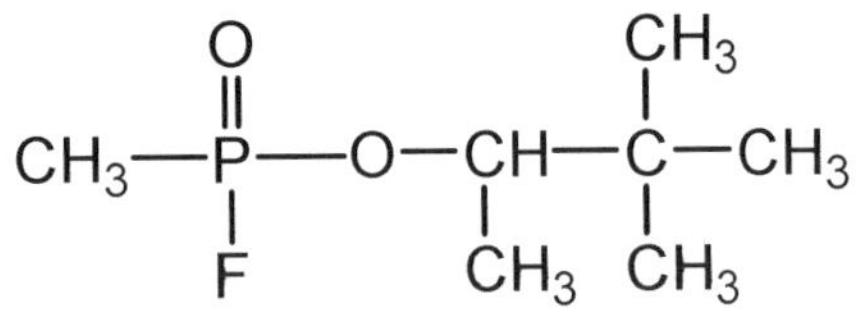

Molecular Formula: $C_7H_{16}FO_2P$

Molecular Weight: 182.17

Physical State	Colorless liquid when pure [1]
Odor	Fruity; impurities give it the odor of camphor [2,3]
Boiling Point	198°C (extrapolated) decomposes [4]
FP/MP	-42°C (MP); [5,6] generally solidifies to a noncrystalline, glasslike material [7]
Liquid Density (g/mL)	1.0222 @ 25°C; 1.0456 @ 0°C (extrapolated) [7]
Vapor Density (relative to air)	6.3 (calculated)
Vapor Pressure (torr)	4.01×10^{-1} @ 25°C; 4.96×10^{-2} @ 0°C [4]
Volatility (mg/m^3)	3.93×10^{3} @ 25°C; 5.31×10^{2} @ 0°C (calculated from vapor pressure) [4]
Latent Heat of Vaporization (kcal/mol)	13.2 @ 25°C; 13.8 @ 0°C (calculated from vapor pressure) [4]
Viscosity (cP)	3.167 @ 25.0°C, 6.789 @ 0°C (extrapolated)[5]
Viscosity of Vapor (cP)	5.90×10^{-3} @ 25.0°C, 5.33×10^{-3} @ 0°C[5]
Surface Tension (dynes/cm)	24.5 @ 25.5°C[5]
Flash Point	121°C (open cup) [5,8]
Decomposition Temperature	Above 150°C,[1] stabilized GD decomposes in 200 hrs @ 130°C; unstabilized GD decomposes in 4 hrs @ 130°C [9]
Solubility	Solubility of GD in water is 2.1 g GD/100g @ 20°C; [9,10] 3.4 g GD/100g solution @ 0°C; [9] very soluble in organic solvents [1]
Rate of Hydrolysis	Varies with pH; using a 0.003 molar solution of GD @ 25°C, $t_{1/2}$ = 3 hr. @ pH 2; $t_{1/2}$ = 45 hrs @ pH 6.65; $t_{1/2}$ = 60 hrs @ pH 10; complete hydrolysis occurs in less than 5 min in a 5% NaOH solution [11]
Hydrolysis Products	Essentially PMPA and HF [11]
Stability in Storage	Relatively stable in glass for 5-1/2 months @ ambient temperature with or without a stabilizer. [12] Stabilized GD can be stored for at least 6 months @ elevated temperatures (71°C) in glass, steel, and aluminum containers. [13]
Action on Metals or Other Materials	Corrosion rate on steel is 0.00001 inch/month @ 65°C [14]

Table II-10. GD (Continued)

Other Data	
Skin and eye toxicity	Eyes: very high toxicity; much greater through eyes than through skin. Skin: highly toxic; decontamination of smallest drop of liquid agent is essential; liquid penetrates skin. [15]
Inhalation toxicity	Most toxic route of exposure [16]
Rate of action	Rapid [17]
Means of detection	M8 paper, M9 paper, M256A1 CADK, M8A1 ACAA, M90 AMAD, M21 ACAA, M22 ACADA, CAM/ICAM, M272 Water Testing Kit, , CAPDS, IPDS, AN/KAS-1 CWDD, M18A3 CADK,[16] M18A2 CADK, MM1
Protection required	MOPP4; liquid nerve agents penetrate ordinary clothing rapidly [15]
Decontamination	Flush eyes with water immediately. Use the M291 SDK to remove any liquid nerve agent on skin or clothing. Use the M295 IEDK for individual equipment.[15] STB is effective on equipment. Water, steam, and absorbents (earth, sawdust, ashes, and rags) are effective for physical removal. [19]
Use	Quick-acting casualty agent

NOTES

[1]Franke, S., *Manual of Chemistry Volume I- Chemistry of Chemical Warfare Agents*, ACSI-J-3890, Chemie der Kampfstoffe, East Berlin, April 1968, UNCLASSIFIED Technical Manual (AD849866).

[2]Welchman, R.M.A., *Preliminary Report on the Potential Value of Nerve Gases as C.W. Agents*, Porton Report No. 2747 (PR 2747), Chemical Defence Experimental Establishment, Porton, England, January 1947, UNCLASSIFIED Report (ADE470188).

[3] TM 3-215/AFM 355-7, *Military Chemistry and Chemical Agents,* December 1963, UNCLASSIFIED Technical Manual (ADA292141).

[4]Savage, J. J., and Fielder, D., *The Vapor Pressure of Chemical Agents GD, VX, EA2223, EA 3547, EA 3580, EA 5365, and EA 5533*, EC-TR-76058, Aberdeen Proving Ground, MD, August 1976, UNCLASSIFIED Report (ADB013164).

[5]Samuel, J.B., et al., *Physical Properties of Standard Agents, Candidate Agents, and Related Compounds at Several Temperatures (U)*, ARCSL-SP-83015, USA Armament Research and Development Command, Aberdeen Proving Ground, MD, June 1983, UNCLASSIFIED Report (ADC033491).

[6]Stern, R.A., USA Chemical Research and Development Laboratories Notebook # NB 7265, p. 45 (C).

[7]Zeffert, B.M., and Coulter, P.B., *Physical Constants of G-Series Compounds: Compounds EA 1210, EA 1211, EA 1212, EA 1213, EA 1214*, Technical Division Memorandum Report 1292, USA Chemical Research and Development Laboratories, Army Chemical Center, MD, July 1947, UNCLASSIFIED Report (ADB964904).

[8]Fielder, D, USA Chemical Warfare Laboratories Notebook # NB 6695, p. 72 (C).

[9]*Chemical Agent Data Sheets Volume* I, Edgewood Arsenal Special Report EO-SR-74001, Edgewood Arsenal, Aberdeen Proving Ground, MD, December 1974, UNCLASSIFIED Report (ADB028222).

[10]Witten, Benjamin, *The Search for Toxic Chemical Agents (U),* EATR 4210, Edgewood Arsenal Research Laboratories, MD, November 1969, UNCLASSIFIED Report (AD507852).

[11]Buckles, L.C., *The Hydrolysis Rate of GD*, TCIR 373, Chemical Corps Technical Command, Army Chemical Center, MD, March 1947, UNCLASSIFIED Report (ADB966291).

[12]Newman, J.H., et al., *A Thickener for GD (U)*, EC-TR-77016, Edgewood Arsenal, Aberdeen Proving Ground, MD, April 1977, CONFIDENTIAL Report (ADC009719).

[13]Grula, R.S., et al., *Storage Stability of GD, GF and EA 1356 (U)*, CRDLR 3342, USA Chemical Research and Development Laboratories, Edgewood Arsenal, MD, December 1965, CONFIDENTIAL Report (AD369299).

[14]Hormats, S., et al., *Storage Stability in Steel at 65°C of Pure GD. Corrosion Rate of Steel at 65°C*, TDMR 1346, Chemical Corps Technical Compound, Army Chemical Center, MD, March 1948, UNCLASSIFIED Report (ADB964759).

[15]FM 8-285/NAVMED P-5041/AFJMAN 44-149/FMFM 11-11, *Treatment of Chemical Agent Casualties and Conventional Military Chemical Injuries*, 22 December 1995.

[16]Sharon Reutter, et al., SBCCOM Report *Review and Recommendations for Human Toxicity Estimates for FM 3-11.9*, ECBC-TR-349, September 2003.

[17]*NIOSH-DOD-OSHA Sponsored Chemical and Biological Respiratory Protection Workshop Report,* February 2000.

[18] DOD Chemical And Biological Defense Program Annual Report to Congress, Volume I, April 2003.

[19]FM 3-5/MCWP 3-37.3, *NBC Decontamination,* 28 July 2000.

i. GD Toxicity Estimates (Table II-11). Note that for an inhalation/ocular expouse the TLE is greater than 1. This means that the effective dosage increases with longer exposure durations and the concentration of the agent decreases.

Table II-11. GD Toxicity Estimates[10]

Endpoint	Toxicity ($mg\text{-}min/m^3$)	MV (L)	Exposure Duration	ROE	Probit Slope	TLE	ROD	DOC
Lethality	LD_{50}: 350 mg [a]	N/A	N/A; 70-kg man	Percutaneous Liquid [b]	6	N/A	Unknown	Low
	LCt_{50}: 35 [a]	15	2 min	Inhalation/ Ocular	12	1.25 [c]	Some	Low
	LCt_{50}: 3000 [a,d]	N/A	30-360 min	Percutaneous Vapor [e]	6	1 [c,f]	Unknown	Low
	LCt_{50}: 1500 [g,h] (provisional)	N/A	30-360 min	Percutaneous Vapor [e]	6	1 [c,f]	Unknown	Low
Severe effects, includes some deaths	ED_{50}: 200 mg [a]	N/A	N/A; 70-kg man	Percutaneous Liquid [b]	6	N/A	Unknown	Low
	ECt_{50}: 25 [a]	15	2 min	Inhalation/ Ocular	12	1.25 [c]	Some	Low
	ECt_{50}: 2000 [a,d]	N/A	30-360 min	Percutaneous Vapor [e]	6	1 [c,f]	Unknown	Low
	ECt_{50}: 1000 [g,h] (provisional)	N/A	30-360 min	Percutaneous Vapor [e]	6	1 [c,f]	Unknown	Low
Threshold effects	ECt_{50}: 300 [a,d]	N/A	30-360 min	Percutaneous Vapor [e]	6	1 [c,f]	Unknown	Low
	ECt_{50}: 150 [g,h] (provisional)	N/A	30-360 min	Percutaneous Vapor [e]	6	1 [c,f]	Unknown	Low
Mild effects (miosis, rhinorrhea)	ECt_{50}: 0.2 [i]	N/A	2 min	Inhalation/ Ocular	10	1.4 [c]	Some	Low

NOTES

[a]Based on Grotte and Yang (2001).
[b]Bare skin.
[c]See Appendix H for supporting toxicity profile estimates.
[d]Moderate temperatures (65-85°F).
[e]Assumes personnel are masked with eye protection and bare skin.
[f]The TLE value is assumed 1 because the concentration time profile is unknown.
[g]Based on recommendations for GB.
[h]Hot temperatures (greater than 85°F).
[i]Based on recommendations for GB and relative potency of GD and GB.

j. Cyclosarin (GF) (see Table II-12 [page II-24]). GF is a colorless and odorless liquid when pure.[13] See Table II-13 (page II-26) for toxicity estimates.

Table II-12. GF

Alternate Designations: EA 1212 (US); T-2139 (British); CMPF	
Chemical Name: Cyclohexyl methylphosphonofluoridate	
Synonyms: Cyclohexyloxyfluoromethylphosphine oxide; Cyclohexyl methylfluorophosphate; Phosphonofluoridic acid, methyl-, cyclohexyl ester; Methyl cyclohexylfluorophosphonate	
CAS Registry Number: 329-99-7	
RTECS Number: TA 8225000	
Physical and Chemical Properties	
Structural Formula: 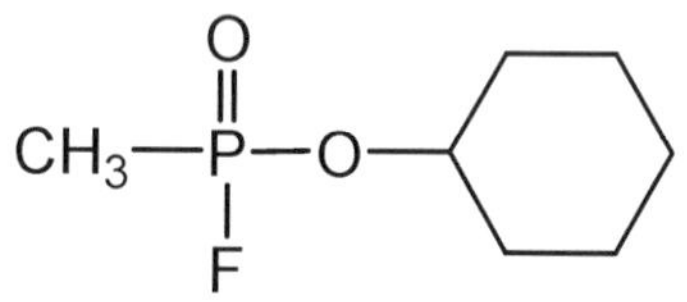 Molecular Formula: $C_7H_{14}FO_2P$ Molecular Weight: 180.16	
Physical State	Colorless liquid [1]
Odor	None if pure [2]
Boiling Point	228°C (extrapolated) [3]
FP/MP	–30 to –50°C (FP); [4] -12°C (MP); below –30°C, a metastable crystalline form of GF is produced which slowly converts into a stable form that melts @ –12°C [5]
Liquid Density (g/mL)	1.1276 @ 25°C; 1.1525 @ 0°C (extrapolated) [4]
Vapor Density (relative to air)	6.2 (calculated)
Vapor Pressure (torr)	9.27×10^{-2} @ 25°C; 9.78×10^{-3} @ 0°C (extrapolated) [3]
Volatility (mg/m^3)	8.98×10^2 @ 25°C; 1.03×10^2 @ 0°C (calculated from vapor pressure) [3]
Latent Heat of Vaporization (kcal/mol)	14.3 @ 25°C; 14.8 @ 0°C (calculated from vapor pressure) [3]
Flash Point	94°C [6]
Viscosity (cP)	5.41 @ 25.0°C, 14.762 @ 0°C (extrapolated)[6]
Viscosity of Vapor (cP)	6.15×10^{-3} @ 25.0°C, 5.5×10^{-3} @ 0°C[6]
Surface Tension (dynes/cm)	32.3 @ 25.5°C[6]
Decomposition Temperature	Completely decomposes within 2 hrs @ 150°C [7]
Solubility	Solubility in water is 3.7 g GF/100g @ 20°C; 5.1 g GF/100 g @ 0°C [6]
Rate of Hydrolysis	$t_{1/2}$ = 42 hrs @ 25°C using a 0.003 M solution of GF in distilled water [8]
Hydrolysis Products	Hydrogen fluoride and cyclohexyl methylphosphonic acid [5]
Stability in Storage	Stabilized GF can be stored @ 71°C for at least 6 months in glass containers and at least 1 year in steel and aluminum containers. [9]
Action on Metals or Other Materials	Corrosion rate on steel is 0.000053 inch/month @ 65°C [10]
Other Data	
Skin and eye toxicity	Eyes: very high toxicity; much greater through eyes than through skin. Skin: highly toxic; decontamination of smallest drop of liquid agent is essential; liquid penetrates skin [11].
Inhalation toxicity	Most toxic route of exposure [12]
Rate of action	Rapid [13]
Means of detection	M8 paper, M9 paper, M256A1 CADK, M8A1 ACAA, M90 AMAD, M21 ACAA, M22 ACADA, CAM/ICAM, M272 Water Testing Kit, , CAPDS, IPDS, AN/KAS-1 CWDD M18A3 CADK, [14] M18A2 CADK, MM1
Protection required	MOPP4; liquid nerve agents penetrate ordinary clothing rapidly [11].
Decontamination	Flush eyes with water immediately. Use the M291 SDK to remove any liquid nerve agent on skin or clothing. Use the M295 IEDK for individual equipment.[11] STB is effective on equipment. Water, steam, and absorbents (earth, sawdust, ashes, and rags) are effective for physical removal. [15]
Use	Quick-acting casualty agent

Table II-12. GF (Continued)

NOTES
[1]Eakle, B.F., *Chemical Agent GF (U)*, Technical Study 69-C4, USA Desert Test Center, Fort Douglas, Utah, January 1969, UNCLASSIFIED Report (AD509689). [2]*Chemical Agent Data Sheets Vol. II,* Edgewood Arsenal Special Report EO-SR-74002, USA Armament Command, Edgewood Arsenal, Aberdeen Proving Ground, MD, December 1974, CONFIDENTIAL Report (AD000020). [3]Tevault, D.E., et al., *Vapor Pressure of GF*, TR-304S, USA ECBC, Aberdeen Proving Ground, MD, submitted for publication 2 May 2003, UNCLASSIFIED Report. [4]Zeffert, B.M., and Coulter, P.B., *Physical Constants of G-Series Compounds: EA1210, EA1211, EA1212, EA1213, EA1214,* Technical Division Memorandum Report 1292), July 1947, USA Chemical Research and Development Laboratories, Army Chemical Center, MD, July 1947, UNCLASSIFIED Report (ADB964904). [5]Chinn, Kenneth, S. K., *Joint CB Technical Data Source Book, Volume III, G Nerve Agents, Part Three: Agents GD and GF (U)*, DPG-TR-82-004, USA Dugway Proving Ground, Utah, August 1983, SECRET Report (ADC032927). [6]Samuel, J.B., et al., *Physical Properties of Standard Agents, Candidate Agents, and Related Compounds at Several Temperatures (U)*, ARCSL-SP-83015, USA Armament Research and Development Command, Aberdeen Proving Ground, MD, June 1983, UNCLASSIFIED Report (ADC033491). [7]Perry, B.J., et al., *The Chemistry of the Alkylfluorophosphonites and Related Compounds*, Porton Technical Paper No. 258, Chemical Defense Experimental Establishment, Porton, England, 31 August 1951, UNCLASSIFIED Report (ADE481528). [8]Buckles, L.C., *The Hydrolysis Rate of G Agents*, TCIR 393, USA Chemical Research and Development Laboratories, Army Chemical Center, MD, December 1947, UNCLASSIFIED Report (ADB966236). [9]Grula, R.S., et al., *Storage Stability of GD, GF and EA 1356 (U),* CRDLR 3342, USA Chemical Research and Development Laboratories, Edgewood Arsenal, MD, December 1965, CONFIDENTIAL Report (AD369299). [10]Kaiser, W.A., *Summary of Information on Agent GF*, CRLR 164, USA Chemical Research and Development Laboratories, Army Chemical Center, MD, March 1954, UNCLASSIFIED Report (ADB969120). [11]FM 8-285/NAVMED P-5041/AFJMAN 44-149/FMFM 11-11, *Treatment of Chemical Agent Casualties and Conventional Military Chemical Injuries*, 22 December 1995. [12]Sharon Reutter, et al., *Review and Recommendations for Human Toxicity Estimates for FM 3-11.9*, ECBC-TR-349, September 2003. [13]*NIOSH-DOD-OSHA Sponsored Chemical and Biological Respiratory Protection Workshop Report*, February 2000. [14]DOD Chemical And Biological Defense Program Annual Report to Congress, Volume I, April 2003. [15]FM 3-5/MCWP 3-37.3, *NBC Decontamination,* 28 July 2000.

k. GF Toxicity Estimates (Table II-13, page II-26). Note that for an inhalation/ocular exposure the TLE is greater than 1. This means that the effective dosage increases with longer exposure durations and the concentration of the agent decreases.

Table II-13. GF Toxicity Estimates[10]

Endpoint	Toxicity (mg-min/m³)	MV (L)	Exposure Duration	ROE	Probit Slope	TLE	ROD	DOC
Lethality	LD_{50}: 350 mg [a]	N/A	N/A; 70-kg man	Percutaneous Liquid [b]	5	N/A	Unknown	Low
	LCt_{50}: 35 [a]	15	2 min	Inhalation/ Ocular	12	1.25 [c]	Some	Moderate
	LCt_{50}: 3000 [a,d]	N/A	30-360 min	Percutaneous Vapor [e]	5	1 [c,f]	Unknown	Low
	LCt_{50}: 1500 [g,h] (provisional)	N/A	30-360 min	Percutaneous Vapor [e]	5	1 [c,f]	Unknown	Low
Severe effects, includes some deaths	ED_{50}: 200 mg [a]	N/A	N/A; 70-kg man	Percutaneous Liquid [b]	5	N/A	Unknown	Low
	ECt_{50} 25 [a]	15	2 min	Inhalation/ Ocular	12	1.25 [c]	Some	Moderate
	ECt_{50}: 2000 [a,d]	N/A	30-360 min	Percutaneous Vapor [e]	5	1 [c,f]	Unknown	Low
	ECt_{50} 1000 [g,h] (provisional)	N/A	30-360 min	Percutaneous Vapor [e]	5	1 [c,f]	Unknown	Low
Threshold effects	ECt_{50}: 300 [a,d]	N/A	30-360 min	Percutaneous Vapor [e]	6	1 [c,f]	Unknown	Low
	ECt_{50}: 150 [g,h] (provisional)	N/A	30-360 min	Percutaneous Vapor [e]	6	1 [c,f]	Unknown	Low
Mild effects (miosis, rhinorrhea)	ECt_{50}: 0.2 [i]	N/A	2 min	Inhalation/ Ocular	10	1.4 c	Some	Low

NOTES

[a]Based on Grotte and Yang (2001).
[b]Bare skin.
[c]See Appendix H for supporting toxicity profile estimates.
[d]Moderate temperatures (65-85°F).
[e]Assumes personnel are masked with eye protection and bare skin.
[f]The TLE value is assumed to be 1 because the Ct profile is unknown.
[g]Hot temperatures (greater than 85°F).
[h]Based on recommendations for GB.
[i]Based on recommendations for GD.

1. O-ethyl methyl phosphonothiolate (VX) (see Table II-14). VX is a colorless and odorless liquid when pure.[13] Although VX is significantly less volatile than the other agents, it does vaporize to some extent and is extremely potent. A significant component of the hazard or airborne VX is percutaneous absorption of the vapor.[10] See Table II-15 (page II-29) for toxicity estimates.

Table II-14. VX

Alternate Designations: EA 1701; TX60

Chemical Name: O-Ethyl-S-(2-diisopropylaminoethyl) methyl phosphonothiolate

Synonyms: S-(2-Diisopropylaminoethyl) O-ethyl methyl phosphonothiolate; Ethyl-S-dimethylaminoethyl methylphosphonothiolate; Phosphonothioic acid, methyl-, S-(2-(diisopropylamino)ethyl) O-ethyl ester; Ethyl S-2-diisopropylaminoethyl methylphosphonothiolate; Ethyl-S-diisopropylaminoethyl methylthiophosphonate; Methylphosphonothioic acid S-(2- (bis(methylethyl)amino)ethyl) O-ethyl ester; O-Ethyl-S-2-diisopropylaminoethylester kyseliny methylthiofosfonove (Czech)

CAS Registry Number: 50782-69-9

RTECS Number: TB1090000

Physical and Chemical Properties

Structural Formula:

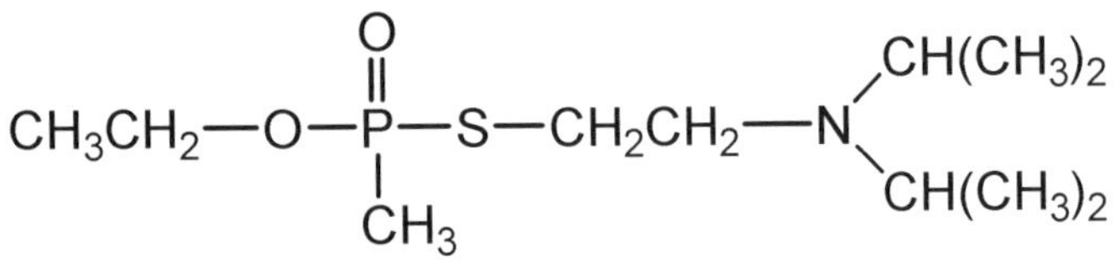

Molecular Formula: $C_{11}H_{26}NO_2PS$

Molecular Weight: 267.37

Physical State	Colorless liquid when pure [1]
Odor	Odorless when pure [1]
Boiling Point	292°C (extrapolated) [2]
FP/MP	Below –51°C and –39 to –60°C (FP) [3-5]
Liquid Density (g/mL)	1.0083 @ 25°C; 1.0209 @ 0°C (extrapolated) [4]
Vapor Density (relative to air)	9.2 (calculated)
Vapor Pressure (torr)	8.78×10^{-4} @ 25°C; 4.22×10^{-5} @ 0°C (extrapolated) [2]
Volatility (mg/m^3)	1.26×10^{1} @ 25°C; 6.62×10^{-1} @ 0°C (calculated from vapor pressure) [2]
Latent Heat of Vaporization (kcal/mol)	19.2 @ 25°C; 20.1 @ 0°C; (calculated from vapor pressure) [2,6]
Viscosity (cP)	10.041 @ 25.0°C, 37.532 @ 0°C (extrapolated)[3]
Viscosity of Vapor (cP)	5.13×10^{-3} @ 25.0°C, 4.63×10^{-3} @ 0°C[3]
Surface Tension (dynes/cm)	31.3 @ 25.0°C, 37.7 @ 0°C[3]
Flash Point	127°C (continuously closed cup method) [7]
Decomposition Temperature	$t_{1/2}$ = 502 days @ 71°C; $t_{1/2}$ = 41 days @ 100°C; $t_{1/2}$ = 34.5 hrs @ 150°C; $t_{1/2}$ = 10 hrs @ 170°C; [8] $t_{1/2}$ = 1.6 hrs @ 200°C; $t_{1/2}$ = 4 min @ 250°C; $t_{1/2}$ = 36 sec @ 295°C [5]
Solubility	Water solubility of VX is 5% @ 21.5°C; [4] miscible with water below 9.4°C; [4] soluble in common organic solvents [5]
Rate of Hydrolysis	Hydrolysis rate of VX varies with temperature and concentration. At 22°C, $t_{1/2}$ = 1.8 min [1.25M NaOH]; $t_{1/2}$ = 10.8 min [0.25M NaOH]; $t_{1/2}$ = 31 min. [0.10M NaOH]; $t_{1/2}$ = 3.3 hrs [0.01M NaOH]; $t_{1/2}$ = 20.8 hrs [0.001M NaOH]; and $t_{1/2}$ = 60 hrs [pure H_2O] [9]
Hydrolysis Products	VX hydrolyzes via three different pathways (P-S, P-O, and C-S), which vary significantly with temperature and pH. At pH below 12, the P-O bond cleavage path produces ethyl methylphosphonate (EMPA) and the toxic S-[2-diisopropylaminoethyl] methylphosphonothiolate ion (EA 2192). At room temperature EA 2192 reacts very slowly with OH^- [EA 2192, $t_{1/2}$ = 7.4 days (1.0M NaOH)] eventually producing less toxic products. [9,10] Using an equimolar ratio of VX and water at elevated temperatures appears to reduce the persistency of EA 2192. [11]
Stability in Storage	Relatively stable @ ambient temperature; unstabilized VX of 95% purity decomposes at a rate of 5% a month @ 71°C. [13] Highly purified VX is stable in both glass and steel. [1]
Action on Metals or Other Materials	Negligible on brass, steel, and aluminum; slight corrosion with copper [12]

Table II-14. VX (Continued)

Other Data	
Skin and eye toxicity	Extremely toxic by skin and eye absorption [14]
Inhalation toxicity	Extremely potent [15]
ROA	Rapid [16]
Means of detection	M8 paper, M9 paper, M256A1 CADK, M8A1 ACAA, M90 AMAD, M22 ACADA, CAM/ICAM, M272 Water Testing Kit, CAPDS, IPDS, AN/KAS-1 CWDD, M18A3 CADK,[17] M18A2 CADK, MM1
Protection required	MOPP4; liquid nerve agents penetrate ordinary clothing rapidly [14]
Decontamination	Flush eyes with water immediately. Use the M291 SDK to remove any liquid nerve agent on skin or clothing. Use the M295 IEDK for individual equipment.[14] STB, HTH, or household bleach are effective on equipment. Water, soaps, detergents, steam, and absorbents (earth, sawdust, ashes, and rags) are effective for physical removal. [18]
Use	Quick-acting casualty agent

NOTES

[1]Witten, B., *The Search for Toxic Chemical Agents (U)*, EATR 4210, Edgewood Arsenal Research Laboratories, MD, November 1969, UNCLASSIFIED Report (AD507852).
[2]Buchanan, J.H., et al., *Vapor Pressure of VX*, ECBC-TR-068, USA Soldier and Biological Chemical Command, Aberdeen Proving Ground, MD, November 1999, UNCLASSIFIED Report (ADA371297).
[3]Samuel, J.B., et al., *Physical Properties of Standard Agents, Candidate Agents, and Related Compounds at Several Temperatures (U)*, ARCSL-SP-83015, June 1983, USA Armament Research and Development Command, Aberdeen Proving Ground, MD, UNCLASSIFIED Report (ADC033491).
[4]Coulter, P.B., et al., *Physical Constants of Thirteen V Agents*, CWLR 2346, USA Chemical Warfare Laboratories, Army Chemical Center, MD, December 1959, UNCLASSIFIED Report (AD314520).
[5]*Chemical Agent Data Sheets Volume I*, Edgewood Arsenal Special Report EO-SR-74001, USA Armament Command, Edgewood Arsenal, Aberdeen Proving Ground, MD, December 1974, UNCLASSIFIED Report (ADB028222).
[6]Abercrombie, P., ECBC Notebook # NB 98-0079, p. 11 (U).
[7]Butrow, B., ECBC Notebook # NB 97-0109 (C).
[8]Rohrbaugh, D.K., et al., *Studies in Support of SUPLECAM (Surveillance Program for Lethal Chemical Agents and Munitions) II, 1. Thermal Decomposition of VX*, CRDEC-TR-88056, USA Chemical Research, Development and Engineering Center, Aberdeen Proving Ground, MD, May 1988, UNCLASSIFIED Report (ADB124301).
[9]Yang, Y, et al., "Hydrolysis of VX: Activation Energies and Autocatalysis," *In Proceedings of the 1994 ERDEC Scientific Conference on Chemical Biological Defense Research 15-18 November1994*, UNCLASSIFIED Paper (ADE479900), ERDEC-SP-036, pp. 375-382, USA Edgewood Research, Development and Engineering Center, Aberdeen Proving Ground, MD, May 1996, UNCLASSFIED Report (ADA313080).
[10]Yang, Y., et al., "Perhydrolysis of Nerve Agent VX," *J. Org. Chem,* Vol. 58, p. 6965, 1993.
[11]Yang, Y., et al., "Hydrolysis of VX with Equimolar Water at Elevated Temperatures: Activation Parameters of VX, CV and EA 2192," *In Proceedings of the 1996 ERDEC Scientific Conference on Chemical Biological Defense Research 19-22 November1996*, UNCLASSIFIED Paper (ADE487572), ERDEC-SP-048, pp. 599-605, USA Edgewood Research, Development and Engineering Center, Aberdeen Proving Ground, MD, October 1997, UNCLASSFIED Report (ADA334105).
[12]Eckhaus, S.R., et al., *Resistance of Various Materials of Construction in Contact with Transester Process*, CWL Technical Memorandum 31-73, USA Chemical Research and Development Laboratories, Army Chemical Center, MD, February 1959, UNCLASSIFIED Report (ADB963125).
[13]Salamon, M.K., *Agent VX*, CWL Special Publication 4-10, USA Chemical Warfare Laboratories, Army Chemical Center, MD, June 1959, UNCLASSIFIED Report (ADE471109).
[14]FM 8-285/NAVMED P-5041/AFJMAN 44-149/FMFM 11-11, *Treatment of Chemical Agent Casualties and Conventional Military Chemical Injuries*, 22 December 1995.
[15]Sharon Reutter, et al., *Review and Recommendations for Human Toxicity Estimates for FM 3-11.9*, ECBC-TR-349, September 2003.
[16]*NIOSH-DOD-OSHA Sponsored Chemical and Biological Respiratory Protection Workshop Report,* February 2000.
[17]DOD Chemical And Biological Defense Program Annual Report to Congress, Volume I, April 2003.
[18]FM 3-5/MCWP 3-37.3, *NBC Decontamination,* 28 July 2000.

m. VX Toxicity Estimates (see Table II-15). VX is unlike the G agents in that the Ct profile for vapor inhalation appears to obey *Haber's Law*. However, some data indicates that the TLE may be less than one. If it does obey the toxic load principle ($C^n t = k$) where the TLE (n) is less than one, it means that the ECt_{50} does not increase with longer exposures at lower concentrations; in fact, it may actually decrease.[9]

Table II-15. VX Toxicity Estimates[10]

Endpoint	Toxicity (mg-min/m³)	MV (L)	Exposure Duration	ROE	Probit Slope	TLE	ROD	DOC
Lethality	LD_{50}: 5 mg [a]	N/A	N/A; 70-kg man	Percutaneous Liquid [b]	6	N/A	Unknown	Low
	LCt_{50}: 15 [a]	15	2-360 min	Inhalation/ Ocular	6	1 [c]	Little, if any	Low [d]
	LCt_{50}: 150 [a,e]	N/A	30-360 min	Percutaneous Vapor [f]	6	1 [c,g]	Little, if any	Low
	LCt_{50}: 75 [h, i] (Provisional)	N/A	30-360 min	Percutaneous Vapor [f]	5	1 [c,g]	Little, if any	Low
Severe effects, includes some deaths	ED_{50}: 2 mg [a]	N/A	N/A; 70-kg man	Percutaneous Liquid [b]	6	N/A	Unknown	Low
	ECt_{50}: 10 [a]	15	2-360 min	Inhalation/ Ocular	6	1 [c]	Little, if any	Low [d]
	ECt_{50}: 25 [a,e]	N/A	30-360 min	Percutaneous Vapor [f]	6	1 [c,g]	Unknown	Moderate
	ECt_{50}: 12 [h,I] (Provisional)	N/A	30-360 min	Percutaneous Vapor [f]	6	1 [c,g]	Unknown	Low
Threshold effects (Slight ChE inhibition)	ECt_{50}: 10 [a,e]	N/A	30-360 min	Percutaneous Vapor [f]	6	1 [c,g]	Unknown	Moderate
	ECt_{50}: 5 [h,i]	N/A	30-360 min	Percutaneous Vapor [f]	6	1 [c,g]	Unknown	Low
Mild effects (miosis, rhinorrhea)	ECt_{50}: 0.1 [a]	N/A	2-360 min	Inhalation/ Ocular	4	1 [c,g,j]	Some	Low

NOTES

[a]Based on Grotte and Yang (2001).
[b]Bare skin.
[c]See Appendix H for supporting toxicity profile estimates.
[d]LCt_{50} /ECt_{50} could be less.
[e]Moderate temperatures (65-85°F).
[f]Assumes personnel are masked with eye protection and bare skin.
[g]The TLE value is assumed to be 1 because the Ct profile is unknown.
[h]Hot temperatures (greater than 85°F).
[i]Based on recommendations for GB and Cummings and Craig (1965).
[j]Estimates should be revised as new data becomes available; human estimate for miosis may go down.

n. V_x (see Table II-16) [page II-30]. Another V agent of interest is V_x, called "V sub x". Information on this agent is limited. Based upon percutaneous liquid exposure, it appears to be less potent than VX. However, it is decidedly more potent than the G agents. It is also noted that it is more volatile than VX, so its potential airborne hazard is greater than that for VX.[10] See Table II-17 (page II-31) for toxicity estimates.

Table II-16. Vx

Alternate Designations: EA 1699; EDMM; Medemo	
Chemical Name: O-ethyl S-(2-dimethylaminoethyl) methylphosphonothiolate	
Synonyms: Phosphonothioic acid, methyl-, S-[2-(dimethylamino)ethyl] O-ethyl ester; O-Aethyl-S-(2-dimethylaminoaethyl)-methylphosphonothioat (German); S-2- Dimethylaminoethyl-O-ethylester kyseliny methylthiofosfonove (Czech); O-Ethyl-S-(dimethylaminoethyl)-methylphosphonothioate	
CAS Registry No: 20820-80-8	
RTECS Number: 51366-09-7	
Physical and Chemical Properties	
Structural Formula: 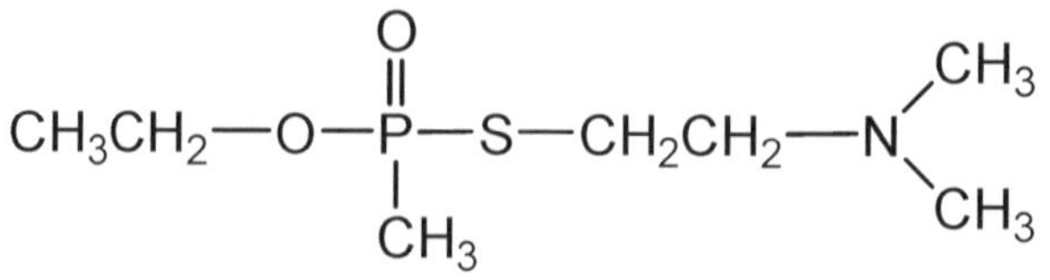 Molecular Formula: $C_7H_{18}NO_2PS$ Molecular Weight: 211.26	
Physical State	Liquid [1]
Odor	Odorless[2]
Boiling Point	256°C (extrapolated) [3, 4]
FP/MP	Data not available
Liquid Density (g/mL)	1.060 @ 25°C; 1.0820 @ 0°C (extrapolated) [1]
Vapor Density (relative to air)	7.3 (calculated)
Vapor Pressure (torr)	6.73×10^{-3} @ 25°C; 5.7×10^{-4} @ 0°C (extrapolated) [3,4]
Volatility (mg/m^3)	7.64×10^{1} @ 25°C; 7.02 @ 0°C (calculated from vapor pressure) [3,4]
Latent Heat of Vaporization (kcal/mol)	16.0 @ 25°C; 16.1 @ 0°C (calculated from vapor pressure) [3,4]
Viscosity (cP)	5.628 @ 25.0°C, 15.335 @ 0°C (extrapolated)[3]
Viscosity of Vapor (cP)	5.56×10^{-3} @ 25.0°C, 5.02×10^{-3} @ 0°C[3]
Surface Tension (dynes/cm)	31.4 @ 25.0°C, 33.7 @ 0°C (extrapolated)[3]
Flash Point	Data not available
Decomposition Temperature	Data not available
Solubility	Soluble in organic solvents; slightly soluble in water (source unidentified)
Rate of Hydrolysis	The rate coefficient for 7.8×10^{-3} hr^{-1} (based on a nonlinear least square fit) [5]
Hydrolysis Products	Ethanol and the toxic product compound S-(2-dialkylamino-ethyl) methylphosphonothioic acid that is very stable in neutral water[3]
Stability in Storage	Data not available
Action on Metals or Other Materials	Data not available
Other Data	
Skin and eye toxicity	Extremely toxic by skin and eye absorption [6]
Inhalation Toxicity	Extremely potent [7]
Rate of action	Rapid [8]
Means of detection	M8 paper, M9 paper, M8A1, IPDS, CAM/ICAM, M18A2 CADK, MM1[7]
Protection required	MOPP4; liquid nerve agents penetrate ordinary clothing rapidly [6]
Decontamination	Flush eyes with water immediately. Use the M291 SDK to remove any liquid nerve agent on skin or clothing. Use the M295 IEDK for individual equipment.[5] STB, HTH, or household bleach is effective on equipment. Water, soaps, detergents, steam, and absorbents (earth, sawdust, ashes, and rags) are effective for physical removal. [10]
Use	Quick-acting casualty agent

Table II-16. Vx (Continued)

NOTES
[1]Coulter, P.B., et al., *Physical Constants of Thirteen V Agents (U)*, CWLR 2346, USA Chemical Warfare Laboratories, Army Chemical Center, MD, December 1959, UNCLASSIFIED Report (AD314520). [2]TM 3-215/AFM, *Military Chemistry and Chemical Agents,* Washington DC, December 1963, UNCLASSIFIED Technical Manual (ADA 292141). [3]Samuel, J.B., et al., *Physical Properties of Standard Agents, Candidate Agents, and Related Compounds at Several Temperatures (U)*, ARCSL-SP-83015, June 1983, USA Armament Research and Development Command, Aberdeen Proving Ground, MD, UNCLASSIFIED Report (ADC033491). [4]Newman, J.H., Edgewood Arsenal Notebook # NB 9298, p. 64 (U). [5]Szafraniec, L.J., et al., *On the Stoichiometry of Phosphonothiolate Ester Hydrolysis*, CRDEC-TR-212, USA Chemical Research Development & Engineering Center, Aberdeen Proving Ground, MD, July 1990, UNCLASSIFIED Report (ADA225952). [6]FM 8-285/NAVMED P-5041/AFJMAN 44-149/FMFM 11-11, *Treatment of Chemical Agent Casualties and Conventional Military Chemical Injuries*, 22 December 1995. [7]Sharon Reutter, et al., Review and Recommendations for Human Toxicity Estimates for FM 3-11.9, ECBC-TR-349, September 2003. [8]*NIOSH-DOD-OSHA Sponsored Chemical and Biological Respiratory Protection Workshop Report*, February 2000. [9]DOD Chemical And Biological Defense Program Annual Report to Congress, Volume I, April 2003. [10]FM 3-5/MCWP 3-37.3, *NBC Decontamination,* 28 July 2000.

Table II-17. Vx Toxicity Estimates[10]

Endpoint	Toxicity ($mg\text{-}min/m^3$)	MV (L)	Exposure Duration	ROE	Probit Slope	TLE	ROD	DOC
N/A	No toxicity estimates are recommended at this time because data are lacking.	N/A	N/A	N/A	N/A	N/A	Unknow n	N/A

6. Blood Agents

Blood agents include AC, CK, and SA. The cyanogen blood agents AC and CK affect the bodily functions by inactivating the cytochrome oxidase system.[18] This poisoning prevents cell respiration and the normal transfer of oxygen from the blood to body tissues.[18] SA causes hemolysis of the red blood cells.[19] Cyanogen agents are highly volatile and, therefore, nonpersistent.[20] Exposure at high concentrations causes effects within seconds and death within minutes in unprotected personnel.[18] The protective mask with fresh filters gives adequate protection against field concentrations.[1] After exposure to AC and CK, filters should be changed.[12] See FM 3-11.4 for filter change criteria.

a. AC (see Table II-18 [page II-32]). Pure AC is a nonpersistent, colorless liquid that is highly volatile. It has a faint odor, similar to bitter almonds, that sometimes cannot be detected even in lethal concentrations.[12] Inhalation of small amounts causes giddiness, headache and faintness, confusion, palpitation and pain in the chest and region of the heart, difficulty breathing, and ultimately unconsciousness.[10] Inhalation of high concentrations can initially cause breathing that is deeper and more rapid than is normal at rest, followed closely by a loss of consciousness. This progresses to respiratory arrest, cessation of cardiac activity, and death.[18] Exposure to AC causes an increase in respiration within a few seconds; a casualty may not be able to hold his breath. The pink color of the casualty's skin suggests AC poisoning.[1] See Table II-19 (page II-33) for toxicity estimates.

Table II-18. AC

Alternate Designations: Cyclone (Russian); Cyclone B; Cyclon; Prussic acid, Forestite (French); Aero Liquid HCN	
Chemical Name: Hydrogen cyanide	
Synonyms: Hydrocyanic acid; Acide cyanhydrique (French); Acido cianidrico (Italian); Blausaeure (German); Blauwzuur (Dutch); Carbon hydride nitride (chn); Cyaanwaterstof (Dutch); Cyanwasserstoff (German); Cyjanowodor (Polish); Evercyn; Formic anammonide; Formonitrile	
CAS Registry Number: 74-90-8	
RTECS Number: MW6825000	
Physical and Chemical Properties	
Structural Formula: $H-C\equiv N$ Molecular Formula: HCN Molecular Weight: 27.03	
Physical State	Colorless liquid [1]
Odor	Bitter almonds or peach kernels [2,3]
Boiling Point	25.5°C [4,1]
FP/BP	-13.3°C (MP) [1]
Liquid Density (g/mL)	0.6797 @ 25°C; 0.7162 @ 0°C [5]
Vapor Density (relative to air)	0.93 (calculated)
Vapor Pressure (torr)	7.60×10^2 @ 25.5°C; 7.46×10^2 @ 25.0°C; 2.65×10^2 @ 0°C [4,1]
Volatility (mg/m^3)	1.10×10^6 @ 25.5°C; 1.08×10^6 @ 25.0°C; 4.20×10^5 @ 0°C (calculated from vapor pressure) [4,1]
Latent Heat of Vaporization (kcal/mol)	6.72 @ 25.5°C; 6.72 @ 25.0°C; 6.71 @ 0°C (calculated from vapor pressure)[4,1]
Viscosity (cP)	Data not available
Viscosity of Vapor (cP)	Data not available
Surface Tension (dynes/cm)	Data not available
Flash Point	-18°C (closed cup); [3] frequently ignites when explosively disseminated[6]
Decomposition Temperature	Above 65.5°C when stabilized; [6] forms explosive polymer on standing; [2,3] stabilized material can be stored up to 65°C [6]
Solubility	Miscible with water and common organic solvents including alcohol and ether [7]
Rate of Hydrolysis	Slow under acidic conditions; rapid with traces of base or basic salts [8]
Hydrolysis Products	Ammonia, formic acid (HCOOH), and amorphous brown solids [9]
Stability in Storage	Pure AC is unstable in storage; forms explosive polymer on long standing; [2,6] with the use of a stabilizer such as phosphoric acid, sulfur dioxide, or powdered copper, AC may be stored in metal containers for long periods of time @ temperatures up to 65°C [2,3,6]
Action on Metals or Other Materials	Corrodes iron, cast iron, chromium steel, and lead [2]
Other Data	
Skin and eye toxicity	None
Inhalation toxicity	Can cause death within minutes. [10]
Rate of action	Rapid [11]
Means of detection	M256A1 CADK, M272 water testing kit, M18A2 CADK,[12] M18A3, MMI
Protection required	Protective mask with fresh filter; MOPP 4 when exposed to or handling liquid AC [13]
Decontamination	Move to fresh air; none required under field conditions [14]
Use	Quick-acting casualty agent

Table II-18. AC (Continued)

NOTES
[1]Giauque, W.F. and Ruehrwein, R.A., "The Entropy of Hydrogen Cyanide. Heat Capacity, Heat of Vaporization and Vapor Pressure. Hydrogen Bond Polymerization of the Gas in Chains of Indefinite Length." *J. Am. Chem. Soc.*, Vol. 61, p. 2626, 1939. [2]Franke, S., *Manual of Military Chemistry Volume I- Chemistry of Chemical Warfare Agents*, ACSI-J-3890, Chemie der Kampfstoffe, East Berlin, April 1968, UNCLASSIFIED Report (AD849866). [3]Lewis, R.J., *Sax's Dangerous Properties of Industrial Materials*, 10th ed., Vol. 3, p. 1992, John Wiley & Sons, Inc., New York, NY, 2001. [4]Abercrombie, P., ECBC Notebook # NB 98-0079, p. 16 (U). [5]Coates, J.E., and Davies, R.H., "Studies on Hydrogen Cyanide. Part XVIII. Some Physical Properties of Anhydrous Hydrogen Cyanide," *J. Chem. Soc.*, p. 1194, 1950. [6]W.R. Kirner, *Summary Technical Report of Division 9, NDRC Volume 1, Chemical Warfare Agents, and Related Chemical Problems Part I-II,* Chapter 2, p. 7, Office of Scientific Research and Development, Washington, DC, 1946, UNCLASSIFIED Report (AD234270). [7]*The Merck Index: An Encyclopedia of Chemicals, Drugs, and Biologicals*, 13th ed., p. 857, Merck & Company, Inc., Whitehouse Station, NJ, 2001. [8]*Properties of War Gases Vol. II: Blood and Nettle Gases (U)*, ETF 100-41/Vol-2, Chemical Corps Board, Army Chemical Center, Maryland, December 1956, CONFIDENTIAL Report (AD108457). [9]Clark, D.N, *Review of Reactions of Chemical Agents in Water, Final Report to USA Biomedical Research and Development Laboratory*, Battelle, Columbus, OH, January 1989, UNCLASSIFIED Report (ADA213287). [10]BG Russ Zajtchuk et al. (eds), *Textbook of Military Medicine: Medical Aspects of Chemical and Biological Warfare,* Office of the Surgeon General, 1997, Chap. 10, "Cyanide Poisoning." [11]*NIOSH-DOD-OSHA Sponsored Chemical and Biological Respiratory Protection Workshop Report,* February 2000. [12]AFMAN 10-2602, *Nuclear, Biological, Chemical, and Conventional (NBCC) Defense Operations and Standards (Operations)*, 29 May 2003. [13]FM 8-285/NAVMED P-5041/AFJMAN 44-149/FMFM 11-11, *Treatment of Chemical Agent Casualties and Conventional Military Chemical Injuries*, 22 December 1995. [14]FM 8-9/NAVMED P-5059/AFJMAN 44-151, *NATO Handbook on the Medical Aspects of NBC Defense Operations AMEDP-6(B),* 1 February 1996.

b. AC Toxicity Estimates (Table II-19). Note that for an inhalation/ocular expouse the TLE is greater than 1. This means that the effective dosage increases with longer exposure durations and the concentration of the agent decreases. No toxicity effects for severe effects is recommended. The existing estimate is not supported by the available data.

Table II-19. AC Toxicity Estimates[10]

Endpoint	Toxicity (mg-min/m^3)	MV (L)	Exposure Duration	ROE	Probit Slope	TLE	ROD	DOC
Lethality	LCt_{50}: 2860 [a] (Provisional)	15	2 min	Inhalation/ Ocular	10	1.85 [b]	Some	Low
Severe effects	ECt_{50}: NR	N/A	N/A	Inhalation/ Ocular	Unknown	Unknown	Some	N/A
Threshold/odor (odor detection)	EC_{50}: 34 mg/m^3 [c]	N/A	Seconds	Inhalation/ Ocular	Unknown	N/A	N/A	Low

NOTES

[a]Based on McNamara (1976).
[b]See Appendix H for supporting toxicity profile estimates.
[c]Based on TM 3-215 (1952) and secondary human data.

c. CK (see Table II-20 [page II-34]). CK is a colorless gas with an irritating odor.[13] It is nonpersistent and is used as a quick-acting casualty agent.[20] It is readily detectable by its immediate lacrimatory effect and its irritant effect on the nasal passage.[20] At high concentrations CK produces effects similar to AC.[10] However, in occasional instances, lung irritation can lead to pulmonary edema.[20] See Table II-21 (page II-35) for CK toxicity estimates.

Table II-20. CK

Alternate Designations: Mauguinite (French); CC; Klortsian	
Chemical Name: Cyanogen chloride	
Synonyms: Chlorcyan; Chlorine cyanide; Chlorocyan; Chlorocyanide; Chlorocyanogen; Chlorure de cyanogene (French)	
CAS Registry Number: 506-77-4	
RTECS Number: GT2275000	
Physical and Chemical Properties	
Structural Formula: Molecular Formula: CNCl Molecular Weight: 61.47	Cl—C≡N
Physical State	Colorless gas [1]
Odor	Lacrimatory and irritating [2]
Boiling Point	12.8°C (calculated) [3-5]
FP/MP	-6.9°C (FP)[5]
Liquid Density (g/mL)	1.202 @ 10°C; 1.222 @ 0°C [4]
Vapor Density (relative to air)	2.1 (calculated)
Vapor Pressure (torr)	7.60×10^2 @ 12.8°C; 6.80×10^2 @ 10°C; 4.48×10^2 @ 0°C [3-5]
Volatility (mg/m^3)	2.62×10^6 @ 12.8°C; 2.37×10^6 @ 10°C; 1.62×10^6 @ 0°C (calculated from vapor pressure) [3-5]
Latent Heat of Vaporization (kcal/mol)	6.40 @ 12.8°C; 6.41 @ 10°C; 6.44 @ 0°C calculated from vapor pressure) [3-5]
Viscosity (cP)	Data not available
Viscosity of Vapor (cP)	Data not available
Surface Tension (dynes/cm)	Data not available
Flash Point	Nonflammable [2]
Decomposition Temperature	Approximately 149°C [6]
Solubility	Solubility of liquefied CK in water is 71.4 g/L @ 20°C, [7] soluble in common organic solvents, sulfur mustard, and AC [1]
Rate of Hydrolysis	The hydrolysis rate of CK with tap water is $t_{1/2}$ = 180 hrs @ ambient temperature and pH 7 [8]
Hydrolysis Products	Hydrogen chloride and cyanic acid (CNOH) [9]
Stability in Storage	CK is stable in glass containers for long periods of time even @ elevated temperatures. Stable in steel containers for at least 1 year @ ambient temperature, but only about 9 weeks @ 60°C, after which time the gas begins to polymerize with formation of the corrosive solid, cyanuric chloride. Impurities have a tendency to promote explosive polymerization. [10,2] When stabilized using 5% anhydrous, powdered sodium pyrophosphate, munitions grade CK with a water content of less than 0.5% can be stored in most common metals for extended periods of time @ temperatures up to 100°C. [2]
Action on Metals or Other Materials	None if CK is dry; slowly polymerizes when stored unstabilized in steel and other common metals @ elevated temperatures (see stability in storage section) [2]
Other Data	
Skin and eye toxicity	Irritation to eyes similar to RCAs [11]
Inhalation toxicity	Can cause death within minutes [11]
Rate of action	Rapid [12]
Means of detection	M256A1 CADK, M272 water testing kit, M18A2 CADK, M18A3 CADK, MMI [13]
Protection required	Protective mask with fresh filters [14]
Decontamination	Move to fresh air; none required under field conditions [15]
Use	Quick-acting casualty agent

Table II-20. CK (Continued)

NOTES
[1]Franke, S., *Manual of Military Chemistry Volume I - Chemistry of Chemical Warfare Agents*, ACSI-J-3890, Chemie der Kampfstoffe, East Berlin, April 1968, UNCLASSIFIED Report (AD849866). [2]W.R. Kirner, *Summary Technical Report of Division 9, NDRC Volume 1, Chemical Warfare Agents, and Related Chemical Problems Part I-II,* Chapter 2, Office of Scientific Research and Development, Washington, DC, 1946, UNCLASSIFIED Report (AD234270). [3]Abercrombie, P., ECBC Notebook # NB 98-0079, p. 16 (U). [4]Cook, R.P., and Robinson, P.L., "Certain Physical Properties of Cyanogen and its Halides," *J. Chem. Soc.*, p. 1001, 1935. [5]Douglas, D.E., and Winkler, C.A., "The Preparation, Purification, Physical Properties and Hydrolysis of Cyanogen Chloride," *Ca. J. Research,* Vol. 25B, p. 381, 1947. [6]Brooks, Marguerite E, et al., *Incineration/Pyrolysis of Several Agents and Related Chemical Materials Contained in Identification Sets*, ARCSL-TR-79040, October 1979, UNCLASSIFIED Report (ADB042888). [7]Carter, R.H., and Knight, H.C., *Fundamental Study of Toxicity: Solubility of Certain Toxics in Water and in Olive Oil*, EACD 445, Chemical Warfare Service, Edgewood Arsenal, MD, May 1928, UNCLASSIFIED Report (ADB955216). [8]Price, C.C., et al., "Hydrolysis and Chlorinolysis of Cyanogen Chloride", *J. Amer. Chem. Soc.*, Vol. 69, p. 1640, 1947. [9]Edwards, J.O., and Sauer, M., *Chemical Reactivity of Cyanogen Chloride in Aqueous Solution, Quarterly Status Report (March through May 1972)*, Report No. III, DAAA15-71-C-0478-QSR 3, USA Chemical Laboratories, Edgewood Arsenal, MD, February 1973, UNCLASSIFIED Report (ADA090556). [10]Henley, F.M., *Surveillance Tests on 75 mm Steel Gas Shell Extending Over a Period of One Year*, EACD 11, Chemical Warfare Service, Edgewood Arsenal, MD, June 1920, UNCLASSIFIED Report (ADB959731). [11]BG Russ Zajtchuk et al. (eds), *Textbook of Military Medicine: Medical Aspects of Chemical and Biological Warfare*, Office of the Surgeon General, 1997, Chap. 10, "Cyanide Poisoning." [12]*NIOSH-DOD-OSHA Sponsored Chemical and Biological Respiratory Protection Workshop Report,* February 2000. [13]AFMAN 10-2602, *Nuclear, Biological, Chemical, and Conventional (NBCC) Defense Operations and Standards (Operations)*, 29 May 2003. [14]FM 8-285/NAVMED P-5041/AFJMAN 44-149/FMFM 11-11, *Treatment of Chemical Agent Casualties and Conventional Military Chemical Injuries*, 22 December 1995. [15]FM 8-9/NAVMED P-5059/AFJMAN 44-151, *NATO Handbook on the Medical Aspects of NBC Defense Operations AMEDP-6(B),* 1 February 1996.

d. CK Toxicity Estimates (Table II-21). Note that for an inhalation/ocular expouse the TLE is greater than 1. This means that the effective dosage increases with longer exposure durations and the concentration of the agent decreases. No toxicity estimates for lethal and severe effects are recommended. The existing estimates are not supported by the available data.

Table II-21. CK Toxicity Estimates[10]

Endpoint	Toxicity (mg-min/m^3)	MV (L)	Exposure Duration	ROE	Probit Slope	TLE	ROD	DOC
Lethality	LCt_{50}: NR	N/A	N/A	Inhalation/ Ocular	Unknown	More than 1	Probably Insignificant	N/A
Severe effects	ECt_{50}: NR	N/A	N/A	Inhalation/ Ocular	Unknown	More than 1	Probably Insignificant	N/A
Threshold (odor detection, tearing)	EC_{50}: 12 mg/m^3 [a]	N/A	Few Seconds	Inhalation/ Ocular	N/A	N/A	N/A	Low
NOTES								
[a]Based on human data and TM 3-215 (1952).								

e. SA (see Table II-22 [page II-36]). SA is a colorless gas with a disagreeable, garlic-like odor.[13] Symptoms from inhalation exposure include abdominal pain, confusion, dizziness, headache, nausea, shortness of breath, vomiting, and weakness. Severe exposure damages blood, causing anemia and kidney damage.[19] Exposure from liquid can cause frostbite.[19] See Table II-23 (page II-37) for toxicity estimates.

Table II-22. SA

Alternate Designations: Arthur	
Chemical Name: Arsenic trihydride	
Synonyms: Hydrogen arsenide, Arseniuretted hydrogen; Arsenic hydride; Arsenous hydride; Arsenowodor (Polish); Aresenwasserstoff (German)	
CAS Registry Number: 7784-42-1	
RTECS Number: CG6475000	
Physical and Chemical Properties	
Structural Formula: 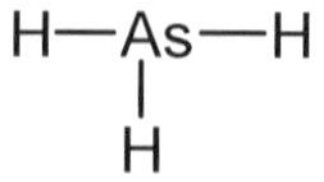 Molecular Formula: AsH_3 Molecular Weight: 77.95	
Physical State	Colorless gas [1]
Odor	Disagreeable, garlic-like [1]
Boiling Point	-62.2°C (extrapolated) [2]
FP/MP	-116°C (MP) [2]
Liquid Density (g/mL)	1.667 @ -75°C; 1.734 @ -100°C [2]
Vapor Density (relative to air)	2.7 (calculated)
Vapor Pressure (torr)	4.00×10^2 @ -75°C and 8.69×10^1 @ -100°C [2]
Volatility (mg/m^3)	2.55×10^6 @ -75°C and 6.27×10^5 @ -100°C (calculated from vapor pressure) [2]
Latent Heat of Vaporization (kcal/mol)	4.17 (calculated from Clausius Clapeyron equation which assumes constant heat of vaporization as a function of temperature) [2]
Viscosity (cP)	Data not available
Viscosity of Vapor (cP)	Data not available
Surface Tension (dynes/cm)	Data not available
Flash Point	Flammable; forms explosive mixtures with air [3]
Decomposition Temperature	300°C [1]
Solubility	Solubility of SA in water is 0.028 g/100 g @ 20°C,[4] soluble in alkalis, halogen alkanes, hydrocarbons, and benzene [3,4]
Rate of Hydrolysis	Rapid in the presence of light.[1] Slow, in the absence of light and air @ 15.5°C and pH ~ 7; 32% of SA is hydrolyzed within 5 hrs and about 66% within 24 hrs [5]
Hydrolysis Products	SA hydrolyzes to produce shiny black arsenic, which is also highly toxic [1]
Stability in Storage	Unstable in most metal containers; metals catalyze decomposition;[6] on exposure to light, moist SA decomposes quickly, depositing shiny black arsenic [1]
Action on Metals or Other Materials	Corrosive to most metals [6]
Other Data	
Skin and eye toxicity	Exposure to liquid causes frostbite [7]
Inhalation toxicity	Acute toxicity is high [7]
Rate of action	1-24 hours (dependent on concentration and exposure duration)[7]
Means of detection	MM1
Protection required	Protective mask with fresh filter; MOPP4 when exposed to or handling liquid SA [7]
Decontamination	Move to fresh air; none required under field conditions[8]
Use	Delayed-action casualty agent

Table II-22. SA (Continued)

NOTES
[1]*The Merck Index: An Encyclopedia of Chemicals, Drugs, and Biologicals*, 13th ed., p. 138, Merck & Company, Inc., Whitehouse Station, NJ, 2001. [2]Johnson, W. and Pechukas, A., "Hydrogen Compounds of Arsenic. I. Preparation of Arsine in Liquid Ammonia Some Physical Properties of Arsine," *J. Am. Chem. Soc.*, Vol. 59, p. 2065, 1937. [3]Franke, S., *Manual of Chemistry Volume I- Chemistry of Chemical Warfare Agents*, ACSI-J-3890, Chemie der Kampfstoffe, East Berlin, April 1968, UNCLASSIFIED Technical Manual (AD849866). [4]Lewis, R.J., *Sax's Dangerous Properties of Industrial Materials*, 10th ed., Vol. 2, p. 309, John Wiley & Sons, Inc., New York, NY, 2001. [5]*Properties of War Gases Volume II: Blood and Nettle Gases (U)*, ETF 100-41/Vol-2, Chemical Corps Board, Army Chemical Center, Maryland, December 1956, CONFIDENTIAL Report (AD108457). [6]TM 3-215/AFM 355-7, *Military Chemistry and Chemical Agents*, Departments of the Army and the Air Force, Washington DC, December 1963, UNCLASSIFIED Technical Manual (ADA292141). [7]L. Fishbein and S. Czerczak, "Concise International Chemical Assessment Document 47: Arsine: Human Health Aspects," WHO, 2002. [8] FM 8-9/NAVMED P-5059/AFJMAN 44-151, *NATO Handbook on the Medical Aspects of NBC Defense Operations AMEDP-6(B)*, 1 February 1996.

Table II-23. SA Toxicity Estimates[10]

Endpoint	Toxicity (mg-min/m^3)	MV (L)	Exposure Duration	ROE	Probit Slope	TLE	ROD	DOC
Lethality	LCt_{50}: 7500 [a] (Provisional)	15	2 min	Inhalation/Ocular	Not Calculated	1.4 [b]	Some	Low
NOTES								
[a]Based on modeling of 6 species. [b]See Appendix H for supporting toxicity profile estimates.								

7. Blister Agents (Vesicants)

Blister agents are used to produce casualties, to degrade fighting efficiency, and to restrict use of terrain and equipment.[1] Blister agents are CW agents that act on the eyes, mucous membranes, lungs, skin, and blood-forming organs.[12] The most toxic route of exposure is inhalation/ocular.[10] The severity of a blister agent burn relates directly to the concentration of the agent, the duration of contact with the skin,[1] and the location on the body. Most blister agents are insidious in action except for lewisite (L) and phosgene oxime (CX), which cause immediate pain on contact. Assume MOPP4 whenever liquid or vaporized agents are known to be present.[1] Decontaminate within 1 or 2 minutes after exposure to help prevent or decrease tissue damage.[21] The blister agents are divided into three groups: mustards, arsenicals, and urticants.

a. Mustards. This group of agents includes the sulfur mustards (H and HD) and the nitrogen mustards (HN-1, HN-2, and HN-3). Because of their physical properties, mustards are persistent under cool conditions; however, evaporation increases as the temperature increases.[21] It is possible to increase their persistency even more by dissolving them in thickeners.[12]

(1) Distilled Mustard (HD) (see Table II-24 [page II-38]). HD is a pale yellow to dark brown oily liquid with a garlic-like odor.[13] The eyes and respiratory tract are the most sensitive target organs.[10] The latency period for ocular effects is shorter than that for pulmonary effects, and acutely, ocular effects are more debilitating.[10] Both mustard vapor and liquid rapidly penetrate the skin. Warm, moist areas with thin skin (perineum, external genitalia, underarms, inside elbow, and neck) are much more sensitive.[21] Sweaty skin absorbs more mustard than dry skin. With an increase in temperature (>85 degrees F) and humidity, the effective dosages decrease and are about half of those for temperatures from 65 to 75 degrees F.[10] Mild symptoms caused from vapor exposure include tearing,

itchy, burning, gritty feeling in the eyes, rhinorrhea, sneezing, hoarseness, hacking cough, and erythema. Severe symptoms include marked lid edema, possible corneal damage, severe pain in the eyes, productive cough, dyspnea, and vesication.[21] Repeated exposures can cause an increase in sensitivity.[21] See Table II-25 (page II-40) for toxicity estimates.

Table II-24. HD

Alternate Designations: EA 1033; HS; G.34; M.O; Kampstoff "Lost"; Mustard HD; Mustard gas; Mustard Sulfur; Mustard vapor; S-Lost; Schewefel-lost; S mustard; Sulfur mustard gas; Sulfur mustard; Sulphur mustard; Sulphur mustard gas; Yellow Cross liquid; Y; Yperite (French & German)	
Chemical Name: Bis (2-chloroethyl) sulfide	
Synonyms: 2, 2'-dichloroethyl sulfide; 1, 1'-Thiobis(2-chloroethane); β- β '-dichlorodiethyl sulphide; β ,β'-dichloroethylsulfide di – (2-chloro-ethyl) sulfide; Sulfide, bis (2-chloroethyl); Bis (beta-chloroethyl)sulfide; Bis (2-chloroethyl)sulfide; Bis (2-chloroethyl) sulphide; 1-Chloro-2-(beta-chloroethylthio)ethane; 2,2'-Dichlorodiethyl sulfide; Di-2-chloroethyl sulfide; beta,beta'-Dichloroethyl sulfide; beta,beta-Dichlor-ethyl-sulphide; 2,2'-Dichloroethyl sulphide; Gelbkreuz (Czech); 1,1'-Thiobis(2-chloroethane)	
CAS Registry Number: 505-60-2	
RTECS Number: WQ0900000	
Physical and Chemical Properties	
Structural Formula: $Cl-CH_2-CH_2-S-CH_2-CH_2-Cl$ Molecular Formula: $C_4H_8Cl_2S$ Molecular Weight: 159.07	
Physical State	Pale yellow to dark brown oily liquid; [1] colorless when pure [1,2]
Odor	Garlic-like [1,3] or horseradish [3]
Boiling Point	218°C (extrapolated); at atmospheric pressure HD starts to decompose below the boiling point [4]
FP/MP	14.45°C (FP) [2]
Solid Density (g/mL)	1.372 @ 0°C; 1.333 @ 10°C [5]
Liquid Density (g/mL)	1.2685 @ 25°C [6]
Vapor Density (relative to air)	5.5 (calculated)
Vapor Pressure (torr)	1.06×10^{-1} @ 25°C [4]
Volatility (mg/m^3)	9.06×10^2 @ 25°C (calculated from vapor pressure) [4]
Latent Heat of Vaporization (kcal/mol)	15.0 @ 25°C (calculated from vapor pressure) [4]
Viscosity (cP)	3.951 @ 25.0°C, 7.746 @ 0°C (extrapolated)[7]
Viscosity of Vapor (cP)	6.65×10^{-3} @ 25.0°C, 6.00×10^{-3} @ 0°C[7]
Surface Tension (dynes/cm)	42.5 @ 25.0°C, 45.9 @ 0°C[7]
Flash Point	105°C [5]
Decomposition Temperature	180°C [8]
Solubility	HD is practically insoluble in water; solubility of HD in distilled water is 0.92g HD/100g solution at 22°C. HD is freely soluble in fats and oils, gasoline, kerosene, most organic solvents, and CW agents. [5]
Rate of Hydrolysis	$t_{1/2}$ = 5 min @ 25°C via a S_n1 mechanism;[9] $t_{1/2}$ = 60 min @ 25°C in salt water.[10] HD on or under water undergoes hydrolysis only if dissolved. The rate of HD hydrolysis is controlled by the rate of mass transfer and is very slow.[11]
Hydrolysis Products	Hydrogen chloride, thiodiglycol, and sulfonium ion aggregates—one of which is also highly toxic[11]
Stability in Storage	A small amount of degradation occurs when stored in steel ton containers for over 50 years.[12] This degradation appears to be caused by the formation of solid deposits "heels" comprised of a six-membered ring cyclic sulfonium ion {1-(2-chloroethyl) -1,4-dithianium chloride}, HD, and Fe, which were detected at the bottom of the containers.[13]
Action on Metals or Other Materials	Very little when pure.[3] The corrosion rate of HD on steel is 0.0001 inch/month @ 65°C using munitions grade HD. [14]

Table II-24. HD (Continued)

Other Data	
Skin and eye toxicity	Eyes are very susceptible to low concentrations; incapacitating effects by skin absorption require higher concentrations. [15]
Inhalation toxicity	Most toxic route of exposure [16]
Rate of action	Delayed—hours to days [17]
Means of detection	M8 paper, M9 paper, M256A1CADK, M90 AMAD, M21 ACAA, M22 ACADA, CAM/ICAM, M272 water testing kit, M18A3 CADK[18], MM1, M18A2 CADK
Protection required	MOPP4 whenever liquid or vaporized agents are present [15]
Decontamination	Flush eyes with water immediately. Use the M291 SDK to remove any liquid nerve agent on skin or clothing. Use the M295 IEDK for individual equipment.[15] HTH or household bleach is effective on equipment. Water, soaps, detergents, steam, and absorbents (earth, sawdust, ashes, and rags) are effective for physical removal. STB does not effectively decontaminate mustard if it has solidified at low temperatures.[19]
Use	Delayed-action casualty agent

NOTES

[1]Franke, S., *Manual of Chemistry Volume I-Chemistry of Chemical Warfare Agents*, ACSI-J-3890, Chemie der Kampfstoffe, East Berlin, April 1968, UNCLASSIFIED Technical Manual (AD849866).
[2]Felsing, W.A., et al., "The Melting Point of Mustard Gas," *J. Amer.Chem. Soc.*, Vol. 70, p. 1966, 1948.
[3]Kibler, A.L., *Data on Chemical Warfare*, Technical Division Memorandum Report 456, Chemical Warfare Center, Edgewood Arsenal, MD, November 1942, UNCLASSIFIED Report (ADB969725).
[4]Penski, E.C., *Properties of Di-(2-Chloroethyl) Sulfide I. Vapor Pressure Data Review and Analysis*, ERDEC-TR-043, USA Edgewood Research, Development and Engineering Center, Aberdeen Proving Ground, MD, April 1993, UNCLASSIFIED Report (ADA 267059).
[5]Buckles, M.F., *CW Vesicants: Selected Values for the Physical Properties of H, T, and Q (U)*, Special Report CRLR 542, Chemical Corps Chemical and Radiological Laboratories, Army chemical Center, MD, May 1956, UNCLASSIFIED Report (AD108272).
[6]Moelwyn-Hughes, E.A., and Owens, R., *The Surface Tension, The Molecular Surface Energy and the Parachor of Toxic Compounds and of Certain Chlorides Used in Their Manufacture, Part XV of the Thermal Decomposition of the Secondary Alkylfluorophosphonites*, Sutton Oak Report 544, Sutton Oak, England, September 1941, UNCLASSIFIED Report.
[7]Samuel, J.B., et al., *Physical Properties of Standard Agents, Candidate Agents, and Related Compounds at Several Temperatures (U)*, ARCSL-SP-83015, USA Armament Research and Development Command, Aberdeen Proving Ground, MD, June 1983, UNCLASSIFIED Report (ADC033491).
[8]Williams, A.H., "The Thermal Decomposition of 2:2'-Dichlorodiethyl Sulphide,," *J. Chem. Soc.*, p. 318, 1947.
[9]Bartlett, P.D., and Swain, C.G., "Kinetics of Hydrolysis and Displacement Reactions of β-β'-(Dichlorodiethyl Sulfide (Mustard Gas) and of β-Chloro-β'-hyroxidediethyl Sulfide (Mustard Chlorohydrin)," *J. Chem. Soc.*, Vol. 71, p. 1406, 1949.
[10]Brookfield, K.J., et al., *The Kinetics of the Hydrolysis of Vesicants Part II-2:2'-Dichlorodiethylsulphide (H)*, SO/R/576, Military Intelligence Division, Great Britain, March 1942, UNCLASSIFIED Report.
[11]Yang, Y., et al., "Decontamination of Chemical Warfare Agents," *Chem. Rev.*, Vol. 92, p. 1729, 1992.
[12]Abercrombie, P.L., and Butrow, A.B., *Selected Physical Properties of Ton Container HD (Mustard) and VX*, ERDEC-TR-450, USA Edgewood Research, Development, and Engineering Center, Aberdeen Proving Ground, MD, July 1998, UNCLASSIFIED Report (ADA350462).
[13]Yang, Y., et al., "Characterization of HD Heels and the Degradation of HD in Ton Containers," *In Proceedings of the 1996 ERDEC Scientifiec Conference on Chemical and Biological Defense Research 19-22 November 1996,* UNCLASSIFIED Paper, ERDEC-SP-048, pp 353-360, USA Edgewood Research, Development and Engineering Center, Aberdeen Proving Ground, MD, October 1997, UNCLASSIFIED Report (ADA334105).
[14]Harris, B.L., et al., *Corrosion by Vesicants: Rate of Corrosion of Steel and Other Metals by H, HQ, HN-3, HN-1 and L. Mostly at 65°C*, Technical Division Memorandum Report 1031, USA Chemical Research and Development Laboratories, Army Chemical Center, MD, April 1945, UNCLASSIFIED Report (ADB963161).
[15]FM 8-285/NAVMED P-5041/AFJMAN 44-149/FMFM 11-11, *Treatment of Chemical Agent Casualties and Conventional Military Chemical Injuries*, 22 December 1995.
[16]Sharon Reutter, et al., Review and Recommendations for Human Toxicity Estimates for FM 3-11.9, ECBC-TR-349, September 2003.
[17]NIOSH-DOD-OSHA Sponsored Chemical and Biological Respiratory Protection Workshop Report, February 2000.
[18]DOD Chemical And Biological Defense Program Annual Report to Congress, Volume I, April 2003.
[19]FM 3-5/MCWP 3-37.3, *NBC Decontamination,* 28 July 2000.

Table II-25. HD Toxicity Estimates[10]

Endpoint	Toxicity (mg-min/m^3)	MV (L)	Exposure Duration	ROE	Probit Slope	TLE	ROD	DOC
Lethality	LD_{50}: 1400 mg [a]	N/A	N/A; 70-kg man	Percutaneous Liquid [b]	7	N/A	Little, if any	Low
	LCt_{50}: 1000 [c]	15	2 min	Inhalation/ Ocular	6	1.5 [d]	Some	Low
	LCt_{50}: 10,000 [a,e]	N/A	30-360 min	Percutaneous Vapor [f]	7	1 [d,g]	Unknown	Low [h]
	LCt_{50}: 5000 [i,j] (Provisional)	N/A	30-360 min	Percutaneous Vapor [f]	7	1 [d,g]	Unknown	Low
Severe effects (vesication)	ED_{50}: 600 mg [a]	N/A	N/A; 70-kg man	Percutaneous Liquid [b]	3	N/A	Little, if any	Low
	ECt_{50}: 500 [e,k]	N/A	30-360 min	Percutaneous Vapor [f]	3	1 [d,g,l]	Little, if any	Moderate
	ECt_{50}: 200 [i,k]	N/A	30-360 min	Percutaneous Vapor [f]	3	1 [d,g,l]	Little, if any	Moderate [h]
Severe effects (eyes)	ECt_{50}: 75 [m]	N/A	2-360 min	Ocular	3	1 [d]	Little, if any	High
Mild effects (erythema, itching, some pain)	ECt_{50}: 50 [a,e]	N/A	30 min	Percutaneous Vapor [f]	3	1 [d]	Little, if any	Moderate
	ECt_{50}: 25 [a,i]	N/A	30 min	Percutaneous Vapor [f]	3	1 [d]	Little, if any	Moderate
Mild effects (eyes)	ECt_{50}: 25 [a]	N/A	2-360 min	Ocular	3	1 [d]	Some	High
Odor detection	EC_{50}: 0.6-1mg/m^3 [n]	N/A	Few seconds	Inhalation	N/A	N/A	Probably insignificant	High

NOTES

[a]Based on Grotte and Yang (2001).
[b]Bare skin.
[c]Based on Grotte and Yang (2001) and Sommerville (2002).
[d]See Appendix H for supporting toxicity profile estimates.
[e]Moderate temperatures (65-85°F).
[f]Assumes personnel are masked with eye protection and clothed skin.
[g]The TLE value is assumed to be 1 because the Ct profile is unknown.
[h]True human LCt_{50}/ECt_{50} values could be lower.
[i]Hot temperatures (greater than 85°F).
[j]Based on temperature factor given in Grotte and Yang (2001) and analysis of lethal data for GB percutaneous vapor exposure.
[k]Based on Grotte and Yang (2001) and Letter (Dec 2001).
[l]Based on human data.
[m]Based on re-analysis of human data.
[n]Based on primary human data.

(2) Levinstein Mustard (H). Levinstein mustard is the original mustard (gas) of World War I vintage. It contains H and about 30 percent impurities. Properties of H are essentially the same as those for HD. The effective dosages of H and HD have been demonstrated to be quite comparable.[10] This manual does not differentiate between H and HD.

(3) Nitrogen Mustard (HN-1) (see Table II-26). HN-1 is a colorless liquid when pure with a faint, fishy or soapy odor.[13] It is used as a delayed-action casualty agent. The most prevalent symptoms in men inadvertently exposed to HN-1 vapor were conjunctivitis, laryngitis, bronchitis, hoarseness, coughing, elevated temperature, nausea, and vomiting. In this accidental exposure, the fact that these men, with all the knowledge available at their command as to precautions, protection against, and physical and chemical properties

of HN-1, were severely affected without knowledge of their exposure, serves to further emphasize the insidious nature of this agent.[10] See Table II-27 (page II-43) for HN-1 toxicity estimates.

Table II-26. HN-1

Alternate Designations: Ethyl S; NH-Lost; NOR nitrogen mustard; Nitrogen mustard gas –1; NSC 10873; TL 329; TL 1149	
Chemical Name: 2,2'-Dichlorotriethylamine	
Synonyms: Bis (2-chloroethyl)ethylamine; Ethylbis(2-chloroethyl)amine; N-ethyl, bis (β -chloroethyl)amine Ethylbis (beta-chloroethyl)amine	
CAS Registry Number: 538-07-8	
RTECS Number: YE1225000	
Physical and Chemical Properties	
Structural Formula: 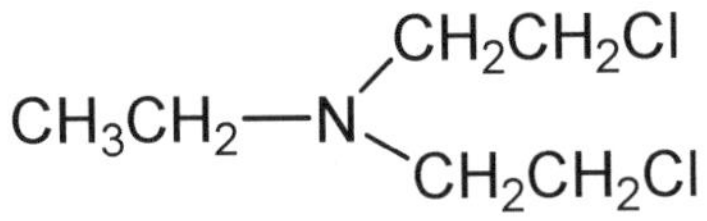 Molecular Formula: $C_6H_{13}Cl_2N$ Molecular Weight: 170.08	
Physical State	Dark oily liquid;[1] colorless when pure [1,2]
Odor	Faint, fishy or soapy [3]
Boiling Point	192°C (extrapolated);[4,2] at atmospheric pressure HN-1 decomposes below the boiling point [5]
FP/MP	-34.2°C (MP) [2]
Liquid Density (g/mL)	1.086 @ 25°C; 1.110 @ 0°C (extrapolated) [2]
Vapor Density (relative to air)	5.9 (calculated)
Vapor Pressure (torr)	2.44×10^{-1} @ 25°C; 3.32×10^{-2}@ 0°C (extrapolated) [4,2]
Volatility (mg/m^3)	2.23×10^3 @ 25°C; 3.31×10^2 @ 0°C (calculated from vapor pressure) [4,2]
Latent Heat of Vaporization (kcal/mol)	13.0 @ 25°C; 12.9 @ 0°C (calculated from vapor pressure) [4,2]
Viscosity (cP)	Data not available
Viscosity of Vapor (cP)	Data not available
Surface Tension (dynes/cm)	Data not available
Flash Point	Data not available; flashing has occurred on static detonation [3]
Decomposition Temperature	For HN-1 · HCl, 12.7% is destroyed @ 149°C and @ 426°C >99% is destroyed. [5]
Solubility	Solubility in water is approximately 4 g HN-1/L solution @ ambient temperature. Miscible with common organic solvents. [3]
Rate of Hydrolysis	$t_{1/2}$ = 1.3 min @ 25°C in aqueous solution [3]
Hydrolysis Products	Complete hydrolysis yields the following: hydrochloric acid and ethyl diethanolamine, $CH_3CH_2N(CH_2CH_2OH)_2$. [1] The process involves a complex series of reactions, with formation of the hydrochloride, cyclic imonium salts, a dimer, etc. [6]
Stability in Storage	Polymerizes with the formation of solid deposits, when stored in steel containers; this amount is slight @ ambient temperature, but increases @ temperatures above 50°C. [7]
Action on Metals or Other Materials	Corrosion of HN-1 on steel @ 65°C is 1×10^{-5} to 5×10^{-5} inch/month [8]
Other Data	
Skin and eye toxicity	Eyes are very susceptible to low concentration; incapacitating effects by skin absorption require higher concentrations. [9]
Inhalation toxicity	Most toxic route of exposure [10]
Rate of action	Delayed: 12 hours or longer [11]
Means of detection	M8 paper, M9 paper, M256A1 CADK, CAM/ICAM, MM1 [12]
Protection required	MOPP4 whenever liquid or vapor is present [9]

Table II-26. HN-1 (Continued)

Decontamination	Liquid on eyes and skin requires immediate decontamination.[9] HTH, household bleach is effective on equipment. Water, soaps, detergents, steam, and absorbents (earth, sawdust, ashes, and rags) are effective for physical removal. STB does not effectively decontaminate mustard if it has solidified at low temperatures.[13]
Use	Delayed-action casualty agent

NOTES

[1]Cheicante, R.L., et al., "Investigation for the Determination of Nitrogen Mustard and Related Compounds in Air by Gas Chromatography Using Solid Sorbent Collection and Thermal Desorption," *In Proceedings of the 1998 ERDEC Scientific Conference on Chemical and Biological Defense Research 17-20 November 1998*, UNCLASSFIED Paper (ADE491775), ERDEC-SP-004, pp 781-792, USA ECBC, Aberdeen Proving Ground, MD, July 1999, UNCLASSFIED Report (ADA375171).

[2]Dawson, T., *A Memorandum Report New Compounds 2,2' Dichlorotriethylamine*, Technical Division Memorandum Report 552, USA Chemical Research and Development Laboratories, Army Chemical Center, MD, February 1943, UNCLASSFIED Report (ADB960467).

[3]W.R. Kirner, *Summary Technical Report of Division 9, NDRC Volume 1, Chemical Warfare Agents, and Related Chemical Problems Part I-II,* Chapter 6, p. 59, Office of Scientific Research and Development, Washington, DC, 1946, UNCLASSIFIED Report (AD234270).

[4]Abercrombie, P., ECBC Notebook # NB 98-0079, p. 24 (U).

[5]Brooks, M. E. and Parker, G.A., et al., *Incineration/Pyrolysis of Several Agents and Related Chemical Materials Contained in Identification Sets*, ARCSL-TR-79040, October 1979, UNCLASSIFIED Report (ADB042888).

[6]W.R. Kirner, *Summary Technical Report of Division 9, NDRC Volume 1, Chemical Warfare Agents, and Related Chemical Problems Part I-II,* Chapter 19, p. 389, Office of Scientific Research and Development, Washington, DC, 1946, UNCLASSIFIED Report (AD234270).

[7]Harris, B.L., et al., *Thickened Vesicants: Storage Stability of Unthickened and Thickened Nitrogen Mustards and Their Mixtures with Levinstein Mustard*, Technical Division Memorandum Report 706, USA Chemical Research and Development Laboratories, Army Chemical Center, MD, July 1943, UNCLASSIFIED Report (ADB962153).

[8]Harris, B.L. andMacy, R., *Corrosion by Vesicants: Rate of Corrosion of Steel and Other Metals by H, HQ, HN-3, HN-1, and L, Mostly at 65°C*, Technical Division Memorandum Report 1031, USA Chemical Research and Development Laboratories, Army Chemical Center, MD, April 1945, UNCLASSIFIED Report (ADB963161).

[9]FM 8-285/NAVMED P-5041/AFJMAN 44-149/FMFM 11-11, *Treatment of Chemical Agent Casualties and Conventional Military Chemical Injuries*, 22 December 1995.

[10]Sharon Reutter, et al., Review and Recommendations for Human Toxicity Estimates for FM 3-11.9, ECBC-TR-349, September 2003.

[11]*NIOSH-DOD-OSHA Sponsored Chemical and Biological Respiratory Protection Workshop Report,* February 2000.

[12]DOD Chemical and Biological Defense Program Annual Report to Congress, Volume I, April 2003.

[13]FM 3-5/MCWP 3-37.3, *NBC Decontamination,* 28 July 2000.

Table II-27. HN-1 Toxicity Estimates[10]

Endpoint	Toxicity (mg-min/m³) [a]	MV (L)	Exposure Duration	ROE	Probit Slope	TLE	ROD	DOC
Lethality	LD_{50}: 1400 mg	N/A	N/A; 70-kg man	Percutaneous Liquid [b]	Unknown	N/A	Unknown	Low
	LCt_{50}: 1000	15	2 min	Inhalation/Ocular	Unknown	1 [c,d]	Unknown	Low
	LCt_{50}: 10,000 [e]	N/A	30 min	Percutaneous Vapor [f]	Unknown	1 [c,d]	Unknown	Low
	LCt_{50}: 5000 [g]	N/A	30 min	Percutaneous Vapor [f]	Unknown	1 [c,d]	Unknown	Low
Severe effects (vesication)	ED_{50}: 600 mg	N/A	N/A; 70-kg man	Percutaneous Liquid	Unknown	N/A	Unknown	Low
	ECt_{50}: 500 [e]	N/A	30 min	Percutaneous Vapor [f]	Unknown	1 [c,d]	Unknown	Low
	ECt_{50}: 200 [g]	N/A	30 min	Percutaneous Vapor [f]	Unknown	1 [c,d]	Unknown	Low
Severe effects (eyes)	ECt_5: 75	N/A	2 min	Ocular	Unknown	1 [c,d]	Unknown	Low
Mild effects (pain, erythema, itching)	ECt_{50}: 50 [e]	N/A	30 min	Percutaneous Vapor [f]	Unknown	1 [c,d]	Unknown	Low
	ECt_{50}: 25 [g]	N/A	30 min	Percutaneous Vapor [f]	Unknown	1 [c,d]	Unknown	Low
Mild effects (eyes)	ECt_{50}: 25	N/A	2 min	Ocular	Unknown	1 [c,d]	Unknown	Low

NOTES

[a]All toxicity values given are provisional and based on recommendations for H/HD.
[b]Bare skin.
[c]See Appendix H for supporting toxicity profile estimate.
[d]The TLE value is assumed to be 1 because the Ct profile is unknown.
[e]Moderate temperatures (65-85°F).
[f]Assumes personnel are masked with eye protection.
[g]Hot temperatures (greater than 85°F).

(4) Nitrogen mustard (HN-2) (see Table II-28). HN-2 is a colorless liquid when pure, and it has a fishy or soapy odor.[13] HN-2 is irritating to the eyes.[20] For other symptoms, see discussions of HN-1. See Table II-29 (page II-45) for HN-2 toxicity estimates.

Table II-28. HN-2

Alternate Designations: Dichloren; N-methyl-Lost (German); Mustine; Mustargen; Mutagen; Nitrogen mustard; NSC 762; S; TL 146; T-1024; ENT-25294; MBA
Chemical Name: Bis-(2-chloroethyl)methylamine
Synonyms: 2,2'-Dichloro-N-methyldiethylamine; N, N-bis(2-chloroethyl)methylamine, N-methyl, bis(β chloroethyl)amine; Bis(beta-chloroethyl)methylamine; Chloramine; Chlorethazine; Chlormethine; 2-Chloro-N-(2-chloroethyl)-N-methylethanamine; beta, beta'-Dichlorodiethyl-N-methylamine; 2,2'-Dichlorodiethyl-methylamine; Di(2-chloroethyl)methylamine; N,N-Di(chloroethyl)methylamine; Ethanamine, 2-chloro-N-(2-chloroethyl)-N-methyl-; Mechlorethamine; Mecloretamina (Italian); Methylbis(beta-chloroethyl)amine; Methylbis(2-chloroethyl)amine; N-Methyl-bis-chloraethylamin (German); N-Methyl-bis(beta-chloroethyl)amine; N-Methyl-bis(2-chloroethyl)amine; N-Methyl-2,2'-dichlorodiethylamine; Methyldi(2-chloroethyl)amine
CAS Registry Number: 51-75-2
RTECS Number: IA750000

Table II-28. HN-2 (Continued)

Physical and Chemical Properties	
Structural Formula: Molecular Formula: $C_5H_{11}Cl_2N$ Molecular Weight: 156.05	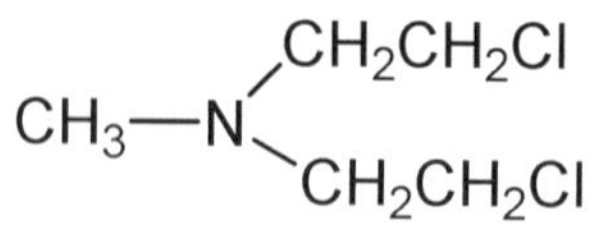
Physical State	Colorless liquid when pure [1]
Odor	Fishy or soapy [2]
Boiling Point	177°C (extrapolated);[3,2,4] at atmospheric pressure, HN-2 decomposes below its boiling point [5]
FP/MP	- 70°C (FP) [2]
Liquid Density (g/mL)	1.118 @ 25°C; 1.1425 @ 0°C (extrapolated) [3,4]
Vapor Density (relative to air)	5.4 (calculated)
Vapor Pressure (torr)	4.16×10^{-1} @ 25°C; 5.70×10^{-2} @ 0°C (extrapolated) [3,2,4]
Volatility (mg/m^3)	3.49×10^3 @ 25°C; 5.22×10^2 @ 0°C (calculated from vapor pressure) [3,2,4]
Latent Heat of Vaporization (kcal/mol)	12.9 @ 25°C; 12.8 @ 0°C (calculated from vapor pressure) [3,2,4]
Viscosity (cP)	Data not available
Viscosity of Vapor (cP)	Data not available
Surface Tension (dynes/cm)	Data not available
Flash Point	Data not available
Decomposition Temperature	Decomposes before boiling point is reached; instability of HN-2 is associated with its tendency to polymerize or condense; the reactions involved could generate enough heat to cause an explosion. [5]
Solubility	Solubility in water is approximately 13 g HN-2/L solution @ ambient temperature. Miscible with common organic solvents. [2]
Rate of Hydrolysis	$t_{1/2}$ = 4 min @ 25°C in an aqueous solution. Slow except where alkali is present; dimerizes fairly rapidly in water. [2]
Hydrolysis Products	The process involves a complex series of reactions, with formation of the hydrochloride, cyclic imonium salts, a dimer, etc. [6]
Stability in Storage	Not stable; dimerizes on storage and deposits crystalline dimers [2]
Action on Metals or Other Materials	None on steel and brass [2]
Other Data	
Skin and eye toxicity	Eyes are very susceptible to low concentration; incapacitating effects by skin absorption require higher concentrations [7]
Inhalation toxicity	Most toxic route of exposure [8]
Rate of action	Delayed: 12 hrs or longer [9]
Means of detection	M8 paper, M9 paper, M256A1 CADK, CAM/ICAM, MM1 [10]
Protection required	MOPP4 whenever liquid or vapor is present [7]
Decontamination	Liquid on eyes and skin requires immediate decontamination. [7] HTH or household bleach is effective on equipment. Water, soaps, detergents, steam, and absorbents (earth, sawdust, ashes, and rags) are effective for physical removal. STB does not effectively decontaminate mustard if it has solidified at low temperatures.[11]
Use	Delayed-action casualty agent

Table II-28. HN-2 (Continued)

NOTES
[1]Witten, Benjamin, *The Search for Toxic Chemical Agents (U)*, EATR 4210, Edgewood Arsenal Research Laboratories, MD, November 1969, UNCLASSIFIED Report (AD507852). [2]W.R. Kirner, *Summary Technical Report of Division 9, NDRC Volume 1, Chemical Warfare Agents, and Related Chemical Problems Part I-II,* Chapter 6, p. 59, Office of Scientific Research and Development, Washington, DC, 1946, UNCLASSIFIED Report (AD234270). [3]Abercrombie, P., ECBC Notebook # NB 98-0079 p. 27 (U). [4]Dawson, T.P., and Witten, B., *Bis (2-chloroethyl) methylamine, Preparation, Decontamination, and Stability*, Technical Division Memorandum Report 442, Chemical Warfare Center, Edgewood Arsenal, MD, September 1942, UNCLASSIFIED Report (ADB960331). [5]TM 3-215/AFM 355-7, *Military Chemistry and Chemical Agents*, December 1963, UNCLASSIFIED Technical Manual (ADA292141). [6]W.R. Kirner, *Summary Technical Report of Division 9, NDRC Volume 1, Chemical Warfare Agents, and Related Chemical Problems Part I-II,* Chapter 19, p. 389, Office of Scientific Research and Development, Washington, DC, 1946, UNCLASSIFIED Report (AD234270). [7]FM 8-285/NAVMED P-5041/AFJMAN 44-149/FMFM 11-11, *Treatment of Chemical Agent Casualties and Conventional Military Chemical Injuries*, 22 December 1995. [8]Sharon Reutter, et al., Review and Recommendations for Human Toxicity Estimates for FM 3-11.9, ECBC-TR-349, September 2003. [9]*NIOSH-DOD-OSHA Sponsored Chemical and Biological Respiratory Protection Workshop Report,* February 2000. [10]DOD Chemical and Biological Defense Program Annual Report to Congress, Volume I, April 2003. [11]FM 3-5/MCWP 3-37.3, *NBC Decontamination,* 28 July 2000.

Table II-29. HN-2 Toxicity Estimates[10]

Endpoint	Toxicity ($mg\text{-}min/m^3$)[a]	MV (L)	Exposure Duration	ROE	Probit Slope	TLE	ROD	DOC
Lethality	LD_{50}: 1400 mg	N/A	N/A; 70-kg man	Percutaneous Liquid[b]	Unknown	N/A	Unknown	Low
	LCt_{50}: 1000	15	2 min	Inhalation/ Ocular	Unknown	Unknown[c]	Unknown	Low
	LCt_{50}: 10,000[d]	N/A	30 min	Percutaneous Vapor[e]	Unknown	1[c,f]	Unknown	Low
	LCt_{50}: 5000[g]	N/A	30 min	Percutaneous Vapor[e]	Unknown	1[c,f]	Unknown	Low
Severe effects (vesication)	ED_{50}: 600 mg	N/A	N/A; 70-kg man	Percutaneous Liquid[b]	Unknown	N/A	Unknown	Low
	ECt_{50}: 500[d]	N/A	30-360 min	Percutaneous Vapor[e]	Unknown	1[c,f]	Unknown	Low
	ECt_{50}: 200[g]	N/A	30-360 min	Percutaneous Vapor[e]	Unknown	1[c,f]	Unknown	Low
Severe effects (eyes)	ECt_{50}: 75	N/A	2 min	Ocular	Unknown	1[c,f]	Unknown	Low
Mild effects (pain, erythema, itching)	ECt_{50}: 50[d]	N/A	30-360 min	Percutaneous Vapor[e]	Unknown	1[c,f]	Unknown	Low
	ECt_{50}: 25[g]	N/A	30-360 min	Percutaneous Vapor[e]	Unknown	1[c,f]	Unknown	Low
Mild effects (eyes)	ECt_{50}: 25	N/A	2 min	Ocular	Unknown	1[c, f]	Unknown	Low

Table II-29. HN-2 Toxicity Estimates[9] (Continued)

NOTES
[a]All toxicity values given are provisional and based on recommendations for H/HD.
[b]Bare skin.
[c]See Appendix H for supporting toxicity profile estimates.
[d]Moderate temperatures (65-85°F).
[e]Assumes personnel are masked with eye protection.
[f]The TLE value is assumed to be 1 because the Ct profile is unknown.
[g]Hot temperatures (greater than 85°F).

(5) Nitrogen Mustard (HN-3) (see Table II-30). HN-3 is a colorless, odorless liquid when pure.[13] It is the most stable in storage of the three nitrogen mustards.[20] Symptoms noted among humans inadvertently exposed to HN-3 vapor include local irritation of eyes and upper respiratory tract, headache, and vomiting.[20] See HN-1 discussion for additional symptoms. See Table II-31 (page II-48) for HN-3 toxicity estimates.

Table II-30. HN-3

Alternate Designations: EA 1053; Nitrogen mustard-3; TO; TL 145; TS 160	
Chemical Name: 2, 2', 2"-Trichlorotriethylamine	
Synonyms: Tri (2-chloroethyl)amine; Tris (2-chloroethyl)amine; Tris (β-chloroethyl)amine	
CAS Registry Number: 555-77-1	
RTECS Number: YE2625000	
Physical and Chemical Properties	
Structural Formula: 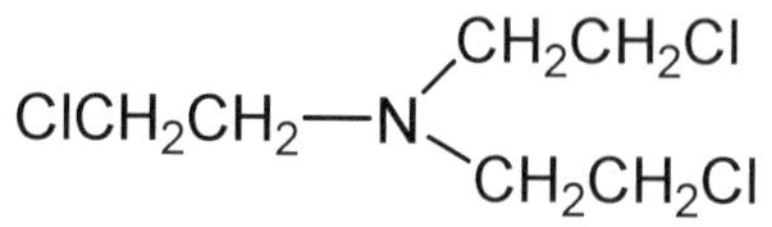 Molecular Formula: $C_6H_{12}Cl_3N$ Molecular Weight: 204.53	
Physical State	Oily dark liquid; colorless when pure [1]
Odor	Geranium-like; none when pure [2]
Boiling Point	257°C (extrapolated); [3] at atmospheric pressure HN-3 decomposes below the boiling point [4]
FP/MP	-3.74°C (MP) [3]
Liquid Density (g/mL)	1.2352 @ 25°C; 1.2596 @ 0°C (extrapolated) [3]
Vapor Density (relative to air)	7.1 (calculated)
Vapor Pressure (torr)	1.1×10^{-2} @ 25°C; 9.2×10^{-4} @ 0°C (extrapolated) [3]
Volatility (mg/m^3)	1.2×10^2 @ 25°C; 1.1×10^1 @ 0°C; (calculated from vapor pressure) [3]
Latent Heat of Vaporization (kcal/mol)	15.8 @ 25°C; 16.0 @ 0°C (calculated from vapor pressure) [3]
Viscosity (cP)	0.073 @ 25.0°C, 0.177 @ 0°C (extrapolated)[3]
Viscosity of Vapor (cP)	5.97×10^{-3} @ 25.0°C, 5.38×10^{-3} @ 0°C[3]
Surface Tension (dynes/cm)	40.9 @ 25.0°C, 44.1 @ 0°C[3]
Flash Point	Data not available
Decomposition Temperature	Above 150°C; [4] remains stable when explosively disseminated [2]
Solubility	Solubility in water is approximately 0.08 g HN-3/L @ ambient temperature. Miscible with common organic solvents. [2]
Rate of Hydrolysis	Very slow; hydrolysis is not complete even after several days unless alkali is present [5]

Table II-30. HN-3 (Continued)

Hydrolysis Products	Complete hydrolysis gives the following products: Hydrochloric acid and triethanolamine (TEA), $N(CH_2CH_2OH)_3$. [1] The process involves a complex series of reactions, with formation of the hydrochloride, cyclic imonium salts, a dimer, etc. [6]
Stability in Storage	In storage HN-3 darkens and forms crystalline deposits. [4] Relatively stable in steel containers, if dry. [2]
Action on Metals or Other Materials	No attack on iron if dry.[4] Corrodes steel @ a rate of 1 x 10^{-5} to 5 x 10^{-5} inch/month @ 65°C.[7]
Other Data	
Skin and eye toxicity	Eyes are very susceptible to low concentration; incapacitating effect by skin absorption require higher concentrations [8]
Inhalation Toxicity	Most toxic route of exposure [9]
Rate of action	Delayed: 12 hours or longer [10]
Means of detection	M8 paper, M9 paper, M256A1 CADK, CAM/ICAM, MM1 [11]
Protection required	MOPP4 whenever liquid or vapor is present[8]
Decontamination	Liquid on eyes and skin requires immediate decontamination. [8] HTH or household bleach is effective on equipment. Water, soaps, detergents, steam, and absorbents (earth, sawdust, ashes, and rags) are effective for physical removal. STB does not effectively decontaminate mustard if it has solidified at low temperatures. [12]
Use	Delayed-action casualty agent

NOTES

[1]Cheicante, R.L., et al., "Investigation for the Determination of Nitrogen Mustard and Related Compounds in Air by Gas Chromatography Using Solid Sorbent Collection and Thermal Desorption," *In Proceedings of the 1998 ERDEC Scientific Conference on Chemical and Biological Defense Research 17-20 November 1998*, UNCLASSFIED Paper, ERDEC-SP-004, pp 781-792, USA ECBC, Aberdeen Proving Ground, MD, July 1999, UNCLASSFIED Report (ADA375171).

[2]W.R. Kirner, *Summary Technical Report of Division 9, NDRC Volume 1, Chemical Warfare Agents, and Related Chemical Problems Part I-II,* Chapter 6, p. 59, Office of Scientific Research and Development, Washington, DC, 1946, UNCLASSIFIED Report (AD234270).

[3]Samuel, J.B., et al., *Physical Properties of Standard Agents, Candidate Agents, and Related Compounds at Several Temperatures (U)*, ARCSL-SP-83015, USA Armament Research and Development Command, Aberdeen Proving Ground, MD, June 1983, UNCLASSIFIED Report (ADC033491).

[4]Kibler, A.L, *Data on Chemical Warfare*, Technical Division Memorandum Report 456, Chemical Warfare Center, Edgewood Arsenal, MD, November 1942,UNCLASSIFIED Report (ADB969725).

[5]Bartlett, P.D., et al., "Kinetics and Mechanism of the Reactions of Tertiary β-Chloroethylamines in Solution. III. β-Chloroethyldiethylamine and tris-β-Chloroethylamine," *J. Am. Chem. Soc.*, Vol. 71, p. 1415, 1949.

[6]W.R. Kirner, *Summary Technical Report of Division 9, NDRC Volume 1, Chemical Warfare Agents, and Related Chemical Problems Part I-II,* Chapter 19, p. 389, Office of Scientific Research and Development, Washington, DC, 1946, UNCLASSIFIED Report (AD234270).

[7]Harris, B.L. and Macy, R, *Corrosion by Vesicants: Rate of Corrosion of Steel and Other Metals by H, HQ, HN-3, HN-1, and L, Mostly at 65°C*, Technical Division Memorandum Report 1031, USA Chemical Research and Development Laboratories, Army Chemical Center, MD, April 1945, UNCLASSIFIED Report (ADB963161).

[8]FM 8-285/NAVMED P-5041/AFJMAN 44-149/FMFM 11-11, *Treatment of Chemical Agent Casualties and Conventional Military Chemical Injuries*, 22 December 1995.

[9]Sharon Reutter, et al., Review and Recommendations for Human Toxicity Estimates for FM 3-11.9, ECBC-TR-349, September 2003.

[10]*NIOSH-DOD-OSHA Sponsored Chemical and Biological Respiratory Protection Workshop Report,* February 2000.

[11]DOD Chemical and Biological Defense Program Annual Report to Congress, Volume I, April 2003.

[12]FM 3-5/MCWP 3-37.3, *NBC Decontamination,* 28 July 2000.

Table II-31. HN-3 Toxicity Estimates[10]

Endpoint	Toxicity (mg-min/m^3)[a]	MV (L)	Exposure Duration	ROE	Probit Slope	TLE	ROD	DOC
Lethality	LD_{50}: 1400 mg	N/A	N/A; 70-kg man	Percutaneous Liquid [b]	Unknown	N/A	Unknown	Low
	LCt_{50}: 1000	15	2-360 min	Inhalation/Ocular	Unknown	1 [c,d]	Unknown	Low
	LCt_{50}: 10,000 [e]	N/A	30-360 min	Percutaneous Vapor [f]	Unknown	1 [c,d]	Unknown	Low
	LCt_{50}: 5000 [g]	N/A	30-360 min	Percutaneous Vapor [f]	Unknown	1 [c,d]	Unknown	Low
Severe effects (vesication)	ED_{50}: 600 mg	N/A	N/A; 70-kg man	Percutaneous Liquid[b]	Unknown	N/A	Unknown	Low
	ECt_{50}: 500 [e]	N/A	30-360 min	Percutaneous Vapor [f]	Unknown	1 [c,d]	Unknown	Low
	ECt_{50}: 200 [g]	N/A	30-360 min	Percutaneous Vapor [f]	Unknown	1 [c,d]	Unknown	Low
Severe effect (eyes)	ECt_{50}: 75	N/A	2-360 min	Ocular	Unknown	1 [c,d]	Unknown	Low
Mild effects (pain, erythema, itching)	ECt_{50}: 50 [e]	N/A	30-360 min	Percutaneous Vapor [f]	Unknown	1 [c,d]	Unknown	Low
	ECt_{50}: 25 [g]	N/A	30-360 min	Percutaneous Vapor [f]	Unknown	1 [c,d]	Unknown	Low
Mild effects (eyes)	ECt_{50}: 25	N/A	2-360 min	Ocular	Unknown	1 [c,d]	Unknown	Low

NOTES

[a]The toxicity values given are provisional and based on recommendations for H/HD.
[b]Bare skin.
[c]The TLE value is assumed to be 1 because the Ct profile is unknown.
[d]See Appendix H for supporting toxicity profile estimates.
[e]Moderate temperatures (65-85°F).
[f]Assumes personnel are masked with eye protection.
[g]Hot temperatures (greater than 85°F).

(6) Mustard-T Mixture (HT) (see Table II-32). HT is a pale yellow to brown liquid with a garlic-like odor. It is a mixture of 60 percent HD and 40 percent T.[13] It is somewhat more vesicant than H on bare skin; however, it is less vesicant through wet or dry clothing and is somewhat less effective than H on the eyes.[10] See Table II-33 (page II-50) for HT toxicity estimates.

Table II-32. HT

Alternate Designations: Distilled mustard and T mixture	
Chemical Name: HD: Bis-(2-chloroethyl) sulfide; T: Bis {2(2-chloroethylthio)ethyl} ether	
Synonyms: N/A	
CAS Registry Number: HD: 505-60-2; T: 63918-89-8	
RTECS Number: HD: WQ0900000; T: KN1400000	
Physical and Chemical Properties	
Structural Formula: 60wt% HD: $Cl-CH_2-CH_2-S-CH_2-CH_2-Cl$ 40 wt% T: $(ClCH_2CH_2SCH_2)_2O$ Molecular Formula: HD: $C_4H_8Cl_2S$; T: $C_8H_{16}Cl_2OS_2$ Molecular Weight: HD: 159.07; T: 263.24; Average: 188.96 (based on 60:40 wt %)	
Physical State	Pale yellow to brown liquid [1]
Odor	Garlic-like; less pronounced than mustard [1]
Boiling Point	No constant boiling point [2]

Table II-32. HT (Continued)

FP/MP	1.3°C (MP) [3]
Liquid Density (g/mL)	1.263 @ 20°C [3]
Vapor Density (relative to air)	6.5 (calculated based on 60:40 HT mixture)
Vapor Pressure (torr)	7.7×10^{-2} @ 25°C (calculated based on Raoult's Law equation) [4]
Volatility (mg/m^3)	7.83×10^{2} @ 25°C (calculated from vapor pressure) [4]
Latent Heat of Vaporization (kcal/mol)	Data not available
Viscosity (cP)	Data not available
Viscosity of Vapor (cP)	Data not available
Surface Tension (dynes/cm)	Data not available
Flash Point	Flash point range 109 to 115°C [5]
Decomposition Temperature	165°C to 180°C [4]
Solubility	Slightly soluble in water; soluble in most organic solvents. [2]
Rate of Hydrolysis	Hydrolyzed by prolonged boiling with water or treatment with caustic alkalis. [4]
Hydrolysis Products	Hydrogen chloride, thiodiglycol, and sulfonium aggregates; based on HD [6]
Stability in Storage	Pressure develops in steel [2]
Action on Metals or Other Materials	Very little when pure. Canadian HT corrodes steel at a rate of 0.00007 inch/ month @ 65°C. [7]
Other Data	
Skin and eye toxicity	Eyes are very susceptible to low concentrations; incapacitating effects by skin absorption require higher concentrations than does eye injury [8]
Inhalation toxicity	Most toxic route of exposure [9]
Rate of action	No data available
Means of detection	M8 paper, M9 paper, M256A1 CADK, M18A3 CADK, CAM/ICAM,[10] M18A2 CADK, MM1
Protection required	MOPP4 whenever liquid or vapor is present. [8]
Decontamination	Liquid on eyes and skin require immediate decontamination. [8] HTH or household bleach is effective on equipment. Water, soaps, detergents, steam, and absorbents (earth, sawdust, ashes, and rags) are effective for physical removal. STB does not effectively decontaminate mustard if it has solidified at low temperatures.[11]
Use	Delayed-action casualty agent

NOTES

[1]Cone, N.M. and Rouiller, C.A., *HQ & HT Review of British & US Literature*, TDMR 575, USA Chemical Research and Development Laboratories, Army Chemical Center, MD, February 1943, UNCLASSIFIED Report.
[2]*Chemical Agent Data Sheets Vol. II*, Edgewood Arsenal Special Report EO-SR-74002, USA Armament Command, Edgewood Arsenal, Aberdeen Proving Ground, MD December 1974, CONFIDENTIAL Report (AD000020).
[3]Dawson, T.P., *A Memorandum Report: New Compounds Bis(B-Chloroethylthioethyl) Ether (T) and its Mixtures with Mustard (HT)*, TDMR 534, USA Chemical Research and Development Laboratories, Army Chemical Center, MD, January 1943, UNCLASSIFIED Report (ADB960651).
[4]*Properties of War Gases Volume IV: Vesicants (U)*, ETF 100-41/Vol-4, Chemical Corps Board, Army Chemical Center, MD, December 1956, CLASSIFIED Report (AD108459).
[5]Butrow, A.B., ECBC Notebook # N 03-0025, p. 50 (U).
[6]Yang, Y., et al., "Decontamination of Chemical Warfare Agents," *Chem. Rev.*, Vol. 92, p. 1729, 1992.
[7]Harris, B.L. and Macy, R., *Corrosion by Vesicants: Rate of Corrosion of Steel and Other Metals by H, HQ, HN-3, HN-1, and L, Mostly @ 65°C*, TDMR 1031, USA Chemical Research and Development Laboratories, Army Chemical Center, MD, April 1945, UNCLASSIFIED Report (ADB963161).
[8]FM 8-285/NAVMED P-5041/AFJMAN 44-149/FMFM 11-11, *Treatment of Chemical Agent Casualties and Conventional Military Chemical Injuries*, 22 December 1995.
[9]Sharon Reutter, et al., SBCCOM Report *Review and Recommendations for Human Toxicity Estimates for FM 3-11.9*, Draft.
[10]DOD Chemical and Biological Defense Program Annual Report to Congress, Volume I, April 2003.
[11]FM 3-5/MCWP 3-37.3, *NBC Decontamination,* 28 July 2000.

Table II-33. HT Toxicity Estimates[10]

Endpoint	Toxicity (mg-min/m^3)[a]	MV (L)	Exposure Duration	ROE	Probit Slope	TLE	ROD	DOC
Lethality	LD_{50}: 1400 mg	N/A	N/A; 70-kg man	Percutaneous Liquid [b]	Unknown	N/A	Unknown	Low
	LCt_{50}: 1000	15	2-360 min	Inhalation/Ocular	Unknown	1 [c,d]	Unknown	Low
	LCt_{50}: 10,000 [e]	N/A	30-360 min	Percutaneous Vapor [f]	Unknown	1 [c,d]	Unknown	Low
	LCt_{50}: 5000 [g]	N/A	30-360 min	Percutaneous Vapor [f]	Unknown	1 [c,d]	Unknown	Low
Severe effects (vesication)	ED_{50}: 600 mg	N/A	N/A; 70-kg man	Percutaneous Liquid [b]	Unknown	N/A	Unknown	Low
	ECt_{50}: 500 [e]	N/A	30-360 min	Percutaneous Vapor [f]	Unknown	1 [c,d]	Unknown	Low
	ECt_{50}: 200 [g]	N/A	30-360 min	Percutaneous Vapor [f]	Unknown	1 [c,d]	Unknown	Low
Severe effect (eyes)	ECt_{50}: 75	N/A	2-360 min	Ocular	Unknown	1 [c,d]	Unknown	Low
Mild effects (pain, erythema, itching)	ECt_{50}: 50 [e]	N/A	30-360 min	Percutaneous Vapor [f]	Unknown	1 [c,d]	Unknown	Low
	ECt_{50}: 25 [g]	N/A	30-360 min	Percutaneous Vapor [f]	Unknown	1 [c,d]	Unknown	Low
Mild effects (eyes)	ECt_{50}: 25	N/A	2-360 min	Ocular	Unknown	1 [c,d]	Unknown	Low

NOTES

[a]The toxicity values given are provisional and based on recommendations for H/HD.
[b]Bare skin.
[c]The TLE value is assumed to be 1 because the concentration time profile is unknown.
[d]See Appendix H for supporting toxicity profile estimates.
[e]Moderate temperatures (65-85°F).
[f]Assumes personnel are masked with eye protection.
[g]Hot temperatures (greater than 85°F).

b. Arsenicals. The arsenical vesicants are organic dichloroarsines. They are respiratory tract irritants and produce lung injury on sufficient exposure. The vapors are irritating to the eyes and the liquid may produce serious eye lesions. Skin damage leading to vesication is produced by sufficient exposure to the vapor or by contact with the liquid. Absorption of vapor or liquid through the skin may lead to systemic intoxication or death.[20]

(1) Lewisite (L) (see Table II-34). L is a brown liquid with a geranium-like odor.[13] L is the principal arsenical of military interest.[12] It is extremely irritating to the eyes and quickly produces copious tearing. Liquid on the skin is immediately painful and is absorbed more promptly than H.[10] Blistering starts within several hours.[21] See Table II-35 (page II-53) for toxicity estimates.

Table II-34. L

Alternate Designations: EA 1034; Lyvizit; LI; M-1; Lewisite (arsenic compound)
Chemical Name: Dichloro(2-chlorovinyl)arsine
Synonyms: Arsonous dichloride, (2-chloroethenyl)-; Chlorovinylarsine dichloride; 2-Chlorovinyldichloroarsine; β-Chlorovinyldichloroarsine; (2-Chloroethenyl) arsonous dichloride; Arsine, dichloro (2-chlorovinyl)-
CAS Registry Number: 541-25-3
RTECS Number: CH2975000

Table II-34. L (Continued)

Physical and Chemical Properties

Structural Formula:

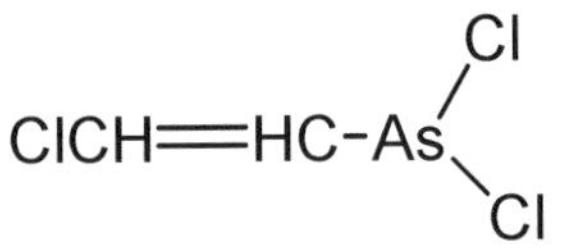

Molecular Formula: $C_2H_2AsCl_3$

Molecular Weight: 207.32

Physical State	Brown liquid; colorless when pure [1]
Odor	Geranium-like; odorless when pure [1]
Boiling Point	196°C (extrapolated); [2,3] decomposes prior to boiling [4]
FP/MP	–44.7 to –1.8°C (FP) (depending on purity and isomers present) [5]
Liquid Density (g/mL)	1.8793 @ 25°C; 1.9210 @ 0°C (extrapolated) [2]
Vapor Density (relative to air)	7.1 (calculated)
Vapor Pressure (torr)	3.46 x 10^{1} @ 25°C (extrapolated); 2.71 x 10^{-2} @ 0°C (extrapolated) [2,3]
Volatility (mg/m^3)	3.86 x 10^{3} @ 25°C; 3.30 x 10^{2} @ 0°C (calculated from vapor pressure) [2,3]
Latent Heat of Vaporization (kcal/mol)	15.5 @ 25°C; 17.5 @ 0°C (calculated from vapor pressure) [2,3]
Viscosity (cP)	2.053 @ 25.0°C, 3.521 @ 0°C (extrapolated)[2]
Viscosity of Vapor (cP)	8.53 x 10^{-3} @ 25.0°C, 7.70 x 10^{-3} @ 0°C[2]
Surface Tension (dynes/cm)	41.1 @ 25.0°C, 44.2 @ 0°C[2]
Flash Point	Nonflammable [6]
Decomposition Temperature	At 149°C, 0.5% of L is destroyed and @ 493°C > 99.99% is destroyed [4]
Solubility	Lewisite on contact with water immediately hydrolyzes to form Lewisite oxide (solid), which dissolves very slowly in water. [1,7] Readily soluble in common organic solvents, oils, and CW agents. [8]
Rate of Hydrolysis	Rapid [1,7]
Hydrolysis Products	2-Chlorovinylarsonous acid (CVAA), 2-chlorovinylarsenious oxide (lewisite oxide), and hydrochloric acid [9]
Stability in Storage	Fairly stable in glass and steel containers, but decomposes considerably upon detonation; alkalis decompose L @ ambient temperatures [1]
Action on Metals or Other Materials	None if L is dry; corrosive penetration on steel is 1 x 10^{-5} to 5 x 10^{-5} inch/month @ 65°C.[10] Extremely corrosive towards aluminum and aluminum alloys. [8]
Other Data	
Skin and eye toxicity	Extremely irritating to the eyes and produces copious tearing,[11] also causes immediate burning sensation on skin.
Inhalation toxicity	Most toxic route of exposure [11]
Rate of action	Rapid [12]
Means of detection	M9 paper, M256A1 CADK, M21 ACAA, M22 ACADA, M272 Water Testing Kit,[13] MM1
Protection required	MOPP4 whenever liquid or vapor is present [14]
Decontamination	Liquid on eyes and skin requires immediate decontamination. [14] STB, HTH, or household bleach is effective on equipment. Water, soaps, detergents, steam, and absorbents (earth, sawdust, ashes, and rags) are effective for physical removal. [15]
Use	Quick-acting casualty agent

Table II-34. L (Continued)

NOTES
[1]W.R. Kirner, *Summary Technical Report of Division 9, NDRC Volume 1, Chemical Warfare Agents, and Related Chemical Problems Part I-II,* Chapter 7, p. 83, Office of Scientific Research and Development, Washington, DC, 1946, UNCLASSIFIED Report (AD234270). [2]Samuel, J.B., et al., *Physical Properties of Standard Agents, Candidate Agents, and Related Compounds at Several Temperatures (U),* ARCSL-SP-83015, USA Armament Research and Development Command, Aberdeen Proving Ground, MD, June 1983, UNCLASSIFIED Report (ADC033491). [3]Sumner, J.F., et al., *The Vapour Pressure of Arsenious Chloride and of Lewisite I,* Sutton Oak Report 561(SO/R/561), Military Intelligence Division, Great Britain, December 1941, UNCLASSIFIED Report. [4]Brooks, M. E. and Parker, G.A., *Incineration/Pyrolysis of Several Agents and Related Chemical Materials Contained in Identification Sets,* ARCSL-TR-79040, October 1979, UNCLASSIFIED Report (ADB042888). [5]Macy, R., *Constants and Physiological Action of Chemical Warfare Agents,* EATR 78, Chemical Warfare Service, Edgewood Arsenal, MD July 1932, UNCLASSIFIED Report (ADB956574). [6]*Chemical Agent Data Sheets Volume I,* Edgewood Arsenal Special Report EO-SR-74001, USA Armament Command, Edgewood Arsenal, Aberdeen Proving Ground, MD December 1974, UNCLASSIFIED Report (ADB028222). [7]Buswell, A.M., et al., *The Chemistry of Certain Arsenical Chemical Warfare Agents as Water Contaminants,* OSRD 4193, Division 9 National Defense Research Committee of the Office of Scientific Research and Development, June 1944, UNCLASSIFIED Report. [8]Franke, S., *Manual of Chemistry Volume I- Chemistry of Chemical Warfare Agents,* ACSI-J-3890, Chemie der Kampfstoffe, East Berlin, April 1968, UNCLASSIFIED Technical Manual, (AD849866). [9]Bossle, P.C., et al., *Determination of Lewisite Contamination in Environmental Waters by High Performance Liquid Chromatography,* CRDEC-TR-042, USA Chemical Research Development and Engineering Center, Aberdeen Proving Ground, MD, January 1989, UNCLASSIFIED Report (ADA206000). [10]Harris, B.L. and Macy, R., *Corrosion by Vesicants: Rate of Corrosion of Steel and Other Metals by H, HQ, HN-3, HN-1, and L, Mostly at 65°C,* Technical Division Memorandum Report 1031, USA Chemical Research and Development Laboratories, Army Chemical Center, MD, April 1945, UNCLASSIFIED Report (ADB963161). [11]Sharon Reutter, et al., *Review and Recommendations for Human Toxicity Estimates for FM 3-11.9,* ECBC-TR-349, September 2003. [12]*NIOSH-DOD-OSHA Sponsored Chemical and Biological Respiratory Protection Workshop Report,* February 2000. [13]DOD Chemical and Biological Defense Program Annual Report to Congress, Volume I, April 2003. [14]FM 8-285/NAVMED P-5041/AFJMAN 44-149/FMFM 11-11, *Treatment of Chemical Agent Casualties and Conventional Military Chemical Injuries,* 22 December 1995. [15]FM 3-5/MCWP 3-37.3, *NBC Decontamination,* 28 July 2000.

Table II-35. L Toxicity Estimates[10]

Endpoint	Toxicity (mg-min/m^3)	MV (L)	Exposure Duration	ROE	Probit Slope	TLE	ROD	DOC
Lethality	LD_{50}: 1400 mg [a] (Provisional)	N/A	N/A; 70-kg man	Percutaneous Liquid [b]	Unknown	N/A	Unknown	Low
	LCt_{50}: 1000 [a] (Provisional)	15	2-360 min	Inhalation/ Ocular	Unknown	1 [c,d]	Unknown	Low
	LCt_{50}: 5000 – 10,000 [e,f] (Provisional)	N/A	30-360 min	Percutaneous Vapor [g]	Unknown	1 [c,d]	Unknown	Low [h]
	LCt_{50}: 2500 – 5000 [f,l] (Provisional)	N/A	30-360 min	Percutaneous Vapor [g]	Unknown	1 [c,d]	Unknown	Low [h]
Severe effects (vesication)	ED_{50}: 600 mg [a] (Provisional)	N/A	N/A; 70-kg man	Percutaneous Liquid [b]	Unknown	N/A	Unknown	Low
	ECt_{50}: 500 [a,e] (Provisional)	N/A	30-360 min	Percutaneous Vapor [g]	Unknown	1 [c,d]	Unknown	Low
	ECt_{50}: 200 [a, i] (Provisional)	N/A	30-360 min	Percutaneous Vapor [g]	Unknown	1 [c,d]	Unknown	Low
Severe effects (eyes)	ECt_{50}: 75 [a] (Provisional)	N/A	2-360 min	Ocular	Unknown	1 [c,d]	Unknown	Low
Mild effects (erythema, pain)	ECt_{50}: 50 [a,e] (Provisional)	N/A	30-360 min	Percutaneous Vapor [g]	Unknown	1 [c,d]	Unknown	Low
	ECt_{50}: 25 [a,l] (Provisional)	N/A	30-360 min	Percutaneous Vapor [g]	Unknown	1 [c,d]	Unknown	Low
Mild effects (eyes)	ECt_{50}: 25 [a] (Provisional)	N/A	2-360 min	Ocular	Unknown	1 [c,d]	Unknown	Low
Threshold/ odor detection	EC_{50}: 8 [j]	N/A	Few Seconds	Inhalation/ Ocular	N/A	N/A	Unknown	Low

NOTES

[a]Based on recommendations for H/HD.
[b]Bare skin.
[c]The TLE value is assumed 1 because the Ct profile is unknown.
[d]See Appendix H for supporting toxicity profile estimates.
[e]Moderate temperatures (65-85°F).
[f]Based on H estimate and animal data.
[g]Assumes personnel are masked with eye protection.
[h]True human values could be lower.
[i]Hot temperatures (greater than 85°F).
[j]Based on TM-215 (1952) and secondary human data.

(2) Mustard-Lewisite Mixture (HL) (see Table II-36 [page II-54]). HL has a garlic-like odor from its HD content.[13] Table II-36 lists the properties for the mixture with 37 percent HD and 63 percent L by weight. Statistical analysis of comparative data for HL, H, and L indicates that HL is equipotent to H and both are statistically less potent than L.[10] See Table II-37 (page II-56) for toxicity estimates.

Table II-36. HL

Alternate Designations: Distilled mustard gas and lewisite mixture	
Chemical Name: HD: Bis-(2-chloroethyl) sulfide; L: Dichloro-(2-chlorovinly)arsine	
Synonyms: N/A	
CAS Registry Number: HD: 505-60-2; L: 541-25-3	
RTECS Number: HD: WQ0900000; L: CH2975000	
Physical and Chemical Properties	
Structural Formula: 37 wt% HD: $Cl\text{-}CH_2\text{-}CH_2\text{-}S\text{-}CH_2\text{-}CH_2\text{-}Cl$; 63 wt% L: $ClCH{=}HC\text{-}AsCl_2$	
Molecular Formula: HD: $C_4H_8Cl_2S$; L: $C_2H_2AsCl_3$	
Molecular Weight: HD: 159.07; L: 207.32; Average: 186.39 (based on 37:63 wt %)	
Physical State	Liquid [1]
Odor	Garlic-like (HD) [2]
Boiling Point	200°C (extrapolated) [3,4]
FP/MP	Munitions grade: -42°C (FP); pure: -25.4°C (FP) [4]
Liquid Density (g/mL)	1.6383 @ 20°C (calculated) (based on 67 wt% L) [1]
Vapor Density (relative to air)	6.4 (calculated)
Vapor Pressure (torr)	3.63×10^{-1} @ 25°C; 4.93×10^{-2} @ 0°C; (calculated based on Raoult's law; actual values are assumed to be somewhat lower than calculated values.) [3,4]
Volatility (mg/m^3)	3.64×10^3 @ 25°C, 5.39×10^2 @ 0°C (calculated based on Raoult's law; actual values are assumed to be somewhat lower than calculated values.) [3,4]
Latent Heat of Vaporization (kcal/mol)	12.8 @ 25°C; 13.1 @ 0°C (calculated from vapor pressure) [3,4]
Flash Point	Data not available for the mixture; HD flashes @ 105°C [5]
Viscosity (cP)	Data not available
Viscosity of Vapor (cP)	Data not available
Surface Tension (dynes/cm)	Data not available
Decomposition Temperature	Above 100°C; based on data which shows that HD decomposes @ 180°C [6] and that L starts to decompose @ 150°C; [7] this might suggest that HL also decomposes in this temperature range
Solubility	Both HD and L are soluble in most organic solvents but only slightly soluble in water, suggesting that HL has a similar degree of solubility towards organic solvents and water.[8]
Rate of Hydrolysis	HD $t_{1/2}$ = 5 min @ 25°C;[9] HD on or under water undergoes hydrolysis only if dissolved. The rate of HD hydrolysis is controlled by the rate of mass transfer and is very slow. [10] Lewisite on contact with water or moist surfaces immediately hydrolyzes to form lewisite oxide (solid), which dissolves very slowly in water. [11,12]
Hydrolysis Products	Hydrogen chloride, thiodiglycol, sulfonium aggregates, (CVAA), and lewisite oxide-based on HD and L [13,14]
Stability in Storage	Stable in lacquered steel containers for approximately 3 months @ 65°C, 6 months @ 50°C, and for a year or more @ ambient temperature when using a 50:50 mixture of HD and L. Less stable in uncoated steel containers @ temperatures above 50°C. Stable in glass @ 65°C. [15]
Action on Metals or Other Materials	Little or none if dry [8]
Other Data	
Skin and eye toxicity	Equal to L in vesication action;[16] both H and L are irritating to the eyes.
Inhalation toxicity	Most toxic route of exposure [16]
Rate of action	Prompt stinging; blistering delayed approximately 12 hours [17]
Means of detection	CAM/ICAM, M256A1 CADK, M18A2 CADK,[18] MM1
Protection required	MOPP4 whenever liquid or vapor is present [19]

Table II-36. HL (Continued)

Decontamination	Liquid on eyes and skin requires immediate decontamination.[19] HTH or household bleach is effective on equipment. Water, soaps, detergents, steam, and absorbents (earth, sawdust, ashes, and rags) are effective for physical removal. STB does not effectively decontaminate mustard if it has solidified at low temperatures.[20]
Use	Delayed-action casualty agent

NOTES

[1]Mumford, S.A. and Parry, G.A., *Report on Physical Properties of Mixtures of H and Lewisite I*, PR-1342, USA Chemical Research and Development Laboratories, Army Chemical Center, MD, March 1935, UNCLASSIFIED Report.
[2]TM 3-215/AFM 355-7, *Military Chemistry and Chemical Agents*, Washington DC, December 1963, UNCLASSIFIED Technical Manual (ADA292141).
[3]Abercrombie, P., ECBC Notebook # NB 98-0079, p. 29 (U).
[4]Macy, R, *Freezing Point and Volatilities of Mustard and Lewisite Mixtures*, TCIR 512, USA Chemical Research and Development Laboratories, Army Chemical Center, MD, March 1935, UNCLASSIFIED Report (ADB9670444).
[5]Buckles, M.F., *CW Vesicants: Selected Values for the Physical Properties of H, T, and Q (U)*, Special Report CRLR 542, Chemical Corps Chemical and Radiological Laboratories, Army Chemical Center, MD, May 1956, UNCLASSIFIED Report (AD108272).
[6]Williams, A.H., "The Thermal Decomposition of 2:2'-Dichlorodiethyl Sulphide," *J. Chem. Soc.*, p. 318, 1947.
[7]Brooks, M. E. and Parker, G.A., *Incineration/Pyrolysis of Several Agents and Related Chemical Materials Contained in Identification Sets*, ARCSL-TR-79040, October 1979, UNCLASSIFIED Report (ADB042888).
[8]*Properties of War Gases Volume IV: Vesicants (U)*, ETF 100-41/Vol-4, Chemical Corps Board, Army Chemical Center, MD, December 1956, CONFIDENTIAL Report (AD108459).
[9]Bartlett, P.D., and Swain, C.G., "Kinetics of Hydrolyisis and Displacement Reactions of β,β'-[Dichlorodiethyl Sulfide (Mustard Gas)] and of β-Chloro- β'-hyroxidediethyl Sulfide (Mustard Chlorohydrin)," *J. Chem. Soc.*, Vol. 71, p. 1406, 1949.
[10]Yang, Y., et al., "Decontamination of Chemical Warfare Agents," *Chem. Rev.,* Vol. 92, p. 1729, 1992.
[11]W.R. Kirner, *Summary Technical Report of Division 9, NDRC Volume 1, Chemical Warfare Agents, and Related Chemical Problems, Part I-II,* Office of Scientific Research and Development, Washington, D.C., 1946, UNCLASSIFIED Report (AD234270).
[12]Buswell, A.M., et al., *The Chemistry of Certain Arsenical Chemical Warfare Agents as Water Contaminants*, OSRD 4193, Division 9 National Defense Research Committee of the Office of Scientific Research and Development, June 1944, UNCLASSIFIED Report.
[13]Yang, Y., et al., "Characterization of HD Heels and the Degradation of HD in Ton Containers," *In Proceedings of the 1996 ERDEC Scientific Conference on Chemical and Biological Defense Research 19-22 November 1996*, UNCLASSFIED Paper (ADE487543), ERDEC-SP-048, pp 353-360, USA Edgewood Research, Development and Engineering Center, Aberdeen Proving Ground, MD, October 1997, UNCLASSFIED Report (ADA334105).
[14]Bossle, P.C., et al., *Determination of Lewisite Contamination in Environmental Waters by High Performance Liquid Chromatography*, CRDEC-TR-042, USA Chemical Research Development and Engineering Center, Aberdeen Proving Ground, MD, January 1989, UNCLASSIFIED Report (ADA206000).
[15]Harris, B.L. and Macy, R., *Memorandum Report Storage Stability of HL, Mixtures of Mustard and Lewisite*, Technical Division Memorandum Report 1302, USA Chemical Research and Development Laboratories, Army Chemical Center, MD, February 1947, UNCLASSIFIED Report (ADB96498).
[16]Sharon Reutter, et al., *Review and Recommendations for Human Toxicity Estimates for FM 3-11.9*, ECBC-TR-349, September 2003.
[17]NIOSH-DOD-OSHA Sponsored Chemical and Biological Respiratory Protection Workshop Report, February 2000.
[18]DOD Chemical and Biological Defense Program Annual Report to Congress, Volume I, April 2003.
[19]FM 8-285/NAVMED P-5041/AFJMAN 44-149/FMFM 11-11, *Treatment of Chemical Agent Casualties and Conventional Military Chemical Injuries*, 22 December 1995.
[20]FM 3-5/MCWP 3-37.3, *NBC Decontamination,* 28 July 2000.

Table II-37. HL Toxicity Estimates[10]

Endpoint	Toxicity (mg-min/m^3)	MV (L)	Exposure Duration	ROE	Probit Slope	TLE	ROD	DOC
Lethality	LD_{50}: 1400 mg [a] (Provisional)	N/A	N/A; 70-kg man	Percutaneous Liquid [b]	Unknown	N/A	Unknown	Low
	LCt_{50}: 1000 [a] (Provisional)	15	2-360 min	Inhalation/ Ocular	Unknown	1 [c,d]	Unknown	Low
	LCt_{50}: 10,000 [a,e] (Provisional)	N/A	30-360 min	Percutaneous Vapor [f]	Unknown	1 [c,d]	Unknown	Low
	LCt_{50}: 5000 [a,g] (Provisional)	N/A	30-360 min	Percutaneous Vapor [f]	Unknown	1 [c,d]	Unknown	Low
Severe effects (vesication)	ED_{50}: 600 mg [a] (Provisional)	N/A	N/A; 70-kg man	Percutaneous Liquid [b]	Unknown	N/A	Unknown	Low
	ECt_{50}: 500 [a,e] (Provisional)	N/A	30-360 min	Percutaneous Vapor [f]	Unknown	1 [c,d]	Unknown	Low
	ECt_{50}: 200 [a,g] (Provisional)	N/A	30-360 min	Percutaneous Vapor [f]	Unknown	1 [c,d]	Unknown	Low
Severe effect (eyes)	ECt_{50}: 75 [a] (Provisional)	N/A	2-360 min	Ocular	Unknown	1 [c,d]	Unknown	Low
Mild effects (erythema, itching, pain)	ECt_{50}: 50 [a,e] (Provisional)	N/A	30-360 min	Percutaneous Vapor [f]	Unknown	1 [c,d]	Unknown	Low
	ECt_{50}: 25 [a,g] (Provisional)	N/A	30-360 min	Percutaneous Vapor [f]	Unknown	1 [c,d]	Unknown	Low
Mild effects (eyes)	ECt_{50}: 25 [a] (Provisional)	N/A	2-360 min	Ocular	Unknown	1 [c,d]	Unknown	Low
Odor Detection	EC_{50}: 2 mg/m^3 [h]	N/A	Few Seconds	Inhalation	N/A	N/A	N/A	Low

NOTES

[a]Based on recommendations for H/HD.
[b]Bare skin.
[c]The TLE value is assumed to be 1 because the Ct profile is unknown.
[d]See Appendix H for supporting toxicity profile estimates.
[e]Moderate temperatures (65-85°F).
[f]Assumes personnel are masked with eye protection.
[g]Hot temperatures (greater than 85°F).
[h]Based on TM 3-215 (1952).

(3) Phenyldichloroarsine (PD) (see Table II-38). PD is an odorless and colorless to yellow liquid.[13] PD does have marked vesicant and sternutatory properties, but the primary action is lung injury. Although PD is somewhat less vesicant than H and L by the percutaneous liquid route of exposure, it is a fairly potent vesicant.[10] See Table II-39 (page II-58) for toxicity estimates.

Table II-38. PD

Alternate Designations: Pfiffikus; DJ; Sternite, M.A., TL 69; FDA	
Chemical Name: Phenyldichloroarsine	
Synonyms: Arsine, dichlorophenyl-; Arsonous dichloride, phenyl-; Dichlorophenylarsine; Phenylarsonous dichloride; Phenylarsinedichloride; Dichlor-fenylarsin (Czech); Fenildicloroarsina (Italian); Fenyldichlorarsin (Czech)	
CAS Registry Number: 696-28-6	
RTECS Number: CH5425000	
Physical and Chemical Properties	
Structural Formula: 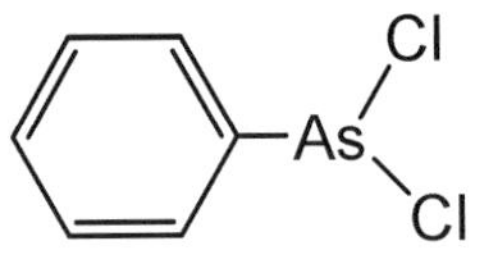 Molecular Formula: $C_6H_5AsCl_2$ Molecular Weight: 222.93	
Physical State	Colorless to yellow liquid [1]
Odor	None [2]
Boiling Point	233°C (extrapolated) [3]
FP/MP	-22.5°C (FP) [3]
Liquid Density (g/mL)	1.645 @ 25°C; 1.677@ 0°C [3]
Vapor Density (relative to air)	7.7 (calculated)
Vapor Pressure (torr)	2.2×10^{-2} @ 25°C; 2.1×10^{-3} @ 0°C (both values are extrapolated) [3]
Volatility (mg/m^3)	2.64×10^{2} @ 25°C; 2.3×10^{1} @ 0°C (calculated from vapor pressure) [3]
Latent Heat of Vaporization (kcal/mol)	15.1 (calculated from Clausius Clapeyron equation which assumes constant heat of vaporization as a function of temperature) [3]
Viscosity (cP)	Data not available
Viscosity of Vapor (cP)	Data not available
Surface Tension (dynes/cm)	Data not available
Flash Point	Data not available
Decomposition Temperature	Stable to the normal boiling point [3]
Solubility	Immediately hydrolyzes in the presence of water,[4] miscible with alcohol, benzene, ether, acetone,[5,6] kerosene, petroleum, and olive oil [6]
Rate of Hydrolysis	Very rapid [4]
Hydrolysis Products	Hydrochloric acid and phenylarsine oxide which are also highly toxic [4]
Stability in Storage	Data not available
Action on Metals or Other Materials	No serious effects on mild steel and cast iron. [7]
Other Data	
Skin and eye toxicity	Fairly potent vesicant [8]
Inhalation toxicity	Toxic lung injurant [8]
Rate of action	Immediate effect on eyes; effects on skin are delayed approximately 1 hour [9]
Means of detection	M18A2 CADK, MM1, M18A3 CADK [10]
Protection required	MOPP4 whenever liquid or vapor is present [11]
Decontamination	Liquid on eyes and skin requires immediate decontamination. [11] Household bleach is effective on equipment. Water, soaps, detergents, steam, and absorbents (earth, sawdust, ashes, and rags) are effective for physical removal. [12]
Use	Delayed-action casualty agent

Table II-38. PD (Continued)

NOTES
[1]Lewis, R.J., *Sax's Dangerous Properties of Industrial Materials*, 10th ed., Volume 3, p. 1215, John Wiley & Sons, Inc., New York, NY, 2001.
[2]TM 3-215/AFM 355-7, *Military Chemistry and Chemical Agents*, December 1963, UNCLASSIFIED Technical Manual (ADA292141).
[3]Owens, R., *Diphenylcyanoarsine: Part V – The Physical Properties of M.A., D.A. T.A., and D.C.*, SO/R 492, Sutton Oak, England, December 1940, UNCLASSIFIED Report.
[4]Buswell, A.M., et al., *The Chemistry of Certain Arsenical Chemical Warfare Agents as Water Contaminants*, OSRD 4193, Division 9 National Defense Research Committee of the Office of Scientific Research and Development, June 1944, UNCLASSIFIED Report.
[5]Lide, D.R., *CRC Handbook of Chemistry and Physics,* 82nd ed., p. 3-15, CRC Press, Washington, DC, 2001.
[6]*Chemical Agent Data Sheets Volume* I, Edgewood Arsenal Special Report EO-SR-74001, Edgewood Arsenal, Aberdeen Proving Ground, MD December 1974, UNCLASSIFIED Report (ADB028222).
[7]Owens, R., *Diphenylcyanoarsine Part III – The Pope-Turner Process*, SO/R 488, Sutton Oak, England, December 1940, UNCLASSIFIED Report.
[8]Sharon Reutter, et al., *Review and Recommendations for Human Toxicity Estimates for FM 3-11.9*, ECBC-TR-349, September 2003.
[9]*NIOSH-DOD-OSHA Sponsored Chemical and Biological Respiratory Protection Workshop Report,* February 2000.
[10]DOD Chemical and Biological Defense Program Annual Report to Congress, Volume I, April 2003.
[11]FM 8-285/NAVMED P-5041/AFJMAN 44-149/FMFM 11-11, *Treatment of Chemical Agent Casualties and Conventional Military Chemical Injuries*, 22 December 1995.
[12]FM 3-5/MCWP 3-37.3, *NBC Decontamination,* 28 July 2000.

(4) PD Toxicity Estimates (Table II-39). No toxicity estimates are recommended for lethal percutaneous, liquid, or vapor exposures or sever percutaneous liquid exposure. The existing estimates are not supported by available data.

Table II-39. PD Toxicity Estimates[10]

Endpoint	Toxicity ($mg\text{-}min/m^3$)	MV (L)	Exposure Duration	ROE	Probit Slope	TLE	ROD	DOC
Lethality	LD_{50}: NR	N/A	N/A	Percutaneous Liquid	Unknown	Unknown	Unknown	N/A
	LCt_{50}: 2600 [a] (Provisional)	15	2-360 min	Inhalation/Ocular	Unknown	1 [b,c]	Unknown	Low
	LCt_{50}: NR	N/A	N/A	Percutaneous Vapor [d]	Unknown	Unknown	Unknown	N/A
Severe effects (vesication)	ED_{50}: NR	N/A	N/A	Percutaneous Liquid	Unknown	Unknown	Unknown	N/A
	ECt_{50}: 200-500 [e] (Provisional)	15	30-50 min	Percutaneous Vapor [d]	Unknown	1 [b,c]	Unknown	Low
Intolerable	ECt_{50}: 16 [f]	15	1 - 2 min	Inhalation/Ocular	Unknown	Unknown	Unknown	Low
Threshold (odor detection)	ECt_{50}: 1 [f]	N/A	1 min	Inhalation/Ocular	Unknown	Unknown	N/A	Low

NOTES

[a]Based on existing estimates.
[b]The TLE value is assumed to be 1 because the Ct profile is unknown.
[c]See Appendix H for supporting toxicity profile estimates.
[d]Assumes personnel are masked with eye protection.
[e]Based on currently accepted estimates for H/HD.
[f]Based on secondary human data.

(5) Ethyldichloroarsine (ED) (see Table II-40). ED is a colorless liquid with a fruity but biting and irritating odor.[13] Although ED is a fairly powerful sternutator and vesicant, it is primarily a toxic lung injurant. ED is a more powerful irritant than L and produces sneezing and lacrimation.[10] See Table II-41 (page II-60) for toxicity estimates.

Table II-40. ED

Alternate Designations: DICK (German); TL 214; Green Cross 3; Yellow Cross 1	
Chemical Name: Ethyldichloroarsine	
Synonyms: Dichloroethylarsine; Arsine, dichloroethyl-; Arsenic dichloroethane; Ethylarsonous dichloride; Arsonous dichloride, ethyl-	
CAS Registry Number: 598-14-1	
RTECS Number: CH3500000	
Physical and Chemical Properties	
Structural Formula: 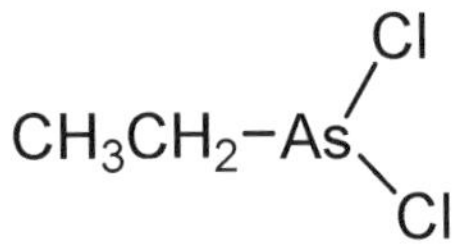 Molecular Formula: $C_2H_5AsCl_2$ Molecular Weight: 174.89	
Physical State	Colorless liquid [1]
Odor	Fruity, biting, and irritating [1]
Boiling Point	156°C; [2] decomposes [3]
FP/MP	Below –65°C (MP) [2]
Liquid Density (g/mL)	1.742 @ 14°C [2]
Vapor Density (relative to air)	6.0 (calculated)
Vapor Pressure (torr)	2.29 @ 21.5°C [2]
Volatility (mg/m^3)	2.19 x 10^4 @ 20°C [2]
Latent Heat of Vaporization (kcal/mol)	9.18 [3]
Viscosity (cP)	Data not available
Viscosity of Vapor (cP)	Data not available
Surface Tension (dynes/cm)	Data not available
Flash Point	Data not available
Decomposition Temperature	Stable to boiling point [3]
Solubility	Immediately hydrolyzes in the presence of water.[4] Soluble in ethyl chloride, alcohol, ether, benzene, acetone, kerosene and cyclohexane.[3]
Rate of Hydrolysis	Very rapid [4]
Hydrolysis Products	Hydrochloric acid and ethylarsine oxide, which are also highly toxic [4]
Stability in Storage	Stable [3]
Action on Metals or Other Materials	None on steel when pure,[2] noncorrosive towards iron @ temperatures up to 50°C, when ED is dry.[5] Attacks brass @ 50°C and is destructive to rubber and plastics [3]
Other Data	
Skin and eye toxicity	Fairly potent vesicant and lacrimator [6]
Inhalation toxicity	Primarily a lung injurant [6]
Rate of action	Immediate irritation; delayed blistering [7]
Means of detection	M18 A3 CADK,[8] M18A2 CADK, MM1
Protection required	MOPP4 whenever liquid or vapor is present [9]
Decontamination	Liquid on eyes and skin requires immediate decontamination. [9] Household bleach is effective on equipment. Water, soaps, detergents, steam, and absorbents (earth, sawdust, ashes, and rags) are effective for physical removal. [10]
Use	Delayed-action casualty agent

Table II-40. ED (Continued)

NOTES
[1]Sax, N.I., *Dangerous Properties of Industrial Materials*, 3rd ed., p. 741, Reinhold Book Corporation, Albany, NY, 1968.
[2]Dawson, T.P., *Ethyldichloroarsine (ED): Preliminary Investigation (1939)*, EATR 325, Chemical Warfare Service, Edgewood Arsenal, MD, November 1941, UNCLASSIFIED Report (ADB957078).
[3]TM 3-215/AFM 355-7, *Military Chemistry and Chemical Agents*, December 1963, UNCLASSIFIED Technical Manual (ADA292141).
[4]Buswell, A.M., et al., *The Chemistry of Certain Arsenical Chemical Warfare Agents as Water Contaminants*, OSRD 4193, Division 9 National Defense Research Committee of the Office of Scientific Research and Development, June 1944, UNCLASSIFIED Report.
[5]Kibler, A.L, *Data on Chemical Warfare*, TDMR 456, Chemical Warfare Center, Edgewood Arsenal, MD, November 1942, UNCLASSIFIED Report (ADB969725).
[6]Sharon Reutter, et al., *Review and Recommendations for Human Toxicity Estimates for FM 3-11.9*, ECBC-TR-349, September 2003.
[7]*NIOSH-DOD-OSHA Sponsored Chemical and Biological Respiratory Protection Workshop Report,* February 2000.
[8]DOD Chemical and Biological Defense Program Annual Report to Congress, Volume I, April 2003.
[9]FM 8-285/NAVMED P-5041/AFJMAN 44-149/FMFM 11-11, *Treatment of Chemical Agent Casualties and Conventional Military Chemical Injuries*, 22 December 1995.
[10]FM 3-5/MCWP 3-37.3, *NBC Decontamination,* 28 July 2000.

(6) ED Toxicity Estimates (Table II-41). No toxicity estimates are recommended for lethal exposure. The existing estimates are not supported by the available data.

Table II-41. ED Toxicity Estimates[10]

Endpoint	Toxicity ($mg\text{-}min/m^3$)	MV (L)	Exposure Duration	ROE	Probit Slope	TLE	ROD	DOC
Lethality	LCt_{50}: NR	N/A	N/A	Inhalation/ Ocular	Unknown	Unknown	Unknown	N/A
	LCt_{50}: NR	N/A	N/A	Percutaneous Vapor	Unknown	Unknown	Unknown	N/A
Severe effects (temporary incapacitation)	ECt_{50}: 5 –10 [a]	N/A	1 min	Inhalation/ Ocular	Unknown	N/A	Unknown	Low
Odor detection	EC_{50}: 1 mg/m^3 [b]	N/A	1 min	Inhalation/ Ocular	Unknown	N/A	Unknown	Low

NOTES

[a]Based on secondary human data and TM 3-215 (1952).
[b]Based on existing estimate.

(7) Methyldichloroarsine (MD) (see Table II-42). MD is a colorless and odorless liquid when pure.[13] It has been described as a vesicant, toxic lung injurant, and respiratory irritant. Inhalation causes a severe irritation in the nose, which produces sneezing and finally extends to the chest where it gives rise to pain.[10] See Table II-43 (page II-62) for toxicity estimates.

Table II-42. MD

Alternate Designations: TL 294; Methyl-dick; Medikus
Chemical Name: Methyldichloroarsine
Synonyms: Arsine, dichloromethyl-; Arsonous dichloride, methyl-; Dichloromethylarsine; Methylarsine dichloride; Methylarsonous dichloride
CAS Registry Number: 593-89-5
RTECS Number: CH4375000

Table II-42. MD (Continued)

Physical and Chemical Properties	
Structural Formula: 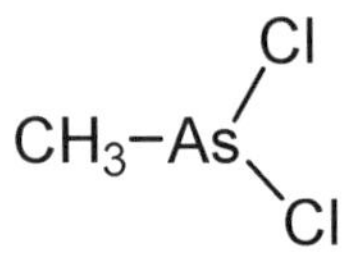 Molecular Formula: CH_3AsCl_2 Molecular Weight: 160.86	
Physical State	Colorless Liquid [1]
Odor	Extremely irritating; none when pure [1]
Boiling Point	132.6°C [2]
FP/MP	-54.8°C (FP) [3]
Liquid Density (g/mL)	1.839 @ 20°C; 1.875 @ 0°C [4]
Vapor Density (relative to air)	5.5 (calculated)
Vapor Pressure (torr)	7.593 @ 20°C; 2.063 @ 0°C [2]
Volatility (mg/m^3)	6.68×10^4 @ 20°C; 1.95×10^4 @ 0°C (calculated from vapor pressure) [2]
Latent Heat of Vaporization (kcal/mol)	10.5 @ 20°C; 10.2 @ 0°C (calculated from vapor pressure) [2]
Viscosity (cP)	Data not available
Viscosity of Vapor (cP)	Data not available
Surface Tension (dynes/cm)	Data not available
Flash Point	Data not available
Decomposition Temperature	Stable up to the boiling point [2]
Solubility	Immediately hydrolyzes in the presence of water. [5] Soluble in common organic solvents @ ambient temperatures.[1]
Rate of Hydrolysis	Very rapid, [5] complete in less than 2 min @ 25°C in dilute solution [6]
Hydrolysis Products	Hydrogen chloride and methylarsenic oxide [5]
Stability in Storage	Stable in steel containers @ 60°C for a period of at least 15 weeks and for at least 1 year @ ambient temperatures [7]
Action on Metals or Other Materials	No appreciable amount of corrosion on steel when MD is pure and acid-free. [8] Satisfactory with steel for at least 1 year @ ambient temperature, but @ elevated temperatures (60°C) crude pitting occurs within 15 weeks. [7]
Other Data	
Skin and eye toxicity	Blistering action less than that of HD and L; eye and skin irritant [9]
Inhalation toxicity	Toxic lung injurant and respiratory irritant [9]
Rate of action	Immediate irritation; delayed blistering [10]
Means of detection	M18A3 CADK,[11] M18A2 CADK, MM1
Protection required	MOPP4 whenever liquid or vapor is present [12]
Decontamination	Liquid on eyes and skin requires immediate decontamination. [12] Household bleach is effective on equipment. Water, soaps, detergents, steam, and absorbents (earth, sawdust, ashes, and rags) are effective for physical removal. [13]
Use	Delayed-action casualty agent

Table II-42. MD (Continued)

NOTES
[1]Macintire, B.G., et al., *Methyldichloroarsine and Methyldifluoroarsine Field Tests*, EACD 410, Chemical Warfare Service, Edgewood Arsenal, MD, March 1931, UNCLASSIFIED Report (ADB955243).
[2]Watson, P.D., *Determination of the Vapor Pressure of Methyldichloroarsine*, EACD 176, Chemical Warfare Service, Edgewood Arsenal, Edgewood, MD, May 1922, UNCLASSIFIED Report (ADB959625).
[3]Mead, W.P. *Freezing Points of Mixtures of Methyldichloroarsine and Mustard Gas and of Lewisite and Mustard Gas*, EACD 170, Chemical Warfare Service, Edgewood Arsenal, Edgewood, MD, May 1922, UNCLASSIFIED Report (ADB95011).
[4]Klosky, S., and Stricker, P.F., *The Physico Chemical Properties of Methyldichloroarsine and Arsenic Trichloride*, EACD 63, Chemical Warfare Service, Edgewood Arsenal, Edgewood, MD, August 1921, UNCLASSIFIED Report (ADB955049).
[5]Buswell, A.M., et al., *The Chemistry of Certain Arsenical Chemical Warfare Agents as Water Contaminants*, OSRD 4193, Division 9 National Defense Research Committee of the Office of Scientific Research and Development, June 1944, UNCLASSIFIED Report.
[6]Beebe, C.H, *Important Constants of Fourteen Common Chemical Warfare Agents*, EACD 328, Chemical Warfare Service, Edgewood Arsenal, Edgewood, MD, December 1924, UNCLASSIFIED Report (ADB958296).
[7]Henley, F.M., *Surveillance Tests on 75 mm Steel Gas Shell Extending Over a Period of One Year*, EACD 11, Chemical Warfare Service, Edgewood Arsenal, MD, June 1920, UNCLASSIFIED Report (ADB959731).
[8]Siegel, M., *The Corrosive Effect of War Gases on Metals and Materials*, EACD 113, Chemical Warfare Service, Edgewood Arsenal, MD, December 1921, UNCLASSIFIED Report (ADB955153).
[9]Sharon Reutter, et al., *Review and Recommendations for Human Toxicity Estimates for FM 3-11.9*, ECBC-TR-349, September 2003.
[10]*NIOSH-DOD-OSHA Sponsored Chemical and Biological Respiratory Protection Workshop Report,* February 2000.
[11]DOD Chemical and Biological Defense Program Annual Report to Congress, Volume I, April 2003.
[12]FM 8-285/NAVMED P-5041/AFJMAN 44-149/FMFM 11-11, *Treatment of Chemical Agent Casualties and Conventional Military Chemical Injuries*, 22 December 1995.
[13]FM 3-5/MCWP 3-37.3, *NBC Decontamination,* 28 July 2000.

(8) MD Toxicity Estimates (Table II-43). No toxicity estimates are recommended for lethal exposure. The existing estimates are not supported by the available data.

Table II-43. MD Toxicity Estimates[10]

Endpoint	Toxicity (mg-min/m³)	MV (L)	Exposure Duration	ROE	Probit Slope	TLE	ROD	DOC
Lethality	LCt_{50}: NR	N/A	N/A	Inhalation/ Ocular	Unknown	Unknown	Unknown	N/A
Severe effects, (temporary incapacitation)	ECt_{50}: 25 [a] (Provisional)	N/A	1 min	Inhalation/ Ocular	Unknown	N/A	Unknown	Low
Odor Detection	EC_{50}: less than 1 mg/m³ [a]	N/A	1 min	Inhalation/ Ocular	Unknown	N/A	Unknown	Low
NOTES								
[a]Based on secondary human data and TM 3-215 (1952).								

c. Urticants. Urticants are not true vesicants because, unlike mustard and L, they do not produce fluid-filled blisters; rather, they produce solid lesions resembling urticaria.[21] CX is the primary urticant of military interest (see Table II-44). It can penetrate garments and rubber much more quickly than other agents.[21] It affects the skin, eyes, and lungs. No other chemical agent produces such an immediately painful onset that is followed by rapid tissue necrosis. The skin lesions, in particular, are similar to those caused by a strong acid. The rapid skin damage renders the skin more susceptible to a second type of agent.[21] Droplets on the skin are potentially lethal. CX has also been classified as a lung poison.[10] See Table II-45 (page II-64) and Appendix H for toxicity estimates.

Table II-44. CX

Alternate Designations: Fosgen Oksim; Phosgen-oxime	
Chemical Name: Dichloroformoxime	
Synonyms: 1,2-Dichloroformoxime; Dichloroformaldoxime; Dichloroximinomethane; Dichlorformaldehyd-oxime; Kohlensaure-dichlorid-oxime; Dichlormethylen-hydroxylamine; Carbonyl chloride oxime	
CAS Registry No: 1794-86-1	
RTECS Number: Data not available	
Physical and Chemical Properties	
Structural Formula: 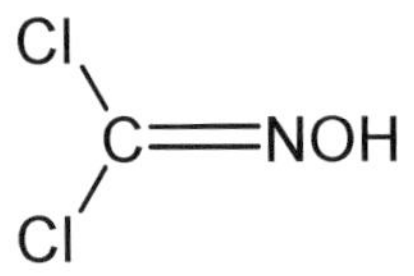	
Molecular Formula: $CHCl_2NO$	
Molecular Weight: 113.93	
Physical State	Colorless, crystalline, deliquescent-solid when pure [1]
Odor	Unpleasant and irritating; [1] resembles new-mown hay at low concentrations [2]
Boiling Point	129°C [3] (with decomposition unless highly pure)
FP/MP	39°C (MP) [3]
Liquid Density (g/mL)	Data not available
Vapor Density (relative to air)	3.9 (calculated)
Vapor Pressure (torr)	2.43×10^{1} @ 50°C [3]
Volatility (mg/m^3)	1.37×10^{5} @ 50°C (calculated from vapor pressure) [3]
Latent Heat of Vaporization (kcal/mol)	11.2 @ 50°C (calculated from vapor pressure) [3]
Viscosity (cP)	Data not available
Viscosity of Vapor (cP)	Data not available
Surface Tension (dynes/cm)	Data not available
Flash Point	Data not available
Decomposition Temperature	Below 129°C [4]
Solubility	Very soluble in both water and common organic solvents [5]
Rate of Hydrolysis	Slow in water @ ambient temperature and pH 7; hydrolyzes 5% within six days @ ambient temperature; dilute acids slow down the hydrolysis rate even further, whereas basic solutions react very violently with CX [2]
Hydrolysis Products	Carbon dioxide, hydrogen chloride, and hydroxylamine [2]
Stability in Storage	Pure, unstabilized CX decomposes on storage @ ambient temperature. If stored @ -20°C it can be kept indefinitely. [1] CX is extremely unstable in the presence of impurities such as metals; even trace amounts of iron chloride may cause explosive decomposition. [2]
Action on Metals or Other Materials	Metals, especially iron, cause rapid decomposition of CX; trace amounts of iron chloride may cause explosive decomposition. [2] Also attacks rubber, especially upon heating. [5]
Other Data	
Skin and eye toxicity	Causes pain, irritation, and severe tissue damage on skin. CX causes pain, conjunctivitis, and inflammation of the cornea of the eye. [6]
Inhalation toxicity	Can cause pulmonary edema [6]
Rate of action	Almost instantaneous [6]
Means of detection	M256A1 CADK, M18A3 CADK, M18A2 CADK, MM1 [7]
Protection required	MOPP4 [8]

Table II-44. CX (Continued)

Decontamination	Because of the rapid reaction of CX with the skin, decontamination will not be entirely effective after pain occurs; nevertheless, decontaminate as rapidly as possible with M291 SDK. If the M291 kit is not available, flush the area with large amounts of water to remove any agent that has not reacted with the skin. [8] Household bleach is effective on equipment. Water, soaps, detergents, steam, and absorbents (earth, sawdust, ashes, and rags) are effective for physical removal. [9]
Use	Rapid-acting casualty agent

NOTES

[1]Witten, B., *The Search for Toxic Chemical Agents (U)*, EATR 4210, Edgewood Arsenal Research Laboratories, MD, November 1969, UNCLASSIFIED Report (AD507852).
[2]Petersen, T.G., *Agent CX (Phosgene Oxime) Summary Report (U)*, CRDL Special Publication 7, USA Chemical Research and Development Laboratories, Edgewood Arsenal, MD, October 1965, CONFIDENTIAL Report (AD367890).
[3]Penski, E.C., *Vapor Pressure Data Analysis of Dichloroformoxime*, ERDEC-TR-042, USA Chemical and Biological Defense Agency, Aberdeen Proving Ground, MD, March 1993, UNCLASSIFIED Report (ADA265873).
[4]*Properties of War Gases Volume II: Blood and Nettle Gases (U)*, ETF 100-41/Vol-2, Chemical Corps Board, Army Chemical Center, MD, December 1956, CONFIDENTIAL Report (AD108457).
[5]Prandtl, W. and Sennewald, K., "Trichloronitrosomethane, Dichloroformoxime (Phosgene Oxime) and Their Derivatives," *Chemische Berichte,* Vol. 62, p. 1766, 1929.
[6]BG Russ Zajtchuk et al. (eds), *Textbook of Military Medicine: Medical Aspects of Chemical and Biological Warfare,* Office of the Surgeon General, 1997, Chap. 7, "Vesicants."
[7]DOD Chemical and Biological Defense Program Annual Report to Congress, Volume I, April 2003.
[8]FM 8-285/NAVMED P-5041/AFJMAN 44-149/FMFM 11-11, Treatment of Chemical Agent Casualties and Conventional Military Chemical Injuries, 22 December 1995.
[9]FM 3-5/MCWP 3-37.3, *NBC Decontamination,* 28 July 2000.

Table II-45. CX Toxicity Estimates[10]

Endpoint	Toxicity ($mg\text{-}min/m^3$)	MV (L)	Exposure Duration	ROE	Probit Slope	TLE	ROD	DOC
Lethality	LCt_{50}: 3200 [a] (Provisional)	15	10 min	Inhalation/ Ocular	Unknown	1 [b,c]	Unknown	Low
Incapacitation/ intolerable	ECt_{50}: 3 [d] (Provisional)	15	1 min	Inhalation/ Ocular	Unknown	Unknown	Unknown	Low
Threshold (odor detection)	ECt_{50}: 1 [d] (Provisional)	15	10 min	Inhalation/ Ocular	Unknown	Unknown	Unknown	Low

NOTES

[a]Based on animal data and existing estimate.
[b]The TLE value is assumed to be 1 because the concentration-time profile is unknown.
[c]See Appendix H for supporting toxicity profile estimates.
[d]Based on secondary human data.

8. Incapacitating Agents

Used in a military context, incapacitation is understood to mean inability to perform one's military mission. Since missions vary, for the purpose of this manual, incapacitation means the inability to perform any military task effectively and implies that the condition was achieved via the deliberate use of a nonlethal weapon.[22] Incapacitating agents differ from other CW agents in that the lethal dose is theoretically many times greater than the incapacitating dose. Thus, they do not seriously endanger life except in cases exceeding many times the effective dose, and they produce no permanent injury.[1] Virtually all drugs whose most prominent effects are psychological or behavioral can be classified into four fairly discrete categories: deliriants, stimulants, depressants, and psychedelics. They interfere with the higher functions of the brain such as attention, orientation, perception, memory, motivation, conceptual thinking, planning, and judgment. [22]

a. Deliriants.[22] These are drugs that produce delirium, which is an incapacitating syndrome involving confusion, hallucinosis, and disorganized speech and behavior. Many drugs can produce delirium; however, the chemicals in the subgroup anticholinergics are regarded as most likely to be used as military incapacitating agents. Of these drugs, 3-Quinuclidinyl benzilate (BZ) is considered the most likely candidate for military use (see Table II-46). BZ is capable of producing delirium at a very low dosage with a high safety margin. Skin absorption is possible with proper solvents. See Table II-47 (page II-66) for toxicity estimates. See Table II-48 (page II-68) for symptoms. BZ intoxication requires from two to three days to reach full recovery.

Table II-46. BZ

Alternate Designation: EA 2277; CS 4030; Oksilidin; QNB	
Chemical Name: 3-Quinuclidinyl benzilate	
Synonym: Benzilic acid, 3-quinuclidinyl ester; 1-Azabicyclo (2.2.2) octan-3-ol, benzilate; Benzeneacetic acid, alpha-hydroxy-alpha-phenyl-,1-azabicyclo (2.2.2)oct-3-yl ester; 3-Chinuclidylbenzilate; 3-(2,2-Diphenyl-2-hydroxyethanoyloxy)-quinuclidine; 3-Quinuclidinol benzilate; 3-Quinuclidyl benzilate	
CAS Registry Number: 6581-06-2	
RTECS Number: DD4638000	
Physical and Chemical Properties	
Structural Formula: 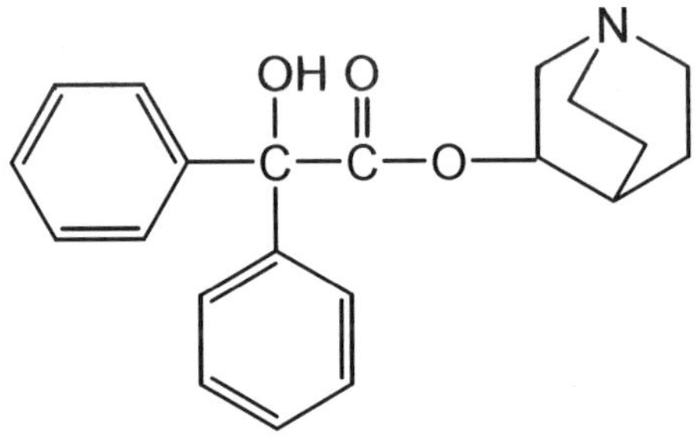 Molecular Formula: $C_{21}H_{23}NO_3$ Molecular Weight: 337.42	
Physical State	White crystalline solid [1]
Odor	None [2]
FP/MP	167.5°C [3] (MP)
Boiling Point	412°C (extrapolated) [4]
Solid Density (g/cm^3)	Bulk: 0.51;[2] Crystal: 1.33 [2]
Vapor Density	11.6 (calculated)
Vapor Pressure (torr)	1.43×10^{-10} @ 25°C (extrapolated); 4.74×10^{-13} @ 0°C (extrapolated)[4]
Volatility (mg/m^3)	2.60×10^{-6} @ 25°C;9.0×10^{-9} @ 0°C (calculated from vapor pressure)[4]
Latent Heat of Vaporization (kcal/mol)	21.2 (calculated from Clausius Clapeyron equation which assumes constant heat of vaporization as a function of temperature) [4]
Flash Point	Pure: 246°C;[2] Munitions grade: 220°C [2]
Decomposition Temperature	Stable up to the melting point. During prolonged heating at temperatures approximately 170°C, BZ begins to decompose, producing carbon dioxide, benzophenone, benzhydrol, and other products. The rate of decomposition is both temperature- and purity-dependent. [1]
Solubility	Solubility in water is approximately 1.18 g/L[5]; slightly soluble in water; soluble in dilute acids and common organic solvents such as alcohol and chloroform; insoluble in aqueous alkali. [1]
Rate of Hydrolysis	$t_{1/2}$ = 6.7 hrs @ 25°C and pH 9.8;[6] $t_{1/2}$ = 1.8 min @ 25°C and pH 13;[6] $t_{1/2}$ = 3 to 4 wks @ 25°C in moist air and pH 7.[6] $t_{1/2}$ = 12 min @ 34° and pH 12;[5] $t_{1/2}$ = 1.4 hr @ 50°C and pH 8.5;[5] $t_{1/2}$ = 9.5 hr @ 100°C and pH 0.[5]
Hydrolysis Products	3-Quinuclidinol and benzylic acid [6]
Stability in Storage	Stable in aluminum and stainless steel @ 71°C for at least 2 years [5]
Action on Metals or Other Materials	Slight pitting of aluminum and stainless steel occurs after 2 years @ 71°C[7]

Table II-46. BZ (Continued)

Other Data	
Skin and eye toxicity	Can cause blurred vision and dilation of pupils [8]
Inhalation toxicity	Primary route of exposure [9]
Rate of action	Delayed action; 1 to 4 hours depending on exposure [10]
Means of detection in field	MM1
Protection required	MOPP4 [11]
Decontamination	Decontaminate skin with soap and water or use M291 kit if soap and water are not available [12]
Use	Delayed-action incapacitating agent

NOTES

[1]Witten, B., *The Search for Toxic Chemical Agents (U)*, EATR 4210, Edgewood Arsenal Research Laboratories, MD, November 1969, UNCLASSIFIED Report (AD507852).
[2]*Chemical Agent Data Sheets Volume I*, Edgewood Arsenal Special Report EO-SR-74001, Edgewood Arsenal, Aberdeen Proving Ground, MD December 1974, UNCLASSIFIED Report (ADB028222).
[3]Lochboehler, C.M., *The Physical Properties of the Glycolates (U),* EASP-100-61; USA Munitions Command, Chemical Research Laboratory; Edgewood Arsenal, MD, 1970, CONFIDENTIAL Report (AD508308).
[4]Cogliano, J.A. and Braude, G.L., *Corrosion,Compatibility, and other Physiocochemical Studies (U), Final Report-Task II*, DA18-108-CML-6602, W.R. Grace and Company, Washington Research Center, Clarksville, MD, 1963, UNCLASSIFIED Report (AD359603).
[5]Rosenblatt, D.H., et al., *Problem Definition Studies on Potential Environmental Pollutants VIII: Chemistry and Toxicology of BZ (3-Quinuclidinyl Benzilate),* USAMBRDL-TR 7710, USA Medical Bioengineering Research and Development Laboratory, Fort Detrick, MD, 1977, UNCLASSIFIED Report (ADB030349).
EA 2277 (U): A Summary Report as of 15 March 1961, CRDL-SP-4-28, USA Chemical Research and Development Laboratories, Army Chemical Center, MD, March 1961, CONFIDENTAL Report.
[6]Sass, S. and Master, I., *Basic Esters of Glycolic Acids (U): Part III Analysis and Chemical Properties of Microgram and Larger Quantities of EA 2277 and Related Compounds*, CRDLR 3088, USA Chemical Research and Development Laboratories, Army Chemical Center, MD, August 1961, UNCLASSIFIED Report (AD325351).
[7]Brooks, M.E., et al., *Corrosion, Compatibility and Other Physicochemical Studies (U)*, DA18-108-CML-6602 (A), Final Report – Task I, W.R. Grace and Company, Washington Research Center, Clarksville, MD, May 1964, UNCLASSIFIED Report (AD350755).
[8]BG Russ Zajtchuk et al. (eds), *Textbook of Military Medicine: Medical Aspects of Chemical and Biological Warfare,* Office of the Surgeon General, 1997, Chap. 11, "Incapacitating Agents."
[9]Sharon Reutter, et al., *Review and Recommendations for Human Toxicity Estimates for FM 3-11.9*, ECBC-TR-349, September 2003.
[10]*NIOSH-DOD-OSHA Sponsored Chemical and Biological Respiratory Protection Workshop Report,* February 2000.
[11]FM 8-9/NAVMED P-5059/AFJMAN 44-151, *NATO Handbook on the Medical Aspects of NBC Defense Operations AMEDP-6(B),* 1 February 1996.
[12]FM 8-285/NAVMED P-5041/AFJMAN 44-149/FMFM 11-11, *Treatment of Chemical Agent Casualties and Conventional Military Chemical Injuries*, 22 December 1995.

b. BZ Toxicity Estimates (Table II-47). No toxicity estimates are recommended for lethal or threshold effects. The existing estimates are not supported by the available data.

Table II-47. BZ Toxicity Estimates[10]

Endpoint	Toxicity (mg-min/m^3)	MV (L)	Exposure Duration	ROE	Probit Slope	TLE	ROD	DOC
Lethality	LCt_{50}: NR	N/A	N/A	Inhalation/ Ocular	Unknown	Unknown	Unknown	N/A
Severe effects (temporary incapacitation)	ECt_{50}: 100 [a] (Provisional)	15	Less than 5 min	Inhalation/ Ocular	Unknown	N/A	Unknown	Low
Threshold effects (odor detection)	ECt_{50}: NR	N/A	N/A	Inhalation/ Ocular	Unknown	N/A	Unknown	N/A

NOTES

[a]Based on human data and "official" estimate (HEC, 1967).

c. Stimulants. Stimulants are drugs that produce a temporary increase of the functional activity or efficiency of an organism or any of its parts.[3] They include amphetamines, cocaine, caffeine, nicotine, and epileptogenic substances such as strychnine and Metrazole. None of the conventional stimulants appears to have sufficient potency to be usable as an airborne incapacitating agent, and low doses could even prove counterproductive, since moderate stimulation might easily lead to a soldier's more energetic and aggressive performance.[22]

d. Depressants. Depressants are drugs that reduce a bodily functional activity or an instinctive desire.[3] They include drugs such as barbiturates, morphine and other opioids, and tranquilizers. The lethal dose for morphine and other opioids is only 10 to 20 fold greater than the incapacitating dose. Major tranquilizers often produce relatively little sedation, although they reduce hyperactivity.[22]

(1) Cannabinols. Cannabinols are another a group of potential incapacitating agents that seem to act basically as CNS depressants. Primary effects of these agents, however, are sedation and destruction of motivation rather than disruption of the ability to think.[1]

(2) Phenothiazine-like compounds have a very high safety index and would not likely involve any special medical care. The onset of action for phenothiazines is about 5 minutes and often lasts about 1 hour.

(3) Fentanyls. On 27 October 2002, Russia used a derivative of fentanyl in a Moscow theatre where Chechen rebels held 800 hostages.[23] Fentanyls are opiates with actions similar to those of morphine; in particular, the capacity to relieve pain. Short-term and long-term effects of the fentanyls are indistinguishable from those of heroin, but they are hundreds of times more potent.[24] Fentanyls depress respiration and heart rate and cause lethargy, sedation, and immobilization.[22] As a potential class of agents, they have a rapid onset of action. Decontamination would involve washing with soap and water.[1]

e. Psychedelics. Psychedelics are drugs capable of producing abnormal psychological effects and sometimes psychological states resembling mental illness.[3] They include D-lysergic acid diethylamide (LSD), MDMA (popularly known as ecstasy), phencyclidine (PCP), and indoles. Psychedelic drugs have been found unsuitable for military use. Affected individuals usually cannot carry out a series of instructions or concentrate on a complex task, but might be capable of isolated, impulsive actions such as firing a weapon accurately enough to be dangerous.[22]

f. Symptoms. Symptoms and possible agent families are shown in Table II-48 (page II-68). See documents such as FM 8-285 for diagnosis and treatment for incapacitating agents.

Table II-48. Correlation of Symptoms and Incapacitating Agent Family [1,22]

Signs and Symptoms	Possible Agent Family
Restless, dizziness, or giddiness; failure to obey orders, confusion, erratic behavior; stumbling or staggering; vomiting	Anticholinergics (BZ), indoles, cannabinoids, anxiety reaction, other intoxications (e.g., alcohol, bromides, barbiturates, lead)
Dryness of mouth, tachycardia at rest, elevated temperature, flushed face, blurred vision, papillary dilation, slurred or nonsensical speech, hallucinatory behavorior, stupor, coma	Anticholinergics (BZ)
Inappropriate smiling or laughing; irrational fear; destructibility; difficulty in expressing self; perceptual distortions; labile increase in pupil size, heart rate, and blood pressure; stomach cramps and vomiting	Indoles (schizophrenic psychosis may mimic in some respects)
Euphoric, relaxed, unconcerned daydreaming attitude; easy laughter; low blood pressure and dizziness on sudden standing	Cannabinols
Tremor, clinging or pleading; crying; decrease in disturbance with reassurance; history of nervousness or immaturity	Anxiety reaction
Respiratory depression; slow pulse; lethargy; sedation; immobilization	Fentanyls

9. Chemical Warfare Agent Precursors[10]

These materials are not CW agents; they are binary agent components and are relatively nontoxic—at least when compared with the CW agents.

a. Methyl Phosphonic Acid (DF) (see Table II-49). DF is an organophosphorous compound that is an intermediate component of binary GB. The Human Estimates Committee of what is now Edgewood Chemical and Biological Center (ECBC) stated that many alkyl organophosphorous compounds produce neurotoxicity in various animal species. The toxic syndrome includes weakness and paralysis, with death resulting from a central action. Deaths that occur within a few hours of exposure, before the appearance of lung damage, are perhaps due to central respiratory failure. Deaths occurring after 24 hours postexposure are thought to result from lung damage, central respiratory failure, or a combination of the two. See Table II-50 (page II-70) for toxicity estimates.

Table II-49. DF

Alternate Designations: Difluoro; EA 1251	
Chemical Name: Methylphosphonic difluoride	
Synonyms: Phosphonic difluoride, methyl-; Difluoromethyl phosphonate; Difluoromethylphosphine oxide; Methyl difluorophosphite; Methylphosphonyldifluoride; Phosphonodifluoridic acid, methyl-	
CAS Registry Number: 676-99-3	
RTECS Number: TA1840700	
Physical and Chemical Properties	
Structural Formula: 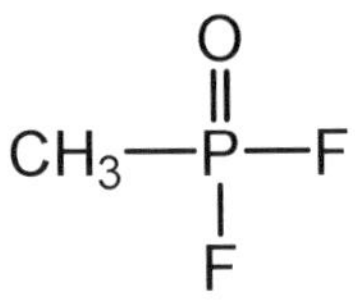 Molecular Formula: CH_3F_2PO Molecular Weight: 100.00	
Physical State	Liquid [1]
Odor	Pungent, acid like[1]
Boiling Point	99.7°C [2]
FP/MP	–36.9°C (FP) [3]
Liquid Density (g/mL)	1.3595 @ 25°C; 1.4060 @ 0°C (extrapolated) [2]
Vapor Density (relative to air)	3.4 (calculated)
Vapor Pressure (torr)	3.6×10^1 @ 25°C; 8.5 @ 0°C (extrapolated) [2]
Volatility (mg/m^3)	1.9×10^5 @ 25°C; 5.0×10^4 @ 0°C (calculated from vapor pressure) [2]
Latent Heat of Vaporization (kcal/mol)	9.2 @ 25°C; 9.5 @ 0°C (calculated from vapor pressure) [2]
Flash Point	Nonflammable [4]
Decomposition Temperature	Data not available
Solubility	Immediately decomposes with the addition of water [5]
Rate of Hydrolysis	Virtually instantaneous to produce methylphosphonofluoridic acid (MF) and hydrogen fluoride (HF) which are also toxic. Further hydrolysis is a slow reaction that produces methylphosphonic acid (MPA); MF $t_{1/2}$ = 162 days @ pH 7, $t_{1/2}$ = 90 days @ pH 4, and $t_{1/2}$ = 47 days @ pH 3. [5]
Hydrolysis Products	Hydrolyzes to give toxic products, MF and HF. Further hydrolysis of MF results in MPA and a second mole of HF. [5]
Stability in Storage	Remains stable for at least 20 years, when stored in high-density polyethylene containers enclosed in steel. [6] Avoid contact with water mist or sprays, metals, alkaline materials, and some organics. [1] Never store DF with alcohols; DF will react with alcohols to form a lethal chemical, such as crude GB. [7]
Action on Metals or Other Materials	Incompatible with water, glass, concrete, [8] most metals, natural rubber, [9] and organic materials like glycols, which is mainly due to the acidic corrosive nature of the hydrolysis products. [1,10-12] HF may react with some metals, to give off hydrogen gas, a potential fire and explosive hazard. [10,13]

Table II-49. DF (Continued)

NOTES
[1]Buchi, K.M., *Environmental Overview of Intermediates, By-Products, and Products in the Production of QL, DC, and DF*, CRDEC-TR-076, USA Chemical Research Development and Engineering Center, Aberdeen Proving Ground, MD, May 1991, UNCLASSIFIED Report (ADB155651).
[2]Zeffert, B.M. et al., "Properties, Interaction and Esterification of Methylphosphonic Dihalides," *J. Am. Chem. Soc.*, Vol. 82, p. 3843, 1960.
[3]Furukawa, G.T., et al., 'Thermodynamic Properties of Some Methylphosphonyl Dihalides From 15 to 335°K," *J. Rsch. NBS Phy. & Chem.*, Vol. 68A, No. 4, p. 367, 1964.
[4]Allan, C.R., *The Relationship Between Oxygen Index and the Flashing Propensity of Explosively Disseminated Liquids*, ARCSL-TR-77061, USA Armament Research and Development Command, Chemical Systems Laboratory, Aberdeen Proving Ground, MD, October 1977, UNCLASSIFIED Report (ADA045976).
[5]Dahl, A.R., et al., *Acute Toxicity of Methylphosphonic Difluoride (DF) Methylphosphonic Dichloride (DC) and their Hydrolysis Products by Inhalation and other Routes in Mice, Rats and Guinea Pigs*, CRDEC-CR-86049, USA Chemical Research Development and Engineering Center, Aberdeen Proving Ground, MD, June 1986, UNCLASSIFIED Report (ADB105158).
[6]Jackson, A.M., and Semiatin, W.J., *Long-Term Storability of the M20 DF Canister Used in the M687 Binary Projectile*, CRDC-TR-84104, USA Chemical Research Development and Engineering Center, Aberdeen Proving Ground, MD, January 1985, UNCLASSIFIED Report (ADB092563).
[7]Hyttinen, L.J., et al., *Mixed Binary Agents New Approach Toward Meeting Expanded Chemical Munitions Effectiveness Requirements*, ARCSL-TR-83080, USA Armament Research and Development Command, APG, MD, June 1983, CONFIDENTIAL Report (ADC033576).
[8]Ellzy, M., et al., *Difluor (DF) – Flooring Compatibility Studies*, CRDEC-TR-229, USA Chemical Research Development and Engineering Center, Aberdeen Proving Ground, MD, May 1991, UNCLASSIFIED Report (ADB154752).
[9]Schweitzer, P.A., *Corrosion Resistance Tables: Metals, Plastics, Nonmetallics, and Rubbers*, 2nd ed., pp. 572- 573, Ed., Marcel Dekker, INC., Chester, NJ, 1986.
[10]Buchi, K.M., *Environmental Overview of Common Industrial Chemicals with Potential Application in the Binary Munitions Program*, CRDEC-TR-87041, USA Chemical Research Development and Engineering Center, Aberdeen Proving Ground, MD, July 1987, UNCLASSIFIED Report (ADA186083).
[11]Thomas, M.T., *Research and Development for Candidate Materials for Use as a DF Containment Vessel*, CRDC-CR-85058, USA Chemical Research Development and Engineering Center, Aberdeen Proving Ground, MD, September 1985, UNCLASSIFIED Report (ADB096058).
[12]Kay Lau, Tony Man, *Glass or Polymer Etching Due to the Reaction of Methylphosphonic Difluoride (DF) with Water (U)*, CRDEC-TR-86074, USA Chemical Research Development and Engineering Center, Aberdeen Proving Ground, MD, August 1986, CONFIDENTIAL Report (ADC039896).
[13]Tarantino, P.A., *Electrochemical Corrosion Study of Miscellaneous Metals/Alloys with Methylphosphonic Difluoride*, CRDEC-TR-88032, USA Chemical Research Development and Engineering Center, Aberdeen Proving Ground, MD, November 1987, UNCLASSIFIED Report (ADB117574).

Table II-50. DF Toxicity Estimates[10]

Endpoint	Toxicity [a] (mg/eye)	MV (L)	Exposure Duration	ROE	Probit Slope	TLE	ROD	DOC
Permanent Corneal Damage	ED_{50}: 10	N/A	N/A	Liquid in eye	Unknown	Unknown	Unknown	Moderate
Temporary Corneal Damage	ED_{50}: 0.2	N/A	N/A	Liquid in eye	Unknown	Unknown	Unknown	Moderate
NOTES								
[a]Based on "official" existing human toxicity estimate.								

b. O-(2 Diisopropylaminoethyl)-O'Ethyl Methyl Phosphonite (QL) (see Table II-51). The only human data for QL are from occupational exposures for which the dosage is unknown. No toxic signs have been observed in workers who have handled large quantities of QL or who have become highly contaminated with it. See Table II-52 (page II-72) for toxicity estimates.

Table II-51. QL

Alternate Designations: EA 1724; EDMP	
Chemical Name: O-(2-Diisopropylaminoethyl) O'-ethyl methylphosphonite	
Synonyms: O-Ethyl-O'-(2-diisopropylaminoethyl) methylphosphonite; Phosphonous acid, methyl-, 2-[bis(1-methylethyl)amino]ethyl ethyl ester	
CAS Registry Number: 57856-11-8	
RTECS Number: Data not available	
Physical and Chemical Properties	
Structural Formula: 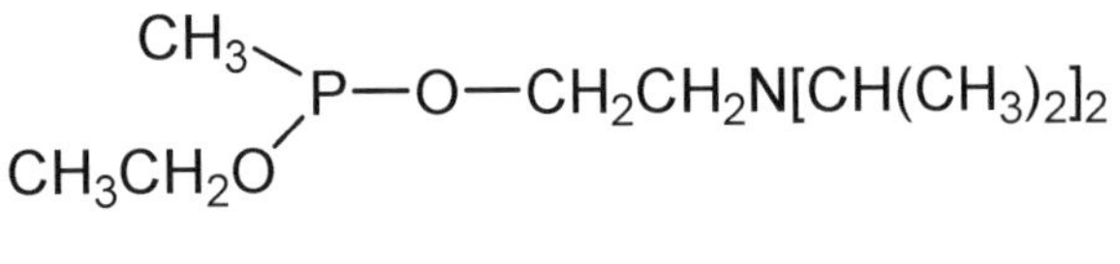 Molecular Formula: $C_{11}H_{26}NO_2P$ Molecular Weight: 235.31	
Physical State	Liquid [1]
Odor	Strong, fishy [1]
Boiling Point	244.8°C (extrapolated) [2]
FP/MP	Data not available
Liquid Density (g/mL)	0.9080 @ 25°C; 0.9307 @ 0°C [2]
Vapor Density (relative to air)	8.1 (calculated)
Vapor Pressure (torr)	1.8×10^{-2} @ 25°C; 7.1×10^{-4} @ 0°C (extrapolated) [2]
Volatility (mg/m^3)	2.3×10^{2} @ 25°C; 9.8 @ 0°C (calculated from vapor pressure) [2]
Latent Heat of Vaporization (kcal/mol)	19.4 @ 25°C; 22.3 @ 0°C (calculated from vapor pressure) [2]
Flash Point	89°C (closed cup). [3] In addition, QL has an autoignition temperature of 129°C. [2] A hydrolysis product, O,O'-diethylmethylphosphonite (TR), has a flash point of 28°C [4] and an autoignition temperature of 40°C. [1]
Decomposition Temperature	Data not available
Solubility	Slightly soluble in water. Soluble in methanol, 2-propanol, acetone, and benzene. [5]
Rate of Hydrolysis	Rapid. QL can be completely hydrolyzed within 5 hrs. [6]
Hydrolysis Products	With excess of water by weight, QL primarily forms O-ethyl methylphosphonic acid (YL) and 2-diisopropylaminoethanol (KB), but also forms O-(2-diisopropylaminoethyl) methylphosphonic acid (QA) and ethanol (ZS) as secondary products. With traces of water or other proton donors, QL will produce O,O'-diethyl methylphosphonite (TR) and O,O'-bis-(2-diisopropylaminoethyl) methylphosphonite (LT). [1] TR has a boiling point of 120°C, [4] a vapor pressure of 10 mm Hg @ 20°C, [1] and is flammable. [4]
Stability in Storage	Stable in aluminum, steel, and stainless steel containers for at least 6 months @ 71°C, if kept dry and pure. [5] Always store QL away from heat or ignition sources and sulfur compounds because of the potential to form highly toxic V-agents. [7]
Action on Metals or Other Materials	Satisfactory against aluminum, steel, and stainless steel, but not glass unless a stabilizer is used. [5] Reacts with sulfur and sulfur compounds to produce highly toxic VX or VX-like compounds. [7] It is incompatible with HTH, many chlorinated hydrocarbons, selenium, selenium compounds, moisture, oxidants, and carbon tetrachloride. [1]

Table II-51. QL (Continued)

NOTES
[1]Buchi, K.M., *Environmental Overview of Intermediates, By-Products, and Products in the Production of QL, DC, and DF*, CRDEC-TR-076, USA Chemical Research Development and Engineering Center, Aberdeen Proving Ground, MD, May 1991, UNCLASSIFIED Report (ADB155651). [2]Samuel, J.B., et al., *Physical Properties of Standard Agents, Candidate Agents, and Related Compounds at Several Temperatures*, ARCSL-SP-83015, June 1983, USA Armament Research and Development Command, Aberdeen Proving Ground, MD, UNCLASSIFIED Report (ADC033491). [3]Butrow, A., Chemical Research and Development Center Notebook #NB 83-0155, p. 45 (U). [4]Kinkead, E.R. *Evaluation of the Acute Toxicity of Four Compounds Associated with the Manufacture of O-Ethyl-O'- (2-Diisopropyaminoethyl) Methylphosphonite*, CRDEC-CR-87077, USA Chemical Research, Development & Engineering Center, Aberdeen Proving Ground, MD, June 1987, UNCLASSIFIED Report (ADB113969). [5]Brooks, M.E. et al, *Final Report – Task VII, Contract DA18-108-CML-6602 (A), Corrosion, Compatibility and Other Physicochemical Studies (U)*, Final Report – Task VII RES-64-86, W. R. Grace & Co., Washington Research Center, Clarksville, Maryland, June 1964, UNCLASSSIFIED Report (AD352753). [6]Rohrbaugh, D.K., *Detection and Identification of QL Impurities by Electron and Chemical Ionization Gas Chromatography/ Mass Spectrometry*, USA Chemical Research, Development, & Engineering Center, Aberdeen Proving Ground, MD, July 1989, UNCLASSIFIED Report (ADB136428). [7]Nowlin, T.E., et al., *A New Binary VX Reaction-Two-Liquid System (U)*, EATR 4700, USA Munitions Command, Edgewood Arsenal, MD, November 1972, UNCLASSIFIED Report (AD524088).

Table II-52. QL Toxicity Estimates[10]

Endpoint	Toxicity (mg-min/m^3)	MV (L)	Exposure Duration	ROE	Probit Slope	TLE	ROD	DOC
N/A	No toxicity estimates are recommended at this time due to lack of data	N/A	N/A	N/A	N/A	N/A	Unknown	N/A

c. Isopropylamine and Isopropyl Alcohol (OPA) (see Table II-53). OPA has an inflammatory and corrosive action when in contact with mucous membranes, tissues, or skin. Death could occur from severe local tissue injury, with secondary complications such as toxemia, shock, perforation, infection, hemorrhage, and obstruction. These effects are the results of the concentration of the material, rather than the total quantity applied. OPA is most likely to be encountered in liquid form. See Table II-54 for toxicity estimates.

Table II-53. OPA

Alternate Designations: N/A
Chemical Name: 2-Propanol (isopropyl alcohol) and Isopropyl amine mixture
Synonyms: N/A
CAS Registry Number: 2-Propanol: 67-63-0; Isopropyl amine: 75-31-0
RTECS Number: 2-Propanol: NT8050000; Isopropyl amine: NT8400000
Physical and Chemical Properties
Structural Formula: 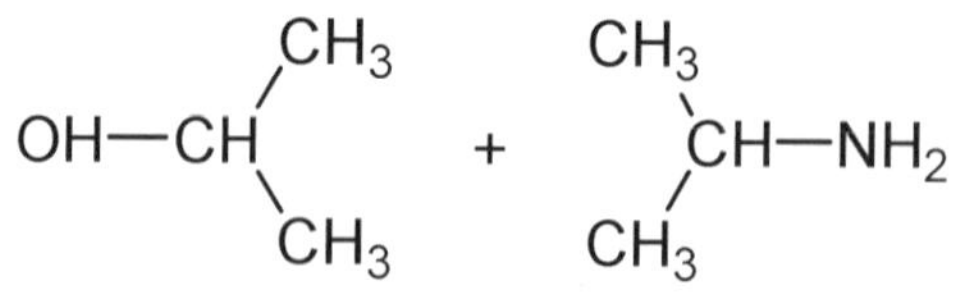 72 wt% 2-Propanol 28 wt% Isopropylamine
Molecular Formula: C_3H_8O and C_3H_9N Molecular Weight: 2-Propanol: 60.10; Isopropyl amine: 59.11; Average: 59.81 (based on 72:28 wt. %)

Table II-53. OPA (Continued)

Physical State	Colorless liquid [1]
Odor	Alcohol and ammonia (based on the two components) [2]
Boiling Point	60.8°C (based on Raoult's law calculation) [3]
FP/MP	Less than –88°C (FP) [1]
Liquid Density (g/mL)	0.7520 @ 25°C [1,4] 0.7759 @ 0°C (extrapolated) [4]
Vapor Density (relative to air)	2.1 (calculated)
Vapor Pressure (torr)	1.955 x 10^2 @ 25°C; 6.128 x 10^1 @ 0°C (based on Raoult's law calculation) [3]
Volatility (mg/m^3)	6.288 x 10^5 @ 25°C; 2.152 x 10^5 @ 0°C (calculated from vapor pressure) [3]
Latent Heat of Vaporization (kcal/mol)	7.51 (calculated from Clausius Clapeyron equation which assumes constant heat of vaporization as a function of temperature)[3]
Flash Point	Less than 0°C [5]
Decomposition Temperature	Data not available
Solubility	Both 2-propanol and isopropyl amine are miscible in water, alcohol, and ether, and soluble in acetone, benzene, and chloroform—suggesting that OPA has a similar degree of solubility. [2,6]
Rate of Hydrolysis	Data not available
Hydrolysis Products	Data not available
Stability in Storage	Relatively stable for at least 5 years at temperatures between ambient and 71°C, if stored in carbon steel containers lined with an ethylenebutylene copolymer. [7] Store OPA away from heat, open flame, and DF because they react to form highly toxic compounds such as crude GB. [1]
Action on Metals or Other Materials	Reacts readily with oxidizing materials and organophosphorus halides, such as DF. Contact with DF can produce extremely toxic compounds such as crude GB. [1]

NOTES

[1]Hyttinen, L.J., et al., *Mixed Binary Agents New Approach Toward Meeting Expanded Chemical Munitions Effectiveness Requirements*, ARCSL-TR-83080, USA Armament Research and Development Command, APG, MD, June 1983, CONFIDENTIAL Report (ADC033576).

[2]*The Merck Index: An Encyclopedia of Chemicals, Drugs, and Biologicals*, 13th ed., p. 5225, Merck & Company, Inc., Whitehouse Station, NJ, 2001.

[3]Abercrombie, P., ECBC Notebook #NB 98-0079, p. 45 (U).

[4]Fielder, D., Chemical Systems Laboratories Notebook #NB-CSL-82-0213, p. 13 (U).

[5]Allan, C.R., *The Relationship Between Oxygen Index and the Flashing Propensity of Explosively Disseminated Liquids*, ARCSL-TR-77061, USA Armament Research and Development Command, Chemical Systems Laboratory, Aberdeen Proving Ground, MD, October 1977, UNCLASSIFIED Report (ADA045976).

[6]Weast, R.C., *CRC Handbook of Chemistry and Physics,* 50th ed., pp. C-453 and C-440, CRC, Cleveland, OH, 1969.

[7]Sze, J.M., and Simak, R.S., *Binary GB: A Compilation of Relevant Data*, ARCSL-TR-82019, USA Armament Research and Development Command, APG, MD, March 1983, CONFIDENTIAL Report (ADC030931).

Table II-54. OPA Toxicity Estimates[10]

Endpoint	Toxicity (mg-min/m^3)	MV (L)	Exposure Duration	ROE	Probit Slope	TLE	ROD	DOC
N/A	No toxicity estimates are recommended at this time due to lack of data	N/A	N/A	N/A	N/A	N/A	Unknown	N/A

d. Sulfur with a Small Amount of Silica Gel (NE) (see Table II-55 [page II-74]). NE is a component of binary VX. It has very low toxicity and may irritate skin, eyes, nose and throat. See Table II-56 (page II-75) for toxicity estimates.

Table II-55. NE

Alternate Designation: Brimstone	
Chemical Name: Sulfur (with small amount of silica aerogel)	
Synonyms: Alpha sulfur; Beta sulfur; Atomic sulfur; Bensulfoid; Brimstone; Colloidal sulfur; Collokit; Colsul; Cosan; Crystex; Elemental sulfur; Flowers of sulfur; Flour sulfur; Ground vocle sulphur; Hexasul; Kocide; Kolofog; Kolospray; Kumulus; Microflotox; Orthorhombic sulfur; Precipitated sulfur; Rhombic sulfur; Sofril; Sperlox-s; Spersul; Spersul thiovit; Sublimed sulfur; Sulfidal; Sulforon; Sulfur flower; Sulkol; Super cosan; Sulphur; Sulsol; Sulfur atom; Sulfur ointment; Sulfur vapor; Tesuloii; Thiolux; Thiovit	
CAS Registry Number: Sulfur: 7704-34-9 and 10544-50-0	
RTECS Number: Data not available	
Physical and Chemical Properties	
Structural Formula: 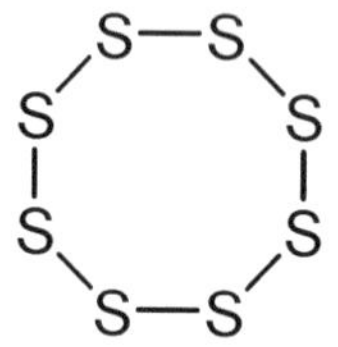 Molecular Formula: S_8 Molecular Weight: 256.48	
Physical State	Rhombic, yellow crystals [1]
Odor	Odorless when pure [2], but many sulfur compounds tend to be vile-smelling
Boiling Point	444.6°C [1]
FP/MP	The rhombic form of sulfur transforms into the monoclinic form @ 95.3°C; the MP of monoclinic sulfur is 115.21°C [1]
Solid Density (g/cc)	2.07 @ 20°C [2]
Vapor Density (relative to air)	8.8 (calculated)
Vapor Pressure (torr)	1 @ 183.8°C [3]
Volatility (mg/m^3)	9.0 x 10^3 @ 184°C (calculated from vapor pressure) [3]
Latent Heat of Vaporization (kcal/mol)	2.21 [4]
Flash Point	207°C (closed cup) [3]
Decomposition Temperature	Data not available
Solubility	Insoluble in water; slightly soluble in alcohol, ether; soluble in carbon disulfide, benzene [1,2] toluene, liquid NH_3, acetone, methylene iodide, and chloroform[1]
Rate of Hydrolysis	Data not available
Stability in Storage	Always store NE away from heat or ignition sources and QL because of the potential to form highly toxic VX [5]
Action on Metals or Other Materials	Reacts with QL to produce extremely toxic VX [5]

NOTES

[1] *The Merck Index: An Encyclopedia of Chemicals, Drugs, and Biologicals*, 13th ed., p. 1599, Merck & Company, Inc., Whitehouse Station, NJ, 2001.

[2] Lide, D.R., *CRC Handbook of Chemistry and Physics,* 82nd ed., CRC Press, Washington, DC, 2001.

[3] Lewis, R.J., *Sax's Dangerous Properties of Industrial Materials*, 10th ed., Vol. 3, p. 3328, John Wiley & Sons, Inc., New York, NY, 2001.

[4] Buchi, K.M., *Environmental Overview of Common Industrial Chemicals with Potential Application in the Binary Munitions Program*, CRDEC-TR-87041, USA Chemical Research Development and Engineering Center, Aberdeen Proving Ground, MD, July 1987, UNCLASSIFIED Report (ADA186083).

[5] Nowlin, T.E., et al., *A New Binary VX Reaction-Two-Liquid System (U)*, EATR 4700, USA Munitions Command, Edgewood Arsenal, MD, November 1972, UNCLASSIFIED Report (AD524088).

Table II-56. NE Toxicity Estimates[10]

Endpoint	Toxicity (mg-min/m^3)	MV (L)	Exposure Duration	ROE	Probit Slope	TLE	ROD	DOC
N/A	No toxicity estimates are recommended at this time due to lack of data	N/A	N/A	N/A	N/A	N/A	Unknown	N/A

e. Dimethylpolysulfide (NM) (see Table II-57). NM is a binary intermediate; as such, it is most likely to be encountered in an occupational setting. No estimates of human toxicity of NM have been derived. Very high vapor concentrations are required to produce toxic effects. The Department of Transportation (DOT) shipping classification is not a class B poison. See Table II-58 (page II-76) for toxicity estimates.

Table II-57. NM (Containing Elemental Sulfur)

Alternate Designation: NM5	
Chemical Name: Dimethyl polysulfide mixture [powdered sulfur + dimethyl disulfide (DMDS)]	
Synonyms: Dimethyl disulfide: 2,3-Dithiabutane; Methyl disulfide; (Methyldithio) methane; Disulfide, dimethyl	
CAS Registry Number: Dimethyl disulfide: 624-92-0	
RTECS Number: Data not available	
Physical and Chemical Properties	
Structural Formula: [1] $CH_3—S—S—S—S—S—CH_3$ Molecular Formula: $C_2H_6S_5$ Molecular Weight: 190.37	
Physical State	Liquid [1]
Odor	Very noxious [2]
Boiling Point	117°C [3]
FP/MP	< -40°C (FP), [1] DMDS: -84.72°C (FP) [4]
Liquid Density (g/mL)	1.3895 @ 25°C [1]
Vapor Density (relative to air)	6.6 (calculated)
Vapor Pressure (torr)	DMDS: 2.864×10^1 @ 25°C, [4] generally vapor pressure decreases with increasing molecular weight, [5] thus, NM will probably have a lower vapor pressure than DMDS since its molecular weight will generally be higher.
Volatility (mg/m^3)	DMDS: 1.45×10^5 @ 25°C (calculated from vapor pressure) [4]
Latent Heat of Vaporization (kcal/mol)	DMDS: 9.21 @ 25°C, [4] generally latent heat of vaporization increases with molecular weight of similar compounds; [5] thus, the latent heat of vaporization for NM will probably be above 9.21 kcal/mol based on DMDS
Flash Point	10^5 to 108°C [6]
Decomposition Temperature	Data not available
Solubility	Dimethyl disulfide is soluble in alcohols, but insoluble in water [7,8]
Rate of Hydrolysis	Data not available
Hydrolysis Products	Data not available
Stability in Storage	Satisfactory in storage for at least 1 year at temperatures between –40 to 71°C. [1] Always store NM away from heat or ignition sources and QL because of the potential to form highly toxic VX. [6]
Action on Metals or Other Materials	Stains the surface of steel and various metal-plated steels when stored in glass vessels for 4 months @ 71°C. [3] Reacts with QL to produce extremely toxic VX. [6]

Table II-57. NM (Containing Elemental Sulfur) (Continued)

NOTES
[1]Brown, H.A., Jr., et al., *Modified NM: An Improved Liquid Binary VX Reactant (U),* EC-TR-76075, USA Armament Command, Edgewood Arsenal, Aberdeen Proving Ground, MD, November 1976, UNCLASSIFIED Report (ADC008561).
[2]Riordan, M.B., *Pilot-Scale Operations of Process for Manufacture of VX Binary Intermediate NM (U),* EM-TR-76055, USA Armament Command, Edgewood Arsenal, Aberdeen Proving Ground, MD, November 1976, CONFIDENTIAL Report (ADC008383).
[3]Grula, R.J., et al., *Compatibility Studies with Candidate Binary VX2 Components*, EC-TM-76009, USA Armament Command, Edgewood Arsenal, Aberdeen Proving Ground, MD, February 1976, UNCLASSIFIED Report (ADE470951).
[4]Scott, D.W., et al., "2,3-Dithiabutane: Low Temperature Heat Capacity, Heat of Fusion, Heat of Vaporization, Vapor Pressure, Entropy and Thermodynamic Functions," *J. Amer. Chem. Soc.*, Vol. 72, p. 2424, 1950.
[5]*Properties of War Gases Volume IV: Vesicants (U),* ETF 100-41/Vol-4, Chemical Corps Board, Army Chemical Center, MD, December 1956, CLASSIFIED Report (AD108459).
[6]Nowlin, T.E., *A New Binary VX Reaction-Two-Liquid System (U)*, EATR 4700, USA Munitions Command, Edgewood Arsenal, MD, November 1972, UNCLASSIFIED Report (AD524088).
[7]Buchi, K.M., *Environmental Overview of Common Industrial Chemicals with Potential Application in the Binary Munitions Program*, CRDEC-TR-87041, USA Chemical Research Development and Engineering Center, Aberdeen Proving Ground, MD, July 1987, UNCLASSIFIED Report (ADA186083).
[8]Lide, D.R., *CRC Handbook of Chemistry and Physics,* 82nd ed., CRC Press, Washington, DC, 2001.

Table II-58. NM Toxicity Estimates[10]

Endpoint	Toxicity ($mg\text{-}min/m^3$)	MV (L)	Exposure Duration	ROE	Probit Slope	TLE	ROD	DOC
N/A	No toxicity estimates are recommended at this time due to lack of data	N/A	N/A	N/A	N/A	N/A	Unknown	N/A

10. Other Chemical Warfare Agents

In the past, other compounds were studied and evaluated to determine their potential as CW agents. Most compounds were found unlikely to be used for various reasons (e.g., unstable in storage). Some CW agents studied but not commonly known are listed in Table II-59.

Table II-59. Other CW Agents

Classification	CW Agent
Choking Agents	Chloropicrin (PS)
Nerve Agents	Ethyl Sarin (GE), VE, Amiton (VG), VS
Blister Agents	Sesqui mustard (Q)

NOTES

[1]FM 8-285/NAVMED P-5041/AFJMAN 44-149/FMFM 11-11, *Treatment of Chemical Agent Casualties and Conventional Military Chemical Injuries,* 22 December 1995.

[2]Henry F. Holtzclaw, Jr., et al., *General Chemistry with Qualitative Analysis,* 9th ed., D.C. Heath and Company, Lexington, MA, 1991.

[3]Stedman's Medical Dictionary, 25th Edition, Williams &Wilkins, Baltimore, MD, 1990.

[4]Holzclaw, Jr. H.F. and Robinson, W.R., *College Chemistry with Qualitative Analysis,* 8th ed., D.C. Heath and Company, Lexington, MA, 1988.

[5]American Society for Testing and Materials (ASTM) Method D 167, Standard Test Method for Apparent and True Specific Gravity and Porosity of Lump Coke.

[6]Matt T. Roberts and Don Etherington, "Vapor Density," Bookbinding and the Conservation of Books: A Dictionary of Descriptive Terminology, 7 January 2002, http://palimpsest.stanford.edu/don/dt/dt3670.html (8 April 2003).

[7]Richard J. Lewis, Sr., *Hawley's Condensed Chemical Dictionary,* 13th ed., John Wiley & Sons, Inc., New York, NY, 1997.

[8]Penski, E.C., *Vapor Pressure Data Analysis Methodology, Statistics, and Applications,* CR DEC-TR-386, Chemical Research, Development, and Engineering Center, Aberdeen Proving Ground, MD, 1992, UNCLASSIFIED Report (AD-A255090).

[9]McGraw Hill Dictionary of Scientific and Technical Terms, 6th ed., McGraw Hill Companies, Inc., New York, 2003.

[10]Sharon Reutter, et al., SBCCOM Report *Review and Recommendations for Human Toxicity Estimates for FM 3-11.9,* Draft.

[11]USACHPPM TG 204, *Glossary of Terms for Nuclear, Biological, and Chemical Agents and Defense Equipment,* December 2001.

[12]FM 8-9/NAVMED P-5059/AFJMAN 44-151, *NATO Handbook on the Medical Aspects of NBC Defense Operations AMEDP-6(B),* 1 February 1996.

[13] Patrice L. Abercrombie, *Physical Property Data Review of Selected Chemical Agents and Related Compounds,* Draft.

[14]BG Russ Zajtchuk, et al. (eds), *Textbook of Military Medicine: Medical Aspects of Chemical and Biological Warfare,* Office of the Surgeon General, 1997, Chap. 9, "Toxic Inhalational Injury."

[15]BG Russ Zajtchuk et al. (eds), *Textbook of Military Medicine: Medical Aspects of Chemical and Biological Warfare,* Office of the Surgeon General, 1997, Chap. 5, "Nerve Agents."

[16]Satu M. Somani and James A. Romano, Jr. (eds), *Chemical Warfare Agents: Toxicity at Low Levels,*" CRC Press, 2001.

[17]BG Russ Zajtchuk et al. (eds), *Textbook of Military Medicine: Medical Aspects of Chemical and Biological Warfare,* Office of the Surgeon General, 1997, Chap. 6, "Pretreatment for Nerve Agent Exposure."

[18]BG Russ Zajtchuk et al. (eds), *Textbook of Military Medicine: Medical Aspects of Chemical and Biological Warfare,* Office of the Surgeon General, 1997, Chap. 10, "Cyanide Poisoning."

[19]L. Fishbein and S. Czerczak, "Concise International Chemical Assessment Document 47: Arsine: Human Health Aspects," WHO, 2002.

[20]W.R. Kirner, *Summary Technical Report of Division 9, NDRC Volume 1, Chemical Warfare Agents, and Related Chemical Problems Part I-II,* Office of Scientific Research and Development, Washington, DC, 1946, UNCLASSIFIED Report (AD234270).

[21]BG Russ Zajtchuk et al. (eds), *Textbook of Military Medicine: Medical Aspects of Chemical and Biological Warfare,* Office of the Surgeon General, 1997, Chap. 7, "Vesicants."

[22]BG Russ Zajtchuk et al. (eds), *Textbook of Military Medicine: Medical Aspects of Chemical and Biological Warfare,* Office of the Surgeon General, 1997, Chap. 11, "Incapacitating Agents."

[23]CNN.com./World, "Russia names Moscow siege gas," 30 October 2002, http://www.cnn.com/2002/WORLD/europe/10/30/moscow.gas/, 27 August 2003.

[24]Scientific Section (Laboratory), Policy Development and Analysis Branch, Division for Operations and Analysis, United Nations Office on Drugs and Crime, *Terminology and Information on Drugs: Part I,* "Opioids: Fentanyls," October 1998, http://www.unodc.org/unodc/report_1998-10-01_1_page016.html, 8 August 2003.

Chapter III

MILITARY CHEMICAL COMPOUNDS AND THEIR PROPERTIES

1. Background

This chapter addresses military chemical compounds including RCAs, respiratory irritants, smoke, obscurants, and incendiaries. See Table III-1 for a list of the selected military chemical compounds discussed.

Table III-1. List of Selected Military Chemical Compounds

Classification	Military Chemical Compounds
RCAs	O-Chlorobenzylidene Malononitrile (CS), CS1, CS2, CSX, Dibenz(b,f)-1:4-oxazepine (CR), Capsaicin (OC)
Respiratory irritants	Adamsite (DM), Diphenylchloroarsine (DA), Diphenylcyanoarsine (DC), Chlorine (Cl_2)
Smoke and obscurants	Hexachloroethane (HC), Phosphorous, Titanium Tetrachloride (FM Smoke), Tantalum (Ta), Synthetic Graphite, Brass, Fog Oil [Smoke Generator Fuels (SGF-2)], DF-1, DF-2, JP-8, Signaling Smoke
Incendiaries	Magnesium Incendiaries, Thermite and Thermate Incendiaries, Oil and metal Incendiary Mixtures

2. Riot Control Agents (Tear-Producing Compounds)

The RCAs are chemicals that rapidly produce sensory irritation or disabling physical effects that disappear within a short time following termination of exposure.[1] The standard tear-producing agents currently in the US inventory for RCAs are o-chlorobenzylidene (CS), other agents in the same family (CS1, CS2, CSX), and dibenz (b,f)-1:4-oxazepine (CR).[2] Generally, they produce a rapid onset of effects (seconds to several minutes) and they have a relatively brief duration of effects (15 to 30 minutes) once the victim has escaped the contaminated atmosphere and has removed the contamination from clothing.[3] Because tear compounds produce only transient casualties, they are widely used for training, riot control, and situations where long-term incapacitation is unacceptable. When used against poorly equipped forces, these compounds have proven extremely effective. When released indoors, they cause serious illness or death.[2]

NOTE: Bleach reacts with RCAs to form a strong irritant.[3] Do not decontaminate RCAs with any form of bleach.[2]

a. Symptoms.[3] Symptoms include an initial burning feeling or irritation to the eyes that progresses to pain accompanied by blepharospasm and lacrimation. The mucous membranes of the mouth have a sensation of discomfort or burning, with excess salivation. Rhinorrhea is accompanied by pain inside the nose. When inhaled, these compounds cause a burning sensation or a feeling of tightness in the chest, with coughing, sneezing, and increased secretions. On unprotected skin, especially if the air is warm and moist, these agents cause tingling or burning.

b. Protection. The protective mask and ordinary field clothing secured at the neck, wrists, and ankles provide protection against field concentrations of RCAs.[4] When handling and loading bulk CS, personnel should wear protective clothing, masks, hoods, and gloves.[2]

c. Toxicity. There are a number of different formulations and methods of dispersion for the RCAs, and the different systems/formulations impact the potency of the materials. The estimates recommended by the former Human Estimates Committee at what is now ECBC were often based upon the most potent formulation.[5]

d. O-Chlorobenzylidene Malononitrile (CS) (see Table III-2). In 1959, the USA adopted CS for combat training and riot control purposes.[2] CS produces an intense burning and irritation of the eyes, with mild to severe conjunctivitis. It also produces a burning sensation in the nose and mucous membranes of the respiratory tract, followed by draining of the nasal sinuses. The chest feels constricted, with a sensation of choking and being unable to breathe. CS is also a primary skin irritant, and can produce erythema, edema, and vesication.[5] CS exists as a family of four forms: CS, CS1, CS2, and CSX. Different forms of CS have different persistence characteristics because of their formulation, dissemination, and rate of hydrolysis.[2] CS has been found to persist in snow for as long as 30 days but its persistency in soil varies, depending on the condition of the soil.[6] See Table III-3 for CS toxicity estimates.

Table III-2. CS

Chemical name: O-Chlorobenzylidene Malononitrile [1]	
Synonym: 2-Chlorbenzalmalonitrile, CS, OCBM [1]	
CAS registry number: 2698-41-1 [1]	
RTECS number: OO3675000 [1]	
Physical and Chemical Properties	
Structural formula: [2] Molecular formula: $C_{10}H_5ClN_2$ Molecular weight: 188.6 [1]	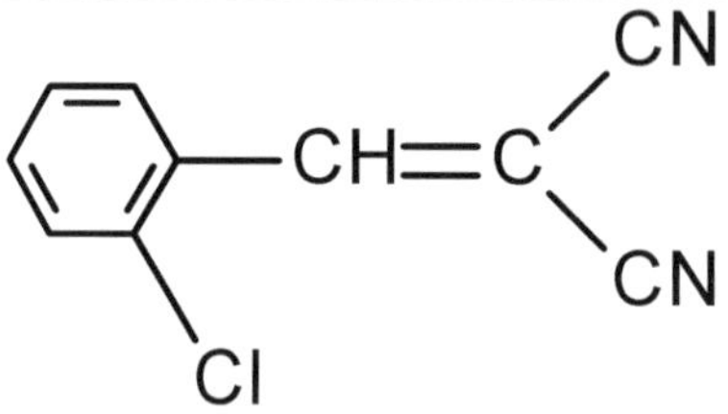
Physical state	White crystalline solid [1]
Odor	Pepper-like [1]
Boiling point	310°C to 315°C (590-599°F)[1]
FP/MP	95°C to 96° C (203-205°F) (MP) [1]
Solid density (g/mL)	Bulk: 0.24-0.26; Crystal: 1.04 [2]
Vapor density (relative to air)	6.5 (calculated)
Vapor pressure (torr)	0.00034 @ 20°C [2]
Volatility (mg/m³)	0.71 @ 25°C [2]
Latent heat of vaporization (kcal/mol)	Data not available
Flash point	Data not available
Decomposition temperature	Data not available
Solubility	Insoluble in water; [1,2] moderate in alcohol; and good in acetone, chloroform, methylene dichloride, ethylacetate, and benzene [3]
Rate of hydrolysis	Data not available
Hydrolysis products	Data not available
Stability in storage	Combustible material; may burn but does not ignite readily. Containers may explode when heated;[4] incompatible with strong oxiders.[1]
Action on metals or other materials	Contact with metals may evolve flammable hydrogen gas [4]

Table III-2. CS (Continued)

Other Data	
Skin and eye toxicity	Burning and irritation to eyes; primary skin irritant [5]
Inhalation toxicity	Causes sensation of choking [5]
Rate of action	Instantaneous [6]
Protection required	A protective mask and dry field clothing secured at neck, wrists, and ankles. Personnel handling and loading bulk CS should wear protective clothing, masks, and gloves. [7]
Decontamination	Move to fresh air. Flush eyes and skin with water. Do not rub eyes. Do not use oil-based lotions. Do not use any form of bleach. Use soap and water on equipment contaminated with CS, CS1, or CS2.[7]
Use	Training and RCA [7]

NOTES

[1]NIOSH Pocket Guide to Chemical Hazards, "o-Chlorobenzylidene malononitrile," CAS 2698-41-1.
[2]BG Russ Zajtchuk et al. (eds), T*extbook of Military Medicine: Medical Aspects of Chemical and Biological Warfare*, Office of the Surgeon General, 1997, Chap. 12, "Riot Control Agents."
[3]FM 8-9/NAVMED P-3059/AFJMAN 49-151, *NATO Handbook on the Medical Aspects of NBC Defense Operations* AMEDP-6(B), 1 February 1996.
[4]*2000 Emergency Response Guidebook*, Guide 153, Substances-Toxic and/or Corrosive (Combustible).
[5]Sharon Reutter, et al., *Review and Recommendations for Human Toxicity Estimates for FM 3-11.9*, ECBC-TR-349, September 2003.
[6]*NIOSH-DOD-OSHA Sponsored Chemical and Biological Respiratory Protection Workshop Report,* February 2000.
[7]FM 3-11.11/MCRP 3-3.7.2, *Flame, Riot Control Agents and Herbicide Operations*, 19 August 1996 (Renumbered from FM 3-11).

Table III-3. CS Toxicity Estimates[5]

Endpoint	Toxicity (mg min/m^3)	MV (L)	Exposure Duration	ROE	Probit Slope	TLE	ROD	DOC
Lethality	LCt_{50}: 52,000-61,000 [a] (Provisional)	15	5-90 min	Inhalation/ Ocular	Unknown	More than 1	Some	Low
Intolerable	ECt_{50}: 7 [b]	15	1 min	Inhalation/ Ocular	Unknown	Unknown	Unknown	Moderate

NOTES

[a]Based on existing human estimates.
[b]Based on "official" estimates (HEC, 1969).

(1) CS1 has been especially formulated to prolong persistency and increase effectiveness. Unlike CS, CS1 is a free-flowing agent powder consisting of 95 percent crystalline CS blended with 5 percent silica aerogel. This formulation reduces agglomeration, increases fluidity, and achieves the desired respiratory effects when dispersed as a solid aerosol. When disturbed, CS1 reaerosolizes and can cause respiratory and eye effects. [2]

(2) CS2 is a siliconized, microencapsulated form of CS1.[3] This treatment improves the physical characteristics of CS by reducing agglomeration and hydrolysis. This form of CS prolongs the effectiveness for both immediate and surface contamination effects.[2]

(3) CSX is a form of CS developed for dissemination as a liquid rather than a powder. One gram of powdered CS is dissolved in 99 grams of trioctyl phosphite (TOF). As with CS, CSX stings and irritates the eyes, skin, nose, throat, and lungs of exposed personnel.[2]

e. Dibenz (b,f)–1:4-oxazepine (CR) (see Table III-4 [page III-4]). In 1974, the USA approved the use of CR in riot control situations.[2] CR is more potent and less toxic than

CS.[3] CR is not used in its pure form but is dissolved in a solution of 80 parts of propylene glycol and 20 parts of water to form a 0.1 percent CR solution. The severity of symptoms increases with the CR solution concentration and in any environment of high temperature and humidity. CR does not degrade in water, and it is quite persistent in the environment. Under suitable conditions, CR can persist on certain surfaces (especially porous) for up to 60 days.[2] (See Table III-5) for CR toxicity estimates

Table III-4. CR

Chemical name: Dibenz(b,f)-1:4-oxazepine [1]	
Synonym: Dibenzoxazepine, CR [2]	
CAS registry number: 257-07-8	
RTECS Number: HQ395000	
Physical and Chemical Properties	
Structural formula: [1] Molecular formula: $C_{13}H_9NO$ Molecular weight: 195.22	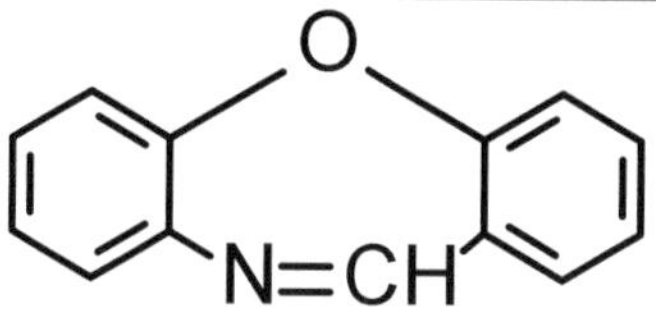
Physical state	Pale yellow crystalline solid [3]
Odor	Pepper-like [3]
Boiling point	Data not available
FP/MP	73°C (163°F) (MP) [3]
Solid density (g/mL)	Data not available
Vapor density (relative to air)	6.7 (calculated)
Vapor pressure (torr)	Data not available
Volatility (mg/m^3)	Data not available
Latent heat of vaporization (kcal/mol)	Data not available
Flash point	Data not available
Decomposition temperature	Data not available
Solubility	Sparingly soluble [1]
Rate of hydrolysis	Not hydrolyzed in aqueous solutions [3]
Hydrolysis products	Data not available
Stability in storage	Stable in organic solutions [3]
Action on metals or other materials	Data not available
Other Data	
Skin and eye toxicity	Irritant; however, CR does not induce inflammatory cell infiltration, vesication, or contact sensitization [1]
Inhalation toxicity	Causes almost no effects in the lower airways and lungs[1]
Rate of action	Instantaneous [4]
Protection required	Personnel exposed to CR should wear protective clothing (secured at neck, wrists, and ankles), masks, hoods, and gloves. [5]
Decontamination	Move to fresh air. Flush eyes with copious amounts of cold water. Do not rub your eyes. Do not use any form of bleach. Soap and water can be used on skin. To decontaminate equipment or surfaces, remove by using towels, rags, absorbent paper, or any other method such as scraping, shoveling, or sweeping.[5]
Use	RCA [5]

Table III-4. CR (Continued)

NOTES
[1]BG Russ Zajtchuk, et al. (eds), *Textbook of Military Medicine: Medical Aspects of Chemical and Biological Warfare*, Office of the Surgeon General, 1997, Chap. 12, "Riot Control Agents."
[2]FM 8-9/NAVMED P-5059/AFJMAN 44-151, *NATO Handbook on the Medical Aspects of NBC Defense Operations AMEDP-6(B),* 1 February 1996.
[3]FM 8-285/NAVMED P-5041/AFJMAN 44-149/FMFM 11-11, *Treatment of Chemical Agent Casualties and Conventional Military Chemical Injuries*, 22 December 1995.
[4]*NIOSH-DOD-OSHA Sponsored Chemical and Biological Respiratory Protection Workshop Report,* February 2000.
[5]FM 3-11.11/MCRP 3-3.7.2, *Flame, Riot Control Agents and Herbicide Operations*, 19 August 1996 (Renumbered from FM 3-11).

Table III-5. CR Toxicity Estimates[5]

Endpoint	Toxicity (mg min/m^3)	MV (L)	Exposure Duration	ROE	Probit Slope	TLE	ROD	DOC
Lethality	LCt_{50}: N/A	N/A	N/A	Inhalation/Ocular	N/A	N/A	N/A	N/A
Severe effects; intolerable	ECt_{50}: 0.15[a]	15	1 min	Inhalation/Ocular	Unknown	N/A	Apparently Rapid	Moderate
Threshold effects; just discernable respiratory tract/ocular symptoms	ECt_{50}: 0.002-0.004[b]	N/A	1 min	Inhalation/Ocular	Unknown	N/A	Apparently Rapid	Moderate
NOTES								
[a]Based on "official" estimates (HEC, 1967). [b]Based on human data.								

f. Capsaicin (OC) (see Table III-6). Capsaicin, also called oleoresin capsicum, is derived from cayenne peppers. OC stimulates sensory nerve endings, causing reflex changes in blood pressure and respiration. It causes pain, edema, and erythema of the tissues with which it makes contact. It also produces bronchoconstriction and edema of the airway mucosa. Contact with the eyes is extremely painful. OC is a powerful irritant and lacrimator. Although it may not have been the proximal cause of death, it has been associated with deaths in humans.[5] See Table III-7 (page III-7) for toxicity estimates.

Table III-6. OC

Alternate Designations:* DC; Pepper Spray.
Chemical Name: Trans-8-methyl-N-vanillyl-6-nonenamide
Synonym: Vanillyl decenamide; 8-Methyl non-6-enoyl vanillylamide; 8 Methyl nonen-6-oxyl-vanillyl amide; N-(4-Hydroxy-3-methoxybenzyl)-8-methylnon-trans-6-enamide; (E)-N-[(4-Hydroxy-3-methoxyphenyl)methyl]-8-methyl-6-nonenamide; Trans-8-methyl-N-vanillyl-6-nonenamide; 6-Nonenamide, N-((4-Hydroxyl-3-methoxyphenyl)methyl)-8-methyl-, (E)-; 6-Nonenamide, 8-methyl-N-vanillyl-, (E)
CAS Registry Number: 404-86-4
RTECS Number: RA8530000
Physical and Chemical Properties
Structural Formula: 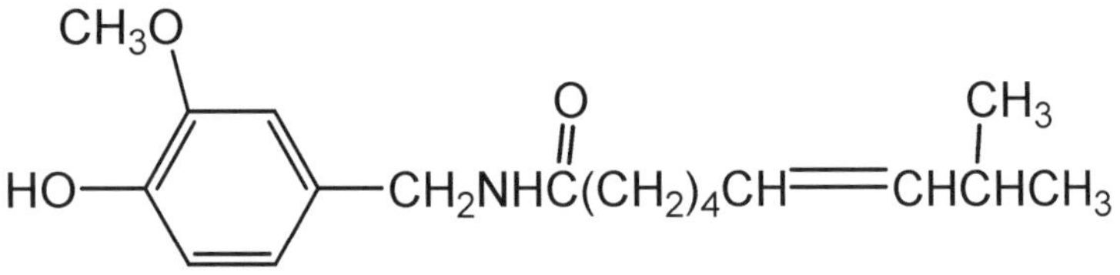
Molecular Formula: $C_{18}H_{27}NO_3$ Molecular Weight: 305.42

Table III-6. OC (Continued)

Physical state	Colorless monoclinic plates [1, 2]
Odor	Pungent, irritating [1]
Boiling point	340.4°C (extrapolated) [3]
FP/MP	65°C (FP) [2,4]
Liquid density (g/mL)	Data not available
Vapor density (relative to air)	10.5 (calculated)
Vapor pressure (torr)	1.5×10^{-7} @ 65°C (extrapolated) [3]
Volatility (mg/m^3)	2.2×10^{-3} @ 65°C (calculated from vapor pressure) [3]
Latent heat of vaporization (kcal/mol)	33.4 (calculated from Clausius Clapeyron equation which assumes constant heat of vaporization of a function of temperature) [3]
Flash point	Capsaicin in the form of pepper spray may be flammable or nonflammable depending on the type of delivery (carrier) system used. [5]
Decomposition temperature	Above 150°C [3]
Solubility	Solubility in water is 0.090 g/L @ 37°C; [6] soluble in alcohol, ether, chloroform, carbon disulfide, concentrated HCl, aromatic solvents, hydrocarbons, ketones, and aqueous alkali [1, 2, 7]
Rate of hydrolysis	Data not available
Hydrolysis products	Alkaline hydrolysis yields vanillylamine and isomeric decenoic acid [7]
Stability in storage	Data not available
Action on metals or other materials	Data not available
Other Data	
Skin and eye toxicity	Powerful irritant and lacrimator [8]
Inhalation toxicity	Causes bronchoconstriction and edema [8]
Rate of action	Almost immediate [9]
Protection required	A protective mask and ordinary field clothing secured at the neck, wrist, and ankles [9]
Decontamination	Move to fresh air. Flush face with cool water; if burning persists, use ice pack. Do not rub area. Decontaminate required areas with soap and water. [9]
Use	Most often used by the MP for incapacitating violent or threatening subjects, also has applications for SF and stability and support operations. [9]

NOTES

*Capsaicin is the principal capsaicinoid compound present in oleoresin capsicum (OC). OC is commonly known as pepper spray. When pepper spray is combined with various carrier systems (i.e., isopropyl alcohol, methylene chloride, water, etc.), it has the potential of being an effective riot control agent.

[1]Sherrill, M.L., *Investigation of the Synthesis of Capsaicin and Related Compounds*, EACD 307, USA Chemical Research Laboratories, Edgewood Arsenal, MD, April 1924, UNCLASSIFIED Report (ADB955292).

[2]Steadman, A., *Isolation of Capsaicin from Capsicum*, EACD 188, USA Chemical Research Laboratories, Edgewood Arsenal, MD, June 1922, UNCLASSIFIED Report (ADB955131).

[3]Watson, P.D., *Determination of the Vapor Pressure of D.C.*, EACD 79, Chemical Warfare Service, Edgewood Arsenal, MD, December 1921, UNCLASSIFIED Report (ADB959611).

[4]Nelson, E.K., and Dawson, L.E., "The Constitution of Capsaicin, The Pungent Principle of Capsicum. III," *J. Am. Chem. Soc.,* Vol. 45, 1923.

[5]Daroff, P.M., et al., *Oleoresin Capsicum: An Effective Less-Than Lethal Riot Control Agent*, DPG/JCP-097-002, Chemical Biological Defense, USA Dugway Proving Ground, UT, January 1997, UNCLASSIFIED Report (ADB225032).

[6]Carter, R.H., and Knight, H.C., *Fundamental Study of Toxicity: Solubility of Certain Toxics in Water and in Olive Oil*, EACD 445, Chemical Warfare Service, Edgewood Arsenal, MD, May 1928, (ADB955216).

[7]Rosenberg, H.R., and Sharp, S.S., *Evaluation and Synthesis of Chemical Compounds Vol. II, DA-18-108-CML-6673 (A), Final Comprehensive Report November 1961 – February 1965*, USA Chemical Research Laboratories, Edgewood Arsenal, MD, February 1965, UNCLASSIFIED Report (ADB253543).

[8]Sharon Reutter, et al., *Review and Recommendations for Human Toxicity Estimates for FM 3-11.9*, ECBC-TR-349, September 2003.

[9]FM 3-11.11/MCRP 3-3.7.2, *Flame, Riot Control Agents and Herbicide Operations*, 19 August 1996 (Renumbered from FM 3-11).

Table III-7. OC Toxicity Estimates[5]

Endpoint	Toxicity ($mg\ min/m^3$)	MV (L)	Exposure Duration	ROE	Probit Slope	TLE	ROD	DOC
Lethality	LCt_{50}: N/A	N/A	N/A	Inhalation/Ocular	N/A	N/A	N/A	N/A

3. Respiratory Irritants

The principal respiratory irritants are diphenylaminochloroarsine (DM) (Adamsite), diphenylchloroarsine (DA), diphenylcyanoarsine (DC) and chlorine (Cl_2). DM, DA, and DC were previously called vomiting agents; however, their primary action is irritation of the respiratory tract. They were originally designed as "mask breakers" during WWI. Their intended purpose was to penetrate the canister, forcing troops to remove their masks and be exposed to more toxic materials.[5] Two characteristics make these compounds different than RCAs. The first characteristic is that the effects do not appear immediately on exposure or seconds afterwards, but several minutes later. In the absence of symptoms, personnel will not mask immediately; by the time they mask, a significant amount of the compound will have been absorbed. The effects may then cause an individual to unmask. The second characteristic of these compounds is that there may be more prolonged systemic effects—such as headache, mental depression, chills, nausea, abdominal cramps, vomiting, and diarrhea—which last for several hours after exposure.[3] They are dispersed as aerosols, and they produce their effects by inhalation or by direct action on the eyes. When released indoors, they can cause serious illness or death.[4] The protective mask gives adequate protection against field concentrations. No protective clothing is required.[4]

a. DM (see Table III-8). Symptoms are decidedly more marked after an exposure than during it. The degree of severity and the duration of effects vary directly with the concentration used. After exposure, there is burning and tightening across the chest with a persistent rasping cough, burning in the nose and throat, and acute general depression. DM has caused a few human deaths.[5] See Table III-9 (page III-9) for DM toxicity estimates.

Table III-8. DM

Alternate Designations: Azine; RI5	
Chemical Name: 10-Chloro-5,10-dihydrophenarsazine	
Synonyms: Diphenylamine chloroarsine; Diphenylamine arsenious chloride; Phenarsazine chloride; 1-Chloro-1,6-dihydrophenarsazine; 10-Chlorophenarsazine; 6-Chlorophenarsazine, Phenarsazine, 10-chloro-5,10-dihydro-; 5-Aza-10-arsenaanthracene chloride; 10-Chloro-5,10-dihydroarsacridine; Diphenylaminechlorarsine; Fenarsazinchlorid (Czech); Phenazsarine chloride; 10-chloro-9,10-dihydro-phenarsazine	
CAS Registry Number: 578-94-9	
RTECS Number: SG0680000	
Physical and Chemical Properties	
Structural Formula: 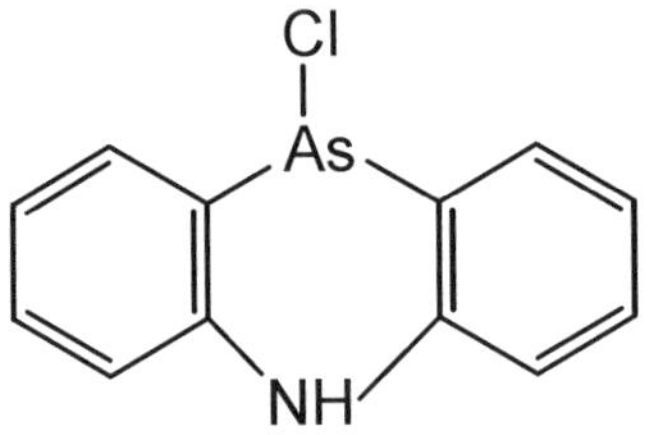 Molecular Formula: $C_{12}H_9AsClN$ Molecular Weight: 277.58	
Physical state	Light yellow to green crystals [1]
Odor	No pronounced odor, but irritating [2]

Table III-8. DM (Continued)

Boiling point	410°C (extrapolated) decomposes [3]
FP/MP	195°C (MP) with slight decomposition [3]
Solid density (g/mL)	1.648 @ 20°C; 1.672 @ 0°C [4]
Vapor density (relative to air)	9.6 (calculated)
Vapor pressure (torr)	Negligible @ ambient temperature [5, 6]
Volatility (mg/m^3)	Negligible @ ambient temperature [5, 6]
Latent heat of vaporization (kcal/mol)	14.2 @ 410°C (calculated from vapor pressure) [5]
Flash point	Nonflammable [2]
Decomposition temperatures	Slight decomposition @ 195°C; [3] 0.02% per min. @ 200°C; and 0.15% per min @ 250°C [5]
Solubility	Solubility in water is 0.044 g/L @ 37°C. [7] Slightly soluble in benzene xylene, carbon tetrachloride, [8, 9] acetone, alcohols, tetrachorethane[9]
Rate of hydrolysis	When solid DM is covered with water, it slowly hydrolyzes and a protective oxide coating forms that hinders further hydrolysis. Finely divided DM hydrolyzes rapidly. [4] Acidic solutions prevent hydrolysis. 0.5% HCl @ room temperature and 0.8% at temperatures between 70-80°C. [4]
Hydrolysis products	Diphenylaminearsenious oxide and hydrochloric acid [4]
Stability in storage	At room temperature, DM is stable for at least 1 year when pure and 6 months with plant grade material. [10] Stable in aluminum and stainless steel for at least 2 years @ 71°C when pure, but the containers are severely pitted. [10]
Action on metals or other material	After 3 months, causes extensive corrosion on aluminum and stainless steel @ 71°C. [10] Also corrodes iron, bronze and brass. [8]
Other Data	
Skin and eye toxicity	Causes eye irritation and burning [11]
Inhalation toxicity	Primary action on upper respiratory tract [11]
Rate of action	Rapid [12]
Protection required	Protective mask [13]
Decontamination	If symptoms persist, the eyes, mouth, and skin may be washed with water. Do not swallow water. [14]
Use	Previously used as an RCA and "mask breaker" [11]

Table III-8. DM (Continued)

NOTES
[1]Lewis, R.J., *Sax's Dangerous Properties of Industrial Materials*, 10th ed., Vol. 3, p. 2875, John Wiley & Sons, Inc., New York, NY, 2001.
[2]Lau, T.M.K., *Brief Evaluation of the Possibilities of Using Arsenicals as Incapacitating Agents*, CRDEC-TR-87061, USA Chemical Research Development and Engineering Center, Aberdeen Proving Ground, MD, July 1987, UNCLASSIFIED Report (ADB114319).
[3]Klosky, S. and Stricker, P.F., *Physico Chemical Constants of Diphenylaminochlorarsine*, EATR 58, Chemical Warfare Service Edgewood Arsenal, MD, July 1921, UNCLASSIFIED Report (ADB955024).
[4]Macy, R., *Constants and Physiological Action of Chemical Warfare Agents*, EATR 78, Chemical Warfare Service Edgewood Arsenal, MD, July 1932, UNCLASSIFIED Report (ADB956574).
[5]Parker, D.H., *Vapor Pressure of D.M. (Diphenylaminechloroarsine)*, EATR 46, Chemical Warfare Service Edgewood Arsenal, MD, June 1921, UNCLASSIFIED Report (ADB955053).
[6]*Properties of War Gases Volume III: Vomiting & Choking Gases & Lacrimators (U),* ETF 100-41/Vol-3, Chemical Corps Board, Army Chemical Center, MD, December 1944, UNCLASSIFIED Report (AD108458).
[7]Carter, R.H., and Knight, H.C., *Fundamental Study of Toxicity: Solubility of Certain Toxics in Water and in Olive Oil*, EACD 445, Chemical Warfare Service, Edgewood Arsenal, MD, May 1928, UNCLASSIFIED Report (ADB955216).
[8]*The Merck Index: An Encyclopedia of Chemicals, Drugs, and Biologicals*, 13th ed., p. 5225, Merck & Company, Inc., Whitehouse Station, NJ, 2001.
[9]Kibler, A.L., *Fundamental Study of Toxicity Miscellaneous Data,* EACD459, Chemical Warfare Service, Edgewood Arsenal, MD, 1928 UNCLASSIFIED Report (ADB955210).
[10]Brooks, M.E., et al., *Corrosion, Compatibility and Other Physicochemical Studies (U)*, DA18-108-CML-6602 (A), Final Report – Task I, W.R. Grace and Company, Washington Research Center, Clarksville, MD, May 1964, UNCLASSIFIED Report (AD350755).
[11]BG Russ Zajtchuk, et al. (eds), T*extbook of Military Medicine: Medical Aspects of Chemical and Biological Warfare*, Office of the Surgeon General, 1997, Chap. 12, "Riot Control Agents."
[12]*NIOSH-DOD-OSHA Sponsored Chemical and Biological Respiratory Protection Workshop Report,* February 2000.
[13]FM 8-285/NAVMED P-5041/AFJMAN 44-149/FMFM 11-11, *Treatment of Chemical Agent Casualties and Conventional Military Chemical Injuries*, 22 December 1995.
[14]FM 8-9/NAVMED P-3059/AFJMAN 49-151, *NATO Handbook on the Medical Aspects of NBC Defense Operations* AMEDP-6(B), 1 February 1996.

Table III-9. DM Toxicity Estimates[5]

Endpoint	Toxicity (mg min/m^3)	MV (L)	Exposure Duration	ROE	Probit Slope	TLE	ROD	DOC
Lethality	LCt_{50}: 11,000[a] (Provisional)	15	2-240 min	Inhalation/Ocular	Unknown	1[b,c]	Unknown	Low
Intolerable	ECt_{50}: 22 – 150[d] (Provisional)	15	1 min	Inhalation/Ocular	Unknown	Unknown	Unknown	Low
Threshold effects, throat irritation	ECt_{50}: less than 1[e]	15	1 min	Inhalation/Ocular	Unknown	Unknown	Unknown	Low

NOTES

[a]Based on existing "official" estimates.
[b]The TLE value is assumed to be 1 because the Ct profile is unknown.
[c]See Appendix H for detailed toxicity profile estimates.
[d]Based on human data.
[e]Based on secondary data.

b. DA (see Table III-10 [page III-10]). DA causes severe irritation of the mucous membranes, coughing, salivation, and nasal discharge. It may lead to serious disturbances of the nervous system from absorption. DA has also been classified as a sternulator.[5] See Table III-11 (page III-11) for toxicity estimates.

Table III-10. DA

Alternate Designations: Clark I (German); Blue Cross (German); Sternite (French); Sneezing gas; DIK	
Chemical Name: Diphenylchloroarsine	
Synonyms: Arsine, chloro, diphenyl; Chlorodiphenylarsine; Diphenylarsenious chloride; Chlorodiphenylarsine; Arsinous chloride, diphenyl-; Diphenyl arsenic chloride; Chlor-difenylarsin (Czech); Chlorodiphenylarsine; Difenylchlorarsin (Czech); Diphenylarsinous chloride; Diphenylchloorarsine (Dutch)	
CAS Registry Number: 712-48-1	
RTECS Number: CG9900000	
Physical and Chemical Properties	
Structural Formula: 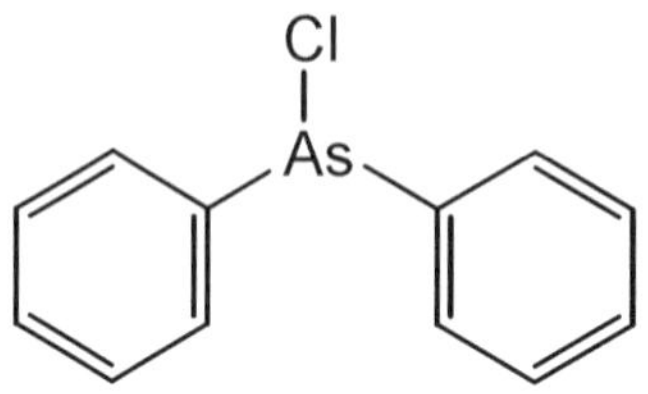 Molecular Formula: $C_{12}H_{10}AsCl$ Molecular Weight: 264.59	
Physical state	Colorless crystalline solid when pure [1]
Odor	None [1]
Boiling point	383°C (extrapolated) decomposes [2]
FP/MP	37.3°C [3]; 39 to 44°C (MP) [2] DA also exists in an unstable modification which melts between 18.2 to 18.4°C [3]
Liquid density (g/mL)	1.3875 @ 50°C [3]
Vapor density (relative to air)	9.1(calculated)
Vapor pressure (torr)	1.79×10^{-2} @ 50°C (extrapolated) [3]
Volatility (mg/m^3)	2.36×10^{2} @ 50°C (calculated from vapor pressure) [3]
Latent heat vaporization (kcal/mol)	15.1 (calculated from Clausius Clapeyron equation which assumes constant heat of vaporization as a function of temperature) [3]
Flash point	350°C [1]
Decomposition temperature	300 to 350°C [2]
Solubility	Solubility in water is 0.078 g/L @ 37°C; [4] soluble in acetone, ether, [5] ethanol, benzene, carbon tetrachloride, ethylene chloride, chloroform, and dichloroethylene [2, 5]
Rate of hydrolysis	Slow in bulk but rapid when finely divided [2]
Hydrolysis products	Diphenylarsenious oxide and hydrogen chloride [2]
Stability in storage	Stable when pure. Stable in steel shells for almost 4 months @ 60°C and for 1 year @ room temperature. [2]
Action on metals or other materials	None when dry [2]
Other Data	
Skin and eye toxicity	Irritant [6]
Inhalation toxicity	Irritant [6]
Rate of action	Rapid [7]
Protection required	Protective mask [8]
Decontamination	If symptoms persist, the eyes, mouth, and skin may be washed with water. Do not swallow water. [9]
Use	Previously used as "mask breaker" [6]

Table III-10. DA (Continued)

NOTES
[1]Lau, T.M.K., *Brief Evaluation of the Possibilities of Using Arsenicals as Incapacitating Agents*, CRDEC-TR-87061, USA Chemical Research, Development and Engineering Center, Aberdeen Proving Ground, MD, July 1987, UNCLASSIFIED Report (ADB114319). [2]Macy, R., *Constants and Physiological Action of Chemical Warfare Agents*, EATR 78, Chemical Warfare Service Edgewood Arsenal, MD, July 1932, UNCLASSIFIED Report (ADB956574). [3]Owens, R., *Diphenylcyanoarsine: Part V – The Physical Properties of M.A., D.A. T.A., and D.C.*, SO/R 492, Sutton Oak, England, December 1940, UNCLASSIFIED Report. [4]Carter, R.H., and Knight, H.C., *Fundamental Study of Toxicity: Solubility of Certain Toxics in Water and in Olive Oil*, EACD 445, Chemical Warfare Service, Edgewood Arsenal, MD, May 1928, UNCLASSIFIED Report (ADB955216). [5]Lide, D.R, *CRC Handbook of Chemistry and Physics*, 82nd, ed., p. 3-15, CRC Press, Washington, DC 2001. [6]Sharon Reutter, et al., *Review and Recommendations for Human Toxicity Estimates for FM 3-11.9*, ECBC-TR-349, September 2003. [7]*NIOSH-DOD-OSHA Sponsored Chemical and Biological Respiratory Protection Workshop Report,* February 2000. [8]FM 8-285/NAVMED P-5041/AFJMAN 44-149/FMFM 11-11, *Treatment of Chemical Agent Casualties and Conventional Military Chemical Injuries*, 22 December 1995. [9]FM 8-9/NAVMED P-3059/AFJMAN 49-151, *NATO Handbook on the Medical Aspects of NBC Defense Operations* AMEDP-6(B), 1 February 1996.

c. DA Toxicity Estimates (Table III-11). No toxicity estimates are recommended for lethal exposure. The existing estimates could not be substantiated.

Table III-11. DA Toxicity Estimates[5]

Endpoint	Toxicity (mg min/m^3)	MV (L)	Exposure Duration	ROE	Probit Slope	TLE	ROD	DOC
Lethality	LCt_{50}: NR	N/A	N/A	Inhalation/Ocular	Unknown	Unknown	Unknown	N/A
Intolerable	ECt_{50}: 12[a]	15	2 min	Inhalation/Ocular	Unknown	Unknown	Unknown	Low
NOTES								
[a]Based on secondary human data.								

d. DC (see Table III-12 [page III-12]). The Germans introduced DC in 1918 as an improvement over DA, which is readily hydrolyzed by water. DC causes irritation of the nose and throat, salivation, and profuse secretion from the eyes and nose. There is a feeling of suffocation and headache. Symptoms last for 30 minutes to 1 hour, except the headache may last for several hours.[5] See Table III-13 (page III-13) for DC toxicity estimates.

Table III-12. DC

Alternate Designations: CLARK II (German); Clark 2 (German); Blue Cross (German), Sternite (French)	
Chemical Name: Diphenylcyanoarsine	
Synonyms: Diphenylarsinous cyanide; Diphenylarsinecarbonitrile; Arsinous cyanide, diphenyl-; Arsinecarbonitrle, diphenyl-	
CAS Registry Number: 23525-22-6	
RTECS Number: Data not available	
Physical and Chemical Properties	
Structural Formula: 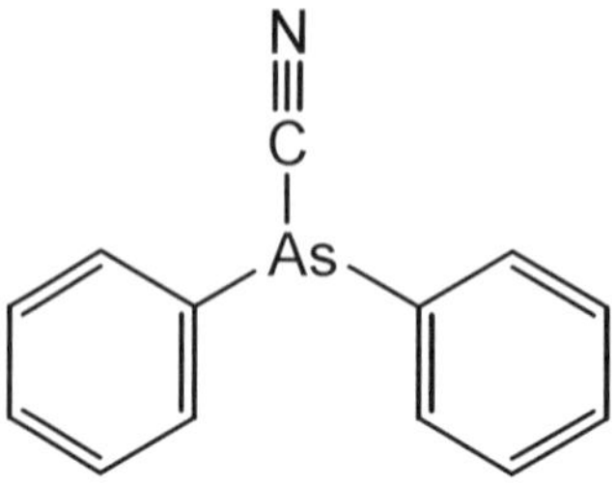 Molecular Formula: $C_{13}H_{10}AsN$ Molecular Weight: 255.15	
Physical state	Colorless crystalline solid [1]
Odor	Similar to garlic and bitter almonds [1]
Boiling point	341°C (extrapolated) decomposes [2]
FP/MP	31.2°C (FP) [2]
Liquid density (g/mL)	1.3338 @ 35°C [2]
Vapor density (relative to air)	8.8 (calculated)
Vapor pressure (torr)	7.2×10^{-4} @ 35°C (extrapolated) [2]
Volatility (mg/m^3)	9.56 @ 35°C (calculated from vapor pressure) [2]
Latent heat of vaporization (kcal/mol)	17.1 (calculated from Clausius Clapeyron equation which assumes constant heat of vaporization as a function of temperature) [2]
Flash point	Low [1]
Decomposition temperatures	Above 240°C [2]
Solubility	Solubility in water is 0.021 mg/L @ 37°C; [3] soluble in chloroform and other organic solvents[4]
Rate of hydrolysis	Very slow [1]
Hydrolysis products	Hydrogen cyanide and diphenylarsenious oxide [1]
Stability in storage	Stable at all ordinary temperatures [1]
Action on metals or other material	None on metals [4]
Other Data	
Skin and eye toxicity	Irritant [5]
Inhalation toxicity	Irritant [5]
Rate of action	Rapid [6]
Protection required	Protective mask [7]
Decontamination	If symptoms persist, the eyes, mouth and skin may be washed with water. Do not swallow water. [8]
Use	Previously used as "mask breaker" [5]

Table III-12. DC (Continued)

NOTES
[1]Lau, T.M.K., *Brief Evaluation of the Possibilities of Using Arsenicals as Incapacitating Agents*, CRDEC-TR-87061, USA Chemical Research Development and Engineering Center, Aberdeen Proving Ground, MD, July 1987, UNCLASSIFIED Report (ADB114319). [2]Owens, R., *Diphenylcyanoarsine: Part V – The Physical Properties of M.A., D.A. T.A., and D.C.*, SO/R 492, Sutton Oak, England, December 1940, UNCLASSIFIED Report. [3]Carter, R.H., and Knight, H.C., *Fundamental Study of Toxicity: Solubility of Certain Toxics in Water and in Olive Oil*, EACD 445, Chemical Warfare Service, Edgewood Arsenal, MD, May 1928, UNCLASSIFIED Report (ADB955216). [4]Franke, S., *Manual of Military Chemistry Volume I- Chemistry of Chemical Warfare Agents*, ACSI-J-3890, Chemie der Kampfstoffe, East Berlin, April 1968, UNCLASSIFIED Technical Manual (AD849866). [5]Sharon Reutter, et al., *Review and Recommendations for Human Toxicity Estimates for FM 3-11.9*, ECBC-TR-349, September 2003. [6]*NIOSH-DOD-OSHA Sponsored Chemical and Biological Respiratory Protection Workshop Report,* February 2000. [7]FM 8-285/NAVMED P-5041/AFJMAN 44-149/FMFM 11-11, *Treatment of Chemical Agent Casualties and Conventional Military Chemical Injuries*, 22 December 1995. [8]FM 8-9/NAVMED P-3059/AFJMAN 49-151, *NATO Handbook on the Medical Aspects of NBC Defense Operations* AMEDP-6(B), 1 February 1996.

Table III-13. DC Toxicity Estimates[5]

Endpoint	Toxicity (mg min/m^3)	MV (L)	Exposure Duration	ROE	Probit Slope	TLE	ROD	DOC
Lethality	LCT_{50}: NR	N/A	N/A	Inhalation/ Ocular	Unknown	Unknown	Unknown	N/A
Severe effects; incapacitation	ECt_{50}: NR	N/A	N/A	Inhalation/ Ocular	Unknown	Unknown	Unknown	N/A

e. Chlorine (Cl_2) (see Table III-14). Chlorine causes spasm of the larynx muscles; burning of the eyes, nose, and throat;, bronchitis; and asphyxiation.[5] Rapid evaporation of the liquid may cause frostbite.[7] Chlorine has been classified as a lung injurant,[5] as a choking agent,[4] and as a TIC.[8] See Table III-15 (page III-14) for toxicity estimates.

Table III-14. Cl_2

Alternate Designations: Bertholite	
Chemical Name: Chlorine	
Synonyms: Chloor (Dutch); Chlor (German); Chlore (French); Chlorine mol.; Cloro (Italian); Molecular chlorine	
CAS Registry Number: 7782-50-5	
RTECS Number: FO2100000	
Physical and Chemical Properties	
Structural Formula: Cl—Cl Molecular Formula: Cl_2 Molecular Weight: 70.91	
Physical state	Greenish-yellow diatomic gas [1]
Odor	Disagreeable and suffocating; irritating to the nose and throat [2]
Boiling point	-34.7°C [3, 4]
MP/FP	-101.6°C (FP) [5]
Liquid density (g/mL)	Liquefied chlorine: 1.393 @ 25°C; 1.468 @ 0°C [4]
Vapor density (relative to air)	2.4 (calculated)
Vapor pressure (torr)	5.75×10^3 @ 25°C; 2.73×10^3 @ 0°C [3, 4]
Volatility (mg/m^3)	2.19×10^7 @ 25°C; 1.14×10^7 @ 0°C (calculated from vapor pressure) [3, 4]
Latent heat of vaporization (kcal/mol)	4.86 @ 25°C; 4.80 @ 0°C (calculated from vapor pressure) [3, 4]
Flash point	Nonflammable [2]
Decomposition temperature	Above 600°C [5]

Table III-14. Cl_2 (Continued)

Solubility	Solubility in water is 0.63 g/100 g water @ 25°C; solubility in carbon tetrachloride is 3.5% @ ambient temperature [5]
Rate of hydrolysis	Slow [6]
Hydrolysis products	HCl and HOCl [5]
Stability in storage	Stable when dry [5]
Action on metals or other materials	None if chlorine is dry. Vigorous action with metals when chlorine is moist due to the presence of hypochlorous acid. [5]
Other Data	
Skin and eye toxicity	Irritant [7]
Inhalation toxicity	Can cause pulmonary edema [7]
Rate of action	Like choking agents, pulmonary edema does not manifest until hours have passed and are aggravated by physical effort [7]
Protection required	Protective mask [8]
Decontamination	Inhalation - fresh air and rest. Eyes and skin - rinse with plenty of water. [7]
Use	Not authorized for military use

NOTES

[1]*The Merck Index: An Encyclopedia of Chemicals, Drugs, and Biologicals*, 13th ed., p. 5225, Merck & Company, Inc., Whitehouse Station, NJ, 2001.
[2]Yaws, C.L., *Matheson Gas Data Book*, 7th ed., p. 162, McGraw-Hill Companies, New York, NY, 2001.
[3]Abercrombie, P., ECBC Notebook # NB 98-0079, p. 32 (U).
[4]Beebe, C.H, *Important Constants of Fourteen Common Chemical Warfare Agents*, EACD 328, Chemical Warfare Service, Edgewood Arsenal, Edgewood, MD, December 1924, UNCLASSIFIED Report (ADB958296).
[5]Macy, R., *Constants and Physiological Action of Chemical Warfare Agents*, EATR 78, Chemical Warfare Service, Edgewood Arsenal, MD July 1932, UNCLASSIFIED Report (ADB956574).
[6] TM 3-215/AFM 355-7, *Military Chemistry and Chemical Agents*, December 1963, UNCLASSIFIED Technical Manual (ADA292141).
[7]ICSC 0126, "Chlorine (Cl_2)."
[8]FM 8-285/NAVMED P-5041/AFJMAN 44-149/FMFM 11-11, *Treatment of Chemical Agent Casualties and Conventional Military Chemical Injuries*, 22 December 1995.

f. Cl_2 Toxicity Estimates. Toxicity estimates for lethal and severe effects are recommended. Existing estimates could not be substantiated.

Table III-15. Cl_2 Toxicity Estimates[5]

Endpoint	Toxicity (mg min/m^3)	MV (L)	Exposure Duration	ROE	Probit Slope	TLE	ROD	DOC
Lethality	LCt_{50}: NR	N/A	N/A	Inhalation/Ocular	Unknown	Unknown	Unknown	N/A
Severe effects, incapacitation	ECt_{50}: NR	N/A	N/A	Inhalation/Ocular	Unknown	Unknown	Unknown	N/A
Threshold effects, odor	ECt_{50}: 10[a]	N/A	Seconds	Inhalation/Ocular	Unknown	N/A	Unknown	Low

NOTES

[a]Based on human secondary data and TM 3-215 (1952).

4. Obsolete Riot Control Agents

The following RCAs are considered obsolete for military employment. The following is primarily of academic and historical interest.

a. Chloroacetophenone (CN) (see Table III-16). The symbol CN identifies the RCA popularly known as tear gas or mace. The USA replaced CN with CS in 1959.[3] Inhalation of CN causes a burning sensation, cough, sore throat, nausea, and shortness of breath. Exposure to skin causes redness and pain. Eye exposure causes redness, pain, and blurred vision. CN can cause pulmonary edema.[9] High concentrations can cause blisters to form, and CN is a potent skin sensitizer. The indiscriminate use of large amounts of CN in confined spaces has caused injuries requiring medical attention and death.[3]

Table III-16. CN

Chemical name: 2-Chloracetophenone [1]	
Synonyms: alpha-Chloroacetophenone, 2-Chloro-1-pheylethanone, Chloromethyl phenyl ketone, Mace®, Phenacyl chloride, Phenyl chloromethyl ketone, Tear Gas [2]	
CAS Registry Number: 532-27-4 [1]	
RTECS Number: AM6300000 [1]	
Physical and Chemical Properties	
Molecular formula	C_8H_7ClO [1]
Molecular weight	154.6 [1]
Physical state	Colorless to gray crystalline solid [3]
Odor	Sharp, irritating odor [3]
Boiling point	244°C (472°F) [3]
FP/MP	57°C (134°F) (MP) [3]
Density (g/mL)	1.3 [1]
Vapor density (relative to air)	5.3 (calculated)
Vapor pressure (torr)	0.005 @ 20°C (68°F) [3]
Volatility (mg/m^3)	Data not available
Latent heat of vaporization (kcal/mol)	Data not available
Flash point	118°C (244°F) [3]
Decomposition temperature	Data not available
Solubility	1.64 g/100 mL @ 25°C, [1] soluble in carbon disulfide, ether, and benzene [2]
Rate of hydrolysis	Slow [2]
Hydrolysis products	HCl [2]
Stability in storage	Sensitive to moisture; incompatible with bases, amines, alcohols, water, and steam [2]
Action on metals and other materials	Reacts slowly with metals, causing mild corrosions [2]
Other Data	
Skin and eye toxicity	Irritant [1]
Inhalation toxicity	Irritant [1]
Rate of action	Instantaneous [4]
Protection required	Protective mask and ordinary field clothing secured at the neck, wrists, and ankles [5]
Decontamination	Move to fresh air. If necessary, wash with water.[6] Skin decontaminants containing bleach should not be used. [7]
Use	Previous RCA still in use by police in some countries [6]
NOTES	
[1]ISCS 0128, "2-Chloroacetophenone." [2]National Toxicity Program, "NTP Chemical Repository: Chloroacetophenone." [3]NIOSH Pocket Guide to Chemical Hazards, "a-Chloroacetophenone," CAS 532-27-4. [4]*NIOSH-DOD-OSHA Sponsored Chemical and Biological Respiratory Protection Workshop Report,* February 2000. [5]FM 8-285/NAVMED P-5041/AFJMAN 44-149/FMFM 11-11, *Treatment of Chemical Agent Casualties and Conventional Military Chemical Injuries*, 22 December 1995. [6]FM 8-9/NAVMED P-5059/AFJMAN 44-151, *NATO Handbook on the Medical Aspects of NBC Defense Operations AMEDP-6(B),* 1 February 1996. [7]BG Russ Zajtchuk, et al. (eds), *Textbook of Military Medicine: Medical Aspects of Chemical and Biological Warfare*, Office of the Surgeon General, 1997, Chap. 12, "Riot Control Agents."	

b. Chloroacetophenone Mixtures. Different CN mixtures were produced to include CN in chloroform (CNC), CN in benzene and carbon tetrachloride (CNB), and CN with PS in chloroform (CNS). However, CS proved more effective and less toxic than any of the CN series and largely has replaced them.

> **CAUTION**
> Benzene is a known carcinogen; carbon tetrachloride and chloroform are suspected carcinogens.[10]

c. Bromobenzylcyanide (CA). CA was the last irritating agent introduced by the Allies in World War I (WWI), and it was the most potent. It corrodes iron and steel, is not chemically stable in storage, and is sensitive to heat—all the characteristics that made it unsuitable for storage and use in artillery shells. CA irritates the eyes and causes lacrimation.[3] CA is too toxic for use as an RCA and is considered obsolete.[11]

5. Smokes, Obscurants, and Incendiaries

Smokes, obscurants, and incendiaries are combat multipliers. Their use provides tactical advantages for offensive and defensive operations. For example, smoke has long been employed as a means of concealing battlefield targets. Fire damage causes casualties and material damage and can impact psychologically.[12] This section contains the physical and chemical properties of selected smokes, obscurants, and incendiaries.

a. Smokes and Obscurants. Smoke is an aerosol that owes its ability to conceal or obscure to its composition of many small particles suspended in the air. These particles scatter or absorb the light, thus reducing visibility. When the density or amount of smoke material between the observer and the object to be screened exceeds a certain minimum threshold value, the object cannot be seen.[12] Many types and combinations of smokes are used, but the three basic types of screening smokes are hexachloroethane (HC) smoke, phosphorous smoke, and fog oil smoke.[6] White phosphorous (WP) and HC are hygroscopic; they absorb water vapor from the atmosphere. This increases their diameters and makes them more efficient at reflecting light rays. Fog oils are nonhygroscopic and depend upon vaporization techniques to produce extremely small diameter droplets to scatter light rays.[12] Most smokes are not hazardous in concentrations that are useful for obscuring purposes. However, any smoke can be hazardous to health if the concentration is sufficient or if the exposure is long enough. The protective mask gives the respiratory tract and the eyes adequate protection against all smokes.[4]

> **CAUTION**
> The protective mask must be worn when operating in or around smoke material.

(1) Hexachloroethane (HC). HC is a pyrotechnic containing an equal amount of hexachloroethane and zinc oxide, with approximately 7 percent grained aluminum. On combustion, the reaction products are zinc chloride and 10 percent phosgene, carbon tetrachloride, ethyl tetrachloride, hexachloroethane, hexachlorobenzene, hydrogen chloride, chlorine, and carbon monoxide. The toxicity of this chemical compound is generally attributed to zinc chloride ($ZnCl_2$) (see Table III-17). The $ZnCl_2$ reacts with the moisture in the air to form a grayish white smoke. The more humid the air, the more dense the HC smoke.[13] Immediately after exposure, symptoms include tightness in the chest, sore throat or hoarseness, and cough.[4] Exposures to very high doses of HC smoke commonly result in sudden, early collapse and death.[13]

CAUTION
Hexachloroethane is a suspected carcinogen.[10]

Table III-17. $ZnCl_2$

Chemical name: Zinc Dichloride	
Synonyms: Zinc Dichloride Flume [1]	
CAS Registry No.: 7646-85-7 [1]	
RTECS Number.: ZH1400000 [1]	
Physical and Chemical Properties	
Structural Formula: **Cl — Zn— Cl** Molecular Formula: $ZnCl_2$ [1] Molecular Weight: 136.3 [1]	
Physical state	Hygroscopic white solid [2]
Odor	Data not available
Boiling point	732°C (1350°F) [1]
FP/MP	290°C (554°F) (MP) [1]
Liquid density (g/mL)	Data not available
Vapor density (relative to air)	4.7 (calculated)
Vapor pressure (torr)	Approximately 0 @ 20°C (68°F) [1]
Volatility (mg/m^3)	Data not available
Specific gravity (water = 1 @ 4°C)	2.91 @ 25°C [1]
Flash point	Data not available
Decomposition temperature	Data not available
Solubility	In water, 432 g/100 mL at 25°C [2]
Stability in storage	Decomposes upon heating, producing toxic fumes of hydrogen chloride and zinc oxide [2]
Action on metals or other materials	The solution in water is a medium-strong acid. [2]
Other Data	
Skin and eye toxicity	Corrosive to the eyes and the skin [2]
Inhalation toxicity	Irritant; inhalation of fumes may cause lung edema [2]
Protection required	Try to remain out of smoke/obscurant clouds, and wear protective mask when making smoke. Personnel can reduce exposure to smoke by rolling down their sleeves and showering after exposure. [3]
Decontamination	Remove contaminated clothing and shoes. Launder clothing following exposure. [3]
Use	See FM 3-50, Smoke Operations
NOTES	
[1]NIOSH Pocket Guide to Chemical Hazards, "Zinc Chloride fume," CAS 7646-85-7. [2]ICSC 1064, "Zinc Chloride." [3]FM 8-285/NAVMED P-5041/AFJMAN 44-149/FMFM 11-11, *Treatment of Chemical Agent Casualties and Conventional Military Chemical Injuries*, 22 December 1995.	

(2) Phosphorous. Phosphorous occurs in three allotropic forms: white, red, and black. The military uses white and red phosphorous.

CAUTION
Phosphorous smoke produces phosphoric acid. Soldiers must wear respiratory protection.[14]

(a) White Phosphorous (WP) (see Table III-18 [page III-18]). WP produces a hot, dense, white smoke composed of particles of phosphorous pentoxide, which are converted by moist air into phosphoric acid.[4] WP is a very active chemical that will readily combine with oxygen in the air, even at room temperature. As oxidation occurs, WP becomes luminous and bursts into flames within minutes. Complete submersion in water is the only way to extinguish the flames.[13] It is used primarily as a smoke agent and can also function as an antipersonnel flame compound capable of causing serious burns.

Table III-18. WP

Chemical name: Phosphorus	
Synonyms: Bonide-Blue-Death-Rat-Killer, Exolit-LPKN; Exolit VPK-n 361, Forforo-Bianco (Italian), Gelber-Phosphor (German), Phosphore-Blanc (French), [1] elemental phosphorus, white phosphorus, [2] yellow phosphorus, WP, WP/F, Willie Peter[3]	
CAS Registry No.: 7723-14-0	
Physical and Chemical Properties	
Chemical formula	P_4 [2]
Molecular weight	124.0 [2]
Physical state	White to yellow; soft waxy solid [2]
Odor	Garlic-like [1]
Boiling point	280°C (536°F) [2]
FP/MP	44°C (111°F) (MP) [2]
Density (g/mL)	1.83 [4]
Vapor density	4.28 (calculated)
Vapor pressure (torr)	0.03 at 20°C [2]
Volatility (relative to air)	Data not available
Specific Gravity (water = 1 @ 4°C)	1.82 @ 20°C [2]
Flash point (mg/m^3)	20°C [4]
Decomposition temperature	Data not available
Solubility	Insoluble in water; soluble in carbon disulfide [1]
Rate of hydrolysis	Data not available
Hydrolysis products	Data not available
Stability in storage	Darkens on exposure to light. This substance spontaneously ignites on contact with air producing toxic fumes. [4]
Action on metals or other materials	Incompatible with sulfur, iodine, oil of turpentine, and potassium chlorate [1]
Other Data	
Skin and eye toxicity	Burns on skin are usually multiple, deep and variable in size. Particles on skin continue to burn unless deprived of atmospheric oxygen. Smoke causes irritation to the eyes, nose, and throat. [5]
Inhalation toxicity	At room temperature may produce a toxic inhalational injury [6]
Protection required	Protective mask. [7] Do not handle the charred wedges on the ground without protective covering. [5]
Decontamination	Flush skin with water. If burning WP strikes the skin, smother the flame with water, a wet cloth, or mud. Keep the WP covered with the wet material to exclude air until the particles can be removed. [5]
Use	See FM 3-50, Smoke Operations
NOTES	
[1]Spectrum Chemical Fact Sheet, "Phosphorus," CAS 7723-14-0. [2]NIOSH Pocket Guide to Chemical Hazards, "Phosphorus (yellow)," CAS 7723-14-0. [3]USA Corps of Engineers, Construction Engineering Research Laboratory, *Methods for Field Studies of the Effects of Military Smokes, Obscurants, and Riot-control Agents on Threatened and Endangered Species, Vol. 4: Chemical Analytical Methods*, USACERL Technical Report 99/56, July 1999. [4]ISCS 0628, "Phosphorus (Yellow)." [5]FM 8-285/NAVMED P-5041/AFJMAN 44-149/FMFM 11-11, *Treatment of Chemical Agent Casualties and Conventional Military Chemical Injuries*, 22 December 1995. [6]BG Russ Zajtchuk, et al. (eds), *Textbook of Military Medicine: Medical Aspects of Chemical and Biological Warfare*, Office of the Surgeon General, 1997, Chapter 9, "Toxic Inhalational Injury." [7]FM 3-50, *Smoke Operations*, 4 December 1990.	

(b) Red Phosphorous (RP). RP smoke is deployed explosively from grenades and mortar shells. The grenades consist of a 95:5 mixture of RP and butyl rubber.[15] RP is produced by heating WP to 270 to 300 degrees C in the absence of air. RP is more dense, has a higher melting point, is much less reactive, is essentially nontoxic, and is easier and safer to handle than WP. When RP is heated, WP molecules sublime from the solid.[16] RP will not ignite spontaneously and, therefore, requires ignition to burn and make smoke.[14]

(3) Titanium Tetrachloride (FM Smoke). FM smoke is a corrosive substance typically dispersed by spray or explosive munitions. It is extremely irritating and corrosive in both liquid and smoke formulations. Exposure to liquid may create burns similar to those of mineral acids on conjunctiva or skin. A dense, white smoke results from the decomposition of FM smoke into hydrochloric acid, titanium oxychloride, and titanium dioxide (TiO_2) (see Table III-19).[13] TiO_2 is the major component of training smoke grenade XM82 and is also used in conjunction with HC in the production of white screening smoke.[6]

Table III-19. TiO_2

Chemical name: Titanium Dioxide[1]	
Synonyms: Titanium (IV) Oxide,[1] Titanium Oxide, Titania, Titanium White, Brookite[2]	
CAS Number: 13463-67-7[1]	
Physical and Chemical Properties	
Molecular formula	TiO_2[1]
Molecular weight	79.87[1]
Physical state	White Powder[1]
Odor	Odorless[1]
Boiling point	2500 - 3000°C (4532 - 5432°F) [1]
FP/MP	1855°C (3371°F) (MP) [1]
Solid density (g/mL)	4.17 [2]
Vapor density (relative to air)	2.75 (calculated)
Vapor pressure (torr)	Data not available
Volatility (mg/m^3)	Data not available
Specific gravity (H_2O = 1)	4.26[1]
Flashpoint	Data not available
Decomposition temperature	Data not available
Solubility	Insoluble in water.[1] Insoluble in hydrochloric acid, nitric acid, and alcohol. It is soluble in hot concentrated sulfuric acid, hydrogen fluoride, and alkali. [3]
Rate of Hydrolysis	Data not available
Hydrolysis Products	Data not available
Stability in storage	Stable under ordinary conditions of use and storage[1]
Action on metals or other materials	Violent reaction with lithium occurs around 200°C with a flash of light; the temperature can reach 900°C. Violent or incandescent reaction may also occur with other metals such as aluminum, calcium, magnesium, potassium, sodium, and zinc. [1]
Other Data	
Skin and eye toxicity	May cause mild irritation and redness [1]
Inhalation toxicity	May cause mild irritation [1]
Protection required	Protective mask [4]
Decontamination	Flush with water [4]
Use	See FM 3-50, *Smoke Operations*

Table III-19. TiO2 (Continued)

NOTES
[1]Mallinckroct Baker, Inc., MSDS Number T3627, "Titanium Dioxide," effective date: 15 February 1998. [2]USA Corps of Engineers, Construction Engineering Research Laboratory, *Methods for Field Studies of the Effects of Military Smokes, Obscurants, and Riot-control Agents on Threatened and Endangered Species, Vol. 4: Chemical Analytical Methods*, USACERL Technical Report 99/56, July 1999. [3]NRC, *Toxicity of Military Smokes and Obscurants,* Vol. 2, National Academy Press, 1999. [4]FM 8-285/NAVMED P-5041/AFJMAN 44-149/FMFM 11-11, *Treatment of Chemical Agent Casualties and Conventional Military Chemical Injuries*, 22 December 1995.

(4) Tantalum (Ta). Ta is a clear, colorless, odorless liquid. It is a mixture of tantalum pentachloride, hydrogen fluoride, and water. It is stable at 68 degrees F when stored and used under proper conditions. On heating to decomposition, it could yield toxic fumes of fluorides. It attacks glass and other silicon-containing compounds. It reacts with silica to produce silicon tetrafluoride, a hazardous, colorless gas. It is incompatible with arsenic trioxide, phosphorus pentoxide, ammonia, calcium oxide, sodium hydroxide, sulfuric acid, vinyl acetate, ethylenediamine, acetic anhydride, alkalis, organic materials, most common metals, rubber, leather, water, strong bases, carbonates, sulfides, cyanides, oxides of silicon, and especially glass, concrete, silica, and fluorine. It is a corrosive poison. It is extremely hazardous in liquid and vapor form. It can cause severe burns that may not be immediately painful or visible. It can be fatal if swallowed or inhaled. Liquid and vapor can burn skin, eyes, and the respiratory tract. It can also cause bone damage.[17]

(5) Synthetic Graphite (see Table III-20). Graphite is used as an obscurant to screen electromagnetic tracking and targeting systems. Graphite flakes perform well in obscuring mid- and far-infrared (IR) bands as well as a combination of visible, near-, mid-, and far-IR bands. Graphite exists naturally and synthetically and is chemically inert.[6] It consists primarily of carbon, with trace impurities totaling less than one percent of the total weight.[18] Inhalation of dust can cause nose and/or throat irritation and shortness of breath.[19]

CAUTION

Graphite is electrically conductive; exercise caution when handling graphite in areas where contact with electrical circuitry is possible. Handle and transfer in a manner that avoids excessive dusting.[19]

Table III-20. Synthetic Graphite

Chemical name: Synthetic Graphite[1]	
Synonyms: Crystalline carbon[1]	
CAS Number: N/A[2]	
Physical and Chemical Properties	
Molecular formula	Carbon
Molecular weight	12
Physical state	Grey to Black[1]
Odor	Odorless[1]
Boiling point	3337.7°C[2]
FP/MP	N/A[1]
Density (g/mL)	2.25 g/mL[2]
Vapor density (relative to air)	N/A[1]
Vapor pressure (torr)	N/A[1]
Volatility (mg/m^3)	Data not available
Specific gravity (H_2O = 1)	2.20 – 2.26[1]

Table III-20. Synthetic Graphite (Continued)

Flashpoint	Data not available
Decomposition temperature	Data not available
Solubility	Insoluble in water[1]
Rate of hydrolysis	Data not available
Hydrolysis products	Data not available
Stability in storage	This material is stable and will not polymerize. Incompatible with oxidizing agents.[1]
Action on metals or other materials	Data not available
Other Data	
Skin and eye toxicity	Minor [1]
Inhalation toxicity	Cases of pulmonary fibrosis, emphysema, and corpulmonale may result from prolonged inhalation of dust[1]
Protection required	Protective mask [1]
Decontamination	Move to fresh air; rinse skin with soap and water; rinse eyes with water[1]
Use	See FM 3-50, *Smoke Operations*
NOTES	
[1]Asbury Graphite Mils, Inc., MSDS, "Synthetic Graphite," CAS No. 7782-42-5, January 2003. [2]Graphite obscurant materials are purchased from manufacturers in accordance with a specification that have specific purity and performance requirements. These specifications do not contain CAS number requirements. The CAS numbers may vary from contract to contract.	

(6) Brass. Brass in flake or powder form is one of the components of screening smoke grenade M76. Smoke munitions containing metal flakes or powders are used to screen against range finders, thermal surveillance systems, and laser target designators as well as to provide protection for armored vehicles. Smoke screens containing brass performed the best against far-IR bands but did poorly on visible light bands. Brass is a metal alloy composed mostly of copper (70 percent) and zinc (30 percent) with approximately 1 percent contamination of trace metals.[6]

(7) Fog Oil (SGF-2). Fog oil is a low-viscosity petroleum oil used to generate screening smokes. The fog is generated by evaporating the hydrocarbons in the oil. Fog oil is composed of many different types of chemicals, but the predominant class of chemicals is aliphatic hydrocarbons with very low levels of noncarcinogenic aromatic hydrocarbons.[6] Fog oil is the overhead fraction of petroleum. It contains no additives and is not refined.[20] Fog oil is commonly called smoke generator fuel number 2 (SGF-2) (see Table III-21).

Table III-21. SGF-2

Chemical name: Mineral Oil [1]	
Synonyms: SGF-2, Fog Oil [1]	
CAS Number: N/A [2]	
Physical and Chemical Properties	
Molecular formula	N/A
Molecular weight	N/A
Physical state	Lube Oil [1]
Odor	Hydrocarbon odor [1]
Boiling point	600°F (316°C) [1]
FP/MP	Data not available
Liquid density (g/mL)	Data not available
Vapor density (relative to air)	> 1[1]
Vapor pressure (torr)	< 1[1]
Volatility (mg/m^3)	Data not available
Specific gravity (H_2O = 1)	0.9
Flashpoint	Data not available

Table III-21. SGF-2 (Continued)

Decomposition temperature	Data not available
Solubility	In water, negligible[1]
Rate of hydrolysis	Data not available
Hydrolysis products	Data not available
Stability in storage	Avoid extreme heat and strong oxidants. No hazardous polymerization. [1]
Action on metals or other materials	Data not available
Other Data	
Skin and eye toxicity	Skin: Prolonged/repeated contact may cause dryness, defatting. Eyes: May cause irritation. [1]
Inhalation toxicity	May cause irritation[1]
Protection required	Protective mask [1]
Decontamination	Move to fresh air [1]
Use	See FM 3-50, *Smoke Operations*
NOTES	
[1]Defense General Supply Center, MSDS MIL-F-12070C, "SGF-2 Type; Fog Oil." [2]Fog oil obscurant materials are purchased from manufacturers in accordance with a specification that has specific purity and performance requirements. These specifications do not contain CAS number requirements. The CAS numbers may vary from contract to contract.	

(8) No. 1 Diesel Fuel (DF-1). DF-1 is clear or straw-colored liquid when undyed and low-sulfur diesel is red. Symptoms of overexposure include eye and skin irritation, dermatitis, upper respiratory tract irritation, nausea, vomiting, diarrhea, lung damage, weakness, headache, confusion, blurred vision, drowsiness, dizziness, slurred speech, flushed face, aortic plaques, heart beat irregularities, convulsions, unconsciousness, and death.[21]

(9) Diesel Fuel No. 2 (DF-2). Symptoms of overexposure include redness, drying to burns or blistering of skin. Overexposure by inhalation can produce symptoms of intoxication such as headache, dizziness, nausea, vomiting, and loss of coordination.[22]

(10) Jet Fuel Grade (JP-8). High vapor concentrations can be irritating to the eyes and respiratory tract, causing headaches, dizziness, anesthesia, drowsiness, unconsciousness, and CNS effects. Chronic exposure of the skin can cause drying, defatting, and dermatitis.[23]

(11) Signaling Smokes. These smokes are produced by explosive dissemination of dyes. There are no reports of ill effects produced by exposure to these smokes.[4]

b. Incendiaries. The purpose of incendiaries is to cause maximum fire damage on flammable materials and objects and to illuminate. The initial action of the incendiary munition may destroy these materials, or the spreading and continuing of fires started by the incendiary may destroy them.[12]

(1) Magnesium Incendiaries. Magnesium is a soft metal, which when raised to its ignition temperature, spontaneously ignites on contact with air or moisture, producing irritating or toxic fumes.[24] Magnesium (Mg) burns at approximately 2000 degrees C with a scattering effect. Its particles produce deep burns. Healing is slow unless these particles are removed quickly. When explosive charges have been added, fragments may be embedded deep in the tissues, causing the localized formation of hydrogen gas and tissue necrosis.[4]

(2) Thermite and Thermate Incendiaries. A thermite reaction occurs when powdered aluminum metal and iron oxide mix. It is an exothermic reaction in which the

temperature rises to about 3000 degrees C, causing the aluminum and iron to become a liquid.[16] Explosive charges are frequently added. Particles of iron that lodge in the skin produce multiple small, deep burns. The particles should be cooled immediately with water and removed.[4]

(3) Oil and Metal Incendiary Mixtures. Lung damage from heat and irritating gases may be complications added to the injuries from incendiaries, especially in confined spaces.[4]

NOTES

[1]USACHPPM TG 204, *Glossary of Terms for Nuclear, Biological, and Chemical Agents and Defense Equipment*, December 2001.

[2]FM 3-11.11/MCRP 3-3.7.2, *Flame, Riot Control Agents and Herbicide Operations*, 19 August 1996 (Renumbered from FM 3-11).

[3]BG Russ Zajtchuk et al (eds), *Textbook of Military Medicine: Medical Aspects of Chemical and Biological Warfare*, Office of the Surgeon General, 1997, Chap. 12, "Riot Control Agents."

[4]FM 8-285/NAVMED P-5041/AFJMAN 44-149/FMFM 11-11, *Treatment of Chemical Agent Casualties and Conventional Military Chemical Injuries*, 22 December 1995.

[5]Sharon Reutter, et al., *Review and Recommendations for Human Toxicity Estimates for FM 3-11.9*, ECBC-TR-349, September 2003.

[6]USA Corps of Engineers, Construction Engineering Research Laboratory, *Methods for Field Studies of the Effects of Military Smokes, Obscurants, and Riot-control Agents on Threatened and Endangered Species, Vol. 4: Chemical Analytical Methods*, USACERL Technical Report 99/56, July 1999.

[7]ICSC 0126, "Chlorine (Cl_2)."

[8]A.K. Steumpfle, et al., *Final Report of International Task Force-25: Hazard From Toxic Industrial Chemicals,* March 18, 1996.

[9]ICSC 0128, "2-Chloroacetophenone."

[10]US HHS, Public Health Service, National Toxicology Program, *10th Report on Carcinogens*, December 2002.

[11]FM 8-9/NAVMED P-5059/AFJMAN 44-151, *NATO Handbook on the Medical Aspects of NBC Defense Operations AMEDP-6(B),* 1 February 1996.

[12]FM 3-6/FMFM 7-11-H/AFM 105-7, *Field Behavior of NBC Agents (Including Smoke and Incendiaries),* 3 November 1986.

[13]BG Russ Zajtchuk, et al. (eds), *Textbook of Military Medicine: Medical Aspects of Chemical and Biological Warfare*, Office of the Surgeon General, 1997, Chap. 9, "Toxic Inhalational Injury."

[14]FM 3-50, *Smoke Operations*, 4 December 1990.

[15]NRC, *Toxicity of Military Smokes and Obscurants,* Vol. 1, National Academy Press, 1997.

[16]Henry F. Holtzclaw, Jr., et al., *General Chemistry with Qualitative Analysis*, 9th ed., D.C. Heath and Company, Lexington, MA, 1991.

[17]Mallinckrodt Baker, Inc., MSDS Number T0080, "Tantalum, 1,000 μg/mL or 10,000 μg/mL," effective date 29 October 2001.

[18]NRC, *Toxicity of Military Smokes and Obscurants,* Vol. 2, National Academy Press, 1999.

[19]Asbury Graphite Mills, Inc., MSDS, "Synthetic Graphite," CAS No. 7782-42-5, January 2003.

[20]Defense General Supply Center, MSDS MIL-F-12070C, "SGF-2 Type; Fog Oil."

[21]Conoco International, Inc., MSDS, "No. 1 Diesel Fuel, No. 1 Fuel Oil."

[22]Costal Corporation, MSDS, "Diesel fuel 2."

[23]Exxon Company, MSDS, "Jet Fuel Grade JP-8, 27020-00079."

[24]ICSC 0701, "Magnesium (Pellets)."

Chapter IV

BIOLOGICAL AGENTS AND THEIR PROPERTIES

1. Background

The military application of biological agents concerns those microorganisms that may be deliberately employed in weapon systems to cause disease or death to man, animals, or plants. Biological agents consist of microorganisms such as pathogens (which include disease-causing bacteria, rickettsiae, and viruses) and toxins.

NOTES:

1. See Table IV-1 (page IV-2) for the list of potential biological agents.

2. See Table IV-2 (page IV-3) for the list of animal and plant pathogens with potential biological agent applications.

3. See Appendix I for the properties of selected biological agents.

4. See Appendix J for more detailed information on animal pathogens.

5. See Appendix K for more detailed information on plant pathogens.

6. See Appendix L for information on the dissemination of biological agents

Table IV-1. List of Potential BW Agents[26]

Viruses	**Bacteria**
Chikungunya virus	*Bacillus anthracis* (Anthrax)
Crimean-Congo hemorrhagic fever virus	*Brucella abortus* (Brucellosis)
Dengue fever virus	*Brucella melitensis* (Brucellosis)
Eastern equine encephalitis virus	*Brucella suis* (Brucellosis)
Ebola virus	*Chlamydia psittaci* (Psittacosis)
Hantaan virus	*Clostridium botulinum* (Botulism)
Junin virus	*Francisella tularensis* (Tularemia)
Lassa virus	*Burkholderia mallei, formerly Pseudomonas mallei* (Glanders)
Lymphocytic choriomeningitis virus	*Burkholderia pseudomallei, formerly Pseudomonas pseudomallei* (Melioidosis)
Machupo virus	*Salmonella typhi* (Typhoid Fever)
Marburg virus	*Shigella dysenteriae* (Shigellosis)
Monkeypox	*Vibrio cholerae* (Cholera and other Vibrioses)
Rift Valley Fever virus	*Yersinia pestis* (Plague)
Tick-borne encephalitis virus *(Russian Spring-Summer encephalitis virus)*	**Toxins**
Variola virus	*Botulinum toxins*
Venezuelan equine encephalitis virus	*Clostidium perfringens toxins*
Western equine encephalitis virus	*Conotoxin*
Yellow fever virus	*Ricin*
Japanese encephalitis virus	*Saxitoxin*
	Shiga toxin
	Staphylococcus aureus toxins
Rickettsia	*Tetrodotoxin*
Coxiella burnetii (Q Fever)	*Verotoxin*
Bart quintana, formerly Rochalimaea quintana or Rickettsia quintana (Trench Fever)	*Microcystin*
Rickettsia prowasecki (Typhus Fever)	*Trichothecene mycotoxin* (see Note)
Rickettsia rickettsii (Rocky Mountain Spotted Fever)	

NOTE: Trichothecene mycotoxin is not on the list of BW agents listed in the reference; however, mycotoxins were alleged to have been used in Southwest Asia in the mid-1980s.

Table IV-2. Animal and Plant Pathogens with Potential BW Applications[26]

Animal Pathogens for Export Control	Plant Pathogens for Export Control
Viruses	**Bacteria**
African swine fever virus	*Xanthomonas albilineans*
Highly pathogenic Avian influenza virus	*Xanthomonas campestris pv. citri*
(synonym: fowl plague)	
Bluetongue virus	**Fungi**
Foot and mouth disease virus	*Colletorichum coffeanum var*
Goat pox virus	*Cochiliobolus miyabeans*
Herpes virus (Aujeszky's disease)	*(Helminthosporium oryzae)*
Hog cholera virus	*Microcylus ulei (syn. Dothidella ulei)*
(synonym: Swine fever virus)	*Puccinia graminis*
Lyssa virus	*(syn. Puccinia graminis fsp. Tritici)*
Newcastle disease virus	*Puccinia striformis (syn. Puccinia glumarium)*
Peste des petits ruminants virus	*Pyricularia grisea/Pyricularia oryzae*
Porcine enterovirus type-9	
(synonym: Swine vesicular disease virus)	**Virus**
Rinderpest virus (synonym: Cattle plague	*Barley Yellow Dwarf Virus*
Sheep pox virus	
Teschen disease virus	**Items for Inclusions in Awareness-Raising Guidelines**
Vesicular stomatitis virus	
	Bacteria
Bacteria	*Xanthomonas campestris pv. oryzae*
Mycoplasma mycoides	*Xylella fastidiosa*
Other Animal Diseases/Pathogens of Concern	**Fungi**
Contagious bovine pleuropneumonia	*Deutrrerophoma tracheiphila*
Contagious Equine Metritis	*(syn. Phoma Tracjeo[jo;a)*
Heartwater (Cowdria)	*Monlia rorei (syn. Moniliophtora rorei)*
Screwworm Myiasis (synonym: Blowfly	
Cochliomyia hominivorax)	**Virus**
Swine vesicular disease	*Banana bunchy top virus*

a. The information presented in this chapter provides descriptions of selected bacteria, rickettsiae, viruses, and toxins.

b. The biological agents addressed in this chapter are presented in a standard format that includes the following information:

(1) Infectious Agent. Identifies the specific pathogen that causes the disease, classifies the pathogen, and may indicate any of its important characteristics.

(2) Occurrence. Provides information on where the disease is known to be prevalent and/or in what population groups it is most likely to occur.

(3) Reservoir. Indicates the ultimate and/or intermediate human, animal, arthropod, plant, soil, or substance (or a combination of these) in which an infectious agent normally lives and multiplies, on which it depends primarily for survival, and where it reproduces itself in such a manner that it can be transmitted to a susceptible host.

(4) Transmission.[1] Describes the mechanisms by which an infectious agent is spread to humans. Such mechanisms include direct, indirect, and airborne.

(a) Direct transmission can be by direct contact (such as touching or biting), or by the direct projection (droplet spread) of droplet spray onto the conjunctiva or onto the mucous membranes of the eye, nose, or mouth during sneezing, coughing, spitting, singing, or talking (usually limited to a distance of about 1 meter or less).

(b) Indirect transmission can be vehicle- or vector-borne. Vehicle-borne transmission is when contaminated inanimate materials or objects (e.g., soiled clothes, cooking or eating utensils, water, or food) serve as an intermediate means by which an infectious agent is transported and introduced into a susceptible host through a suitable portal of entry. Vector-borne transmission can be mechanical or biological. Mechanical includes simple mechanical carriage by a crawling or flying insect through soiling of its feet or proboscis, or by passage of organisms through its GI tract. This does not require multiplication or development of the organism. Biological means that the propagation (multiplication), cyclic development, or a combination of these is required before the arthropod can transmit the infective form of the agent to humans. An incubation period is required following infection before the arthropod becomes infective.

(c) Airborne transmission is the dissemination of microbial aerosols to a suitable portal of entry—usually the respiratory tract. Microbial aerosols are suspensions of particles in the air consisting partially or wholly of microorganisms. They may remain suspended in the air for long periods of time.

(5) Symptoms. Describes the symptoms of the disease.

(6) Incubation period. Identifies the interval (in hours, days, or weeks) between initial, effective exposure to an infectious organism and the first appearance of symptoms of the infection.

(7) Communicability. Describes the time (days, weeks, or months) during which an infectious agent may be transmitted, directly or indirectly, from an infected person to another person or from an infected animal to humans.

(8) Prevention. Describes prophylaxis measures.

(9) Delivery. Describes the most likely means to disseminate the agent.

2. Bacterial Agents of Potential Concern

Bacteria are single-celled, microscopic organisms. These unicellular forms outnumber all other forms of microorganisms. However, few bacteria are of military significance. Of several thousand identified species, only about 100 are known to be pathogenic.[2] Some bacteria that infect man are selective human parasites, but many are zoonotic, transmittable from animals to humans. Zoonotic organisms include anthrax, tularemia, and brucellosis. Bacteria have primary military potential as antipersonnel agents.

a. Anthrax.

(1) Infectious Agent. *Bacillus anthracis*. This organism forms a protective spore under adverse environmental conditions or upon exposure to air. When conditions improve, the spores germinate to produce vegetative bacteria. The spores are extremely hardy and can survive extremes of temperature, dryness, and flooding.[3]

(2) Occurrence. Worldwide.[3]

(3) Reservoir. The soil and domestic and wild animals (primarily herbivores, including goats, sheep, cattle, horses, and swine) serve as reservoirs.[4]

(4) Transmission. Humans can contract anthrax from the inhalation of aerosolized spores, contact with infected animals (their hides, wool, or other products), or ingestion of contaminated meat. Usually, humans do not contract anthrax directly from the soil.[3]

(5) Symptoms. Symptoms of anthrax depend upon the method of transmission.

(a) Cutaneous anthrax features a painless, necrotic ulcer with a black scab and local swelling.[3] Untreated cutaneous anthrax has a case-fatality rate between 5 and 20 percent.[1]

(b) Ingestion causes oropharyngeal and GI anthrax. Initial symptoms of oropharyngeal anthrax are fever, sore throat, and difficulty swallowing. Acute symptoms include ulcer or scab involving the hard palate or tonsils, swelling of neck tissues, and abnormal enlargement of the lymph nodes. Initial symptoms of GI anthrax include fever, loss of appetite, nausea, and vomiting. Abdominal pain, bloody vomiting, bloody diarrhea, and possibly massive abdominal swelling may follow. In both cases, septic shock and death may follow.[3]

(c) Initial symptoms of inhalation anthrax are mild and nonspecific and may include fever, malaise, fatigue, and mild cough or chest discomfort; acute symptoms of respiratory distress, fever, and shock follow, with death occurring shortly thereafter.[1]

(6) Incubation period. Hours to 7 days; although most cases occur within 48 hours post-exposure.[2] Incubation periods up to 60 days are possible.[1]

(7) Communicability. Transmission from person to person is very rare. Articles and soil contaminated with spores may remain infective for decades.[1]

(8) Prevention. Anthrax vaccine is available.[2]

(9) Delivery. Missiles, bomblets, artillery fires, point release, or airborne line release may deliver aerosolized spores. Contamination of food and water could also be used.[3]

b. Brucellosis.

(1) Infectious Agent. *Brucella abortus*, cattle; *Brucella melitensis*, sheep, goats, camels; *Brucella suis*, swine, *Brucella canis*, dogs and coyotes.[3]

(2) Occurrence. Worldwide.[1]

(3) Reservoir. Cattle, swine, goats, and sheep serve as reservoirs. Infection may occur in bison, elk, caribou, and some species of deer.[1] Dogs and coyotes have been found to be infected.

(4) Transmission. The disease is transmitted by the inhalation of aerosols or dusts that contain organisms, ingestion of unpasteurized dairy products and contaminated meat, or inoculation of abraded skin or mucosal surfaces.[3]

(5) Symptoms. A bacterial disease with sudden or insidious onset, characterized by continued, intermittent, or irregular fever of variable duration; headache; weakness; profuse sweating; chills; severe pain in a joint, especially one not inflammatory in character; depression; weight loss; and generalized aching. The disease may last for

several days, months, or occasionally a year or more if not adequately treated. The untreated case-fatality rate is 2 percent or less. Some of the original syndrome may reappear as relapses.[1]

(6) Incubation period. Varies from 5 days to 8 weeks, usually 2 to 8 weeks.[3]

(7) Communicability. This disease is not communicable from person to person.[3]

(8) Prevention. Vaccines are not currently available for human use. Personnel must avoid consuming unpasteurized dairy products or uncooked foods containing the dairy products and avoid contact with suspect infected animals.[3]

(9) Delivery. The primary threat is by aerosol release. A food-borne brucellosis attack is unlikely, but could be executed.[3]

c. Cholera and other Vibrioses.

(1) Infectious Agent. *Vibrio cholerae*[1]

(2) Occurrence. Worldwide.[1]

(3) Reservoir. Humans and some strains of bacteria found in aquatic environments, particularly brackish waters.[1]

(4) Transmission. Through ingestion of food or water contaminated directly or indirectly with feces or vomit of infected persons. Ingestion of raw or undercooked seafood from polluted waters has caused outbreaks.[1]

(5) Symptoms. Symptoms include profuse painless, watery stools; nausea and vomiting; and (in untreated cases) rapid dehydration, acidosis, and circulatory collapse. In severe untreated cases death may occur within a few hours and the case-fatality rate may exceed 50 percent.[1]

(6) Incubation period. From a few hours to 5 days, usually 2 to 3 days.[1]

(7) Communicability. Transmission is negligible from person to person.[6] Occasionally the carrier state may persist for several months, but usually only a few days after recovery.[1]

(8) Prevention. Following proper field sanitation and personal hygiene are useful preventive measures. Do not eat raw seafood. A vaccine is available.[1]

(9) Delivery. The primary threat is from contamination of food and water supplies and secondly by aerosol spray.[5]

d. Glanders.

(1) Infectious Agent. *Burkholderia mallei* (formerly *Pseudomonas mallei).*[3]

(2) Occurrence. The disease is not widespread. The cases have been among workers with jobs that involve horses, donkeys, or mules and laboratory workers.[3] Cases continue to occur in Asia, Africa, the Middle East, and Central and South America.[6]

(3) Reservoir. Horses, mules, and donkeys serve as reservoirs.[3]

(4) Transmission. The organism is transmitted from animals to humans by invading the nasal, oral, and mucous membranes around the eyes; by inhaling into the lungs; and by invading abraded or lacerated skin.[3]

(5) Symptoms. Symptoms include fever, rigors, sweating, muscle pain, headache, pleuritis, chest pain, and generalized papular/pustular eruptions.[3] Mortality rate is over 50 percent despite antibiotic treatment.[7]

(6) Incubation period. 10 to 14 days after inhalation.[3]

(7) Communicability. Cases of person-to-person transmission have been reported.[6]

(8) Prevention. No vaccine is available.[3]

(9) Delivery. The primary threat is aerosol release.[3]

e. Melioidosis.

(1) Infectious Agent. *Burkholderia pseudomallei* (formerly *Pseudomonas pseudomallei).*[3]

(2) Occurrence. In countries between 20 degrees north and south latitudes.[3]

(3) Reservoir. Soil and water are the reservoirs. Several animals including sheep, goats, horses, swine, monkeys, and rodents, can become infected. There is no evidence that animals are important reservoirs except that they spread the agent to new soil and water.[3]

(4) Transmission. Contact with contaminated soil or water through gross or unapparent skin lesions, aspiration or ingestion of contaminated water, or inhalation of contaminated dust.[3]

(5) Symptoms. Melioidosis will most likely present as an acute pulmonary infection ranging from mild bronchitis to severe pneumonia. Other symptoms may include fever (over 102 degrees F), headache, loss of appetite, and pain in one or more muscles. Acute pulmonary disease can progress into a rapidly fatal septicemic disease. The case-fatality rate for acute septicemic disease exceeds 90 percent.[3]

(6) Incubation period. 10 to 14 days following inhalation.[3]

(7) Communicability. Person-to-person transmission has not been proven. Laboratory acquired infections are uncommon, but do occur, especially if procedures produce aerosols.[3]

(8) Prevention. No vaccines are available.[3]

(9) Delivery. The primary threat is aerosol release.[3]

f. Plague.

(1) Infectious Agent. *Yersinia pestis.*[1]

(2) Occurrence. Worldwide.[8] Plague continues to be a threat because of vast areas of persistent wild rodent infection.[1]

(3) Reservoir. The primary reservoir is wild rodents (especially ground squirrels). Domestic cats, wild carnivores, rabbits, and hares can also transmit plague to humans.[1]

(4) Transmission. The disease is transmitted from infected fleas—either from rodents to humans, from dogs or cats to humans, or from person to person—and by handling tissues of infected animals.[1] In the most likely BW scenario, plague would be

spread via aerosol. A rapidly person-to-person spread of fulminant pneumonia would occur.[8] Cat bites or scratches may also transmit plague.[1]

(5) Symptoms. Initial signs and symptoms may be nonspecific with fever, chills, malaise, muscular pain, nausea, exhaustion, sore throat, and headache. This is bubonic plague, and it occurs more often in lymph nodes. The involved nodes become swollen, inflamed, and tender and may form or discharge pus. Fever is usually present. Untreated bubonic plague has a case-fatality rate of about 50 to 60 percent. Untreated primary septicemic plague and pneumonic plague are invariably fatal. Modern therapy markedly reduces fatality from bubonic plague; pneumonic and septicemic plagues also respond if recognized and treated early. However, patients who do not receive adequate therapy for primary pneumonic plague within 18 hours after onset of respiratory symptoms are not likely to survive.[1]

(6) Incubation period. From 1 to 7 days, maybe a few days longer for an immunized individual. For primary plague pneumonia, 1 to 4 days, usually short.[1]

(7) Communicability. Fleas may remain infective for months under suitable conditions of temperature and humidity. Bubonic plague is not usually transmitted directly from person to person unless there is contact with pus from suppurating buboes. Pneumonic plague may be highly communicable under appropriate climatic conditions; overcrowding facilitates transmission.[1] Fleas flourish at humidity just above 65 percent and temperatures between 20 to 26 degrees C, and can survive 6 months without a feeding.[8]

(8) Prevention. Use of insect repellents, approved for human use, will provide a level of protection from bites by infected fleas. A vaccine is available to prevent bubonic plague.[3]

(9) Delivery. The primary threat is by aerosol release or by contamination of food and water.[3]

g. Psittacosis.

(1) Infectious Agent. *Chlamydia psittaci (c. psittaci).*[1]

(2) Occurrence. Worldwide.[1]

(3) Reservoir. Parakeets, parrots, pigeons, turkeys, ducks, and other birds serve as reservoirs. Apparently healthy birds can be carriers and occasionally shed the infectious agent, particularly when subjected to the stresses of crowding and shipping.[1]

(4) Transmission. Infection is acquired by inhaling dried droppings, secretions, and dust from infected feathers of birds. Household birds, usually psittacine birds, are the most frequent source of exposure, followed by turkey, squab, and duck farms and in poultry processing plants.[1]

(5) Symptoms. In humans, fever, headache, rash, muscle pain, chills, and upper or lower respiratory tract disease are common. Respiratory symptoms are mild when compared with the extensive pneumonia confirmed by X-ray.[1] Before antimicrobial agents were available, 15 to 20 percent of persons with *C. psittaci* infection were reported to have died.[9]

(6) Incubation period. From 1 to 4 weeks.[1]

(7) Communicability. Rare person-to-person transmission has been reported. Diseased as well as seemingly healthy birds may shed the agent intermittently, and sometimes continuously, for weeks or months.[1]

(8) Prevention. No vaccine is available.[5]

(9) Delivery. The primary threat is from aerosol release.[5]

h. Shigellosis (Bacillary dysentery).

(1) Infectious Agent. The genus Shigella (S.) is comprised of four species: Group A, *S. dysenteriae*; Group B, *S. flexneri*, Group C, *S. boydii*, and Group D, *S. sonnei*.[1]

(2) Occurrence. Worldwide.[1]

(3) Reservoir. Man; however, some outbreaks have occurred in primate colonies.[1]

(4) Transmission. Transmission is mainly by direct or indirect fecal-oral transmission from a patient or carrier. Infection may occur after the ingestion of very few (10 to 100) organisms. Individuals primarily responsible for transmission are those who fail to clean hands and under fingernails thoroughly after defecation. They may then spread infection to others directly by physical contact or indirectly by contaminating food. Water and milk transmission may occur as the result of direct fecal contamination; flies can transfer organisms from latrines to uncovered food items.[1]

(5) Symptoms. Symptoms involve the large and distal small intestine, characterized by diarrhea accompanied by fever, nausea, and sometimes toxemia, vomiting, cramps, and tenesmus. In typical cases, the stools contain blood, and mucus (dysentery); however, many cases present with a watery diarrhea. *S. dysenteriae 1* is often associated with serious disease. Case-fatality rates have been as high as 20 percent among hospitalized cases even in recent years.[1]

(6) Incubation period. From 12 to 96 hours (usually 1 to 3 days) up to 1 week for *S. Dysenteriae 1*.[1]

(7) Communicability. During acute infection and until the infectious agent is no longer present in the feces, usually within 4 weeks after illness.[1]

(8) Prevention. Follow proper personal hygiene and field sanitation procedures.[1]

(9) Delivery. The primary threat would be contamination of food and water supplies.

i. Tularemia.

(1) Infectious Agent. *Francisella tularensis (F. tularensis)*.[3] There are two biovars: Jellison type A (*F. tularensis* biovar *tularensis)* and type B (*F. tularensis* biovar *palaerctica*).[1] The type A strain is highly virulent. Ten virulent organisms injected subcutaneously and 10 to 50 organisms given by aerosol can cause infection in humans.[10]

(2) Occurrence. Tularemia occurs throughout North America and in many parts of continental Europe, the former Soviet Union, China, and Japan.[1]

(3) Reservoir. This organism is maintained in numerous and diverse mammalian (rabbits, hares, and rodents) and tick reservoirs.[3] In addition, a rodent-mosquito cycle has been described.[1]

(4) Transmission. Transmission is through the bite of arthropods (ticks and deerflies, also mosquitoes in Sweden, Finland, and Russia), direct contact with infected animals, aerosols generated by skinning/processing infected animals, and ingestion of contaminated food or water.[3]

(5) Symptoms. This bacterial disease has a variety of clinical manifestations related to the route of introduction and the virulence of the disease agent. Most often it presents itself as an ulcer at the site of introduction of the organism, together with swelling of the regional lymph nodes. There may be no apparent primary ulcer, but only one or more enlarged and painful lymph nodes. Ingestion of organisms in contaminated food or water may produce painful pharyngitis, abdominal pain, diarrhea, and vomiting. Inhalation of infectious material may be followed by pneumonic involvement with a 30 to 60 percent case-fatality rate if untreated. Type A has a 5 to 15 percent untreated case-fatality rate, and type B produces few fatalities even without treatment.[1]

(6) Incubation period. Related to virulence of infecting strain and to size of inoculum; the range is 1 to 14 days (usually 3 to 5 days).[1]

(7) Communicability. Not directly transmitted from person to person. Unless treated, the infectious agent may be found in the blood during the first 2 weeks of disease and in lesions for a month, sometimes longer. Flies can be infective for 14 days and ticks throughout their lifetime (about 2 years). Rabbit meat frozen at –15 degrees C (5 degrees F) has remained infective longer than 3 years.[1]

(8) Prevention. The protective mask provides protection of the respiratory tract from exposure to aerosol organisms. All food must be thoroughly cooked to kill any organisms before consumption. Water must be thoroughly disinfected before consumption. A live attenuated vaccine is available.[3]

(9) Delivery. The primary threat is by aerosol release or by contamination of food or water supplies.[3]

j. Typhoid Fever.

(1) Infectious Agent. *Salmonella typhi*.[1]

(2) Occurrence. Worldwide. The annual incidence of typhoid fever is estimated at about 17 million cases with approximately 600 thousand deaths.[1]

(3) Reservoir. Man.[1]

(4) Transmission. Transmission is by food and water contaminated by feces and urine of patients and carriers. Important vehicles in some parts of the world include shellfish taken from sewage-contaminated beds (particularly oysters), raw fruits, vegetables fertilized by human excrement and eaten raw, and contaminated milk and milk products. Flies may infect foods where the organisms then multiply and achieve an infective dose.[1]

(5) Symptoms. Systemic bacterial diseases characterized by insidious onset of sustained fever, severe headache, malaise, loss of appetite, a relative bradycardia, enlargement of the spleen, rose spots on the trunk, nonproductive cough, and constipation. Prior to antibiotics the case-fatality rate was 10 to 20 percent.[1]

(6) Incubation period. The incubation period depends on the size of the infecting dose; range is from 3 days to 1 month and usually between 8 to 14 days.[1]

(7) Communicability. Person-to-person transmission is possible. The period of communicability is as long as the bacilli appear in excreta, usually from the first week throughout convalescence; variable thereafter. About 10 percent of untreated typhoid fever patients will discharge bacilli for 3 months after the onset of symptoms, and 2 to 5 percent become permanent carriers.[1]

(8) Prevention. Following proper field sanitation and personal hygiene are useful preventive measures. A vaccine is available.[1]

(9) Delivery. The primary threat is from the sabotage of food and water supplies and, secondly, by aerosol.[5]

3. Rickettsiae of Potential Concern

The rickettsiae are intracellular, parasitic microorganisms that are considered intermediate in size between the bacteria and viruses. They resemble the bacteria in their shape and resemble the viruses in their strict growth requirements for living host cells. Most rickettsiae are parasites, primarily of lower animals and arthropods. Rickettsiae are transmitted to man and animals by vectors such as ticks, lice, fleas, and mites. The rickettsiae have a selective affinity for specific types of cells of the human and animal bodies. As they require living tissue for reproduction, they are considered parasites.[2] Usually, they are easily killed by heat, dehydration, or disinfectants.

a. Query (Q) Fever.

(1) Infectious Agent. *Coxiella burnetii (C. burnetii).* Despite the fact that *C. burnetii* is unable to grow or replicate outside host cells, there is a sporelike form of the organism that is extremely resistant to heat, pressure, desiccation, and many standard antiseptic compounds; this allows the organism to persist in the environment for long periods (weeks or months) under harsh conditions.[11]

(2) Occurrence. Worldwide.[3]

(3) Reservoir. Sheep, goats, cattle, dogs, cats, some wild mammals, birds, and ticks serve as reservoirs. Infected animals usually do not develop the disease, but shed large numbers of organism in placental tissues and body fluids.[1]

(4) Transmission. The organism is highly communicable by aerosol. A single viable organism is enough to cause infection in humans.[3] It is commonly transmitted by airborne dissemination of coxiellae in dust from premises contaminated by placental tissues, birth fluids, and excreta of infected animals; in establishments processing infected animals or their byproducts; and in necropsy rooms. Airborne particles containing organisms may be carried downwind for a considerable distance (one-half mile or more); also by direct contact with infected animals and other contaminated materials such as wool, straw, fertilizer, and laundry. Raw milk from infected cows contains organisms and may be responsible for some cases, but this has not been proven. Direct transmission by blood or marrow transfusion has been reported.[1]

(5) Symptoms. Symptoms include fever, fatigue, chills, sweats, muscular pain, and severe headache in 75 percent of cases.[3] The case-fatality rate in untreated acute cases is usually less than 1 percent, but has been reported as high as 2.4 percent.[1]

(6) Incubation period. Varies from 10 to 40 days and depends on the size of the infecting dose.[11]

(7) Communicability. Direct transmission from person to person occurs rarely, if ever. However, contaminated clothing may be a source of infection.[1] *Coxiella* organisms may persist in the environment and produce infection for weeks or months.[11]

(8) Prevention. The military protective mask provides protection from aerosols. Consume only pasteurized dairy products, and heat all foods sufficiently to destroy the organisms (see TB MED 530 for food preparation guidelines). An investigational vaccine is being tested.[11]

(9) Delivery. The primary threat is by aerosol or through contamination of food.[3]

b. Rocky Mountain Spotted Fever.

(1) Infectious Agent. *Rickettsia rickettsii.*[1]

(2) Occurrence. Infection has been documented throughout the US, in Canada, western and central Mexico, Panama, Costa Rica, Colombia, Argentina, and Brazil.[1]

(3) Reservoir. Maintained in nature in ticks. The rickettsiae can be transmitted to dogs, various rodents, and other animals. Disease in rodents and dogs have been observed.[1]

(4) Transmission. Transmission is ordinarily from the bite of an infected tick. At least 4 to 6 hours of attachment and feeding on blood by the tick are required before the rickettsiae become reactivated and infectious for people. Contamination of breaks in the skin or mucous membranes with crushed tissues or feces of the tick may also lead to infection.[1]

(5) Symptoms. This disease is characterized by sudden onset of moderate to high fever (which ordinarily persists 2 to 3 weeks in untreated cases), significant malaise, deep muscle pain, severe headache, and chills. A rash generally appears on the extremities on about the third to fifth day; this soon includes the palms and soles and spreads rapidly to much of the body. The fatality rate ranges between 13 to 25 percent in the absence of specific therapy.[1]

(6) Incubation period. From 3 to about 14 days.[1]

(7) Communicability. Not directly transmitted from person to person. The tick remains infective for life, commonly as long as 18 months.[1]

(8) Prevention. Measures that can be taken to reduce the opportunity for exposure include good field sanitation and use of insect repellents. No vaccine is available.[1]

(9) Delivery. The primary threat would be from aerosol or from infected vectors.[5]

c. Trench Fever.

(1) Infectious Agent. *Bartonella quintana* (formerly *Rochalimaea Quintana).*[1]

(2) Occurrence. The organism probably can be found wherever the human body louse exists. Epidemics occurred in Europe during World Wars I and II among troops and prisoners of war living in crowded, unhygienic conditions.[1]

(3) Reservoir. Humans serve as reservoirs; the intermediate host and vector is the body louse, *Pediculus humanus corporis.* The organism multiplies extracellularly in the gut lumen for the duration of the insect's life, which is approximately 5 weeks after hatching.[1]

(4) Transmission. The body louse is infected by feeding on the blood of an infected patient. The infected louse excrete rickettsiae in their feces and usually defecate at the time of feeding; people are infected by rubbing feces or crushed lice into the bite or into superficial abrasions.[1]

(5) Symptoms. Trench Fever is typically a nonfatal, bacterial disease varying in manifestations and severity. It is characterized by headache, malaise, and pain and tenderness—especially on the shins. Onset is either sudden or slow, with a fever that may be relapsing, typhoid-like, or limited to a single episode lasting for several days.[1]

(6) Incubation period. Generally 7 to 30 days.[1]

(7) Communicability. Not directly transmitted from person to person. Organisms may circulate in the blood (by which lice are infected) for weeks, months, or years and may recur with or without symptoms.[1]

(8) Prevention. Delousing procedures will destroy the vector and prevent transmission to man. Dust clothing and body with an effective insecticide.[1] Washing sheets and clothing in hot water will eliminate the louse. Good hygiene will help prevent this condition.

(9) Delivery. The primary threat would be by aerosol release or infected vector.

d. Typhus Fever (Epidemic Louse-Borne Typhus Fever).

(1) Infectious Agent. *Rickettsia prowazekii.*[1]

(2) Occurrence. In colder areas where people may live under unhygienic conditions and are louse-infested; enormous and explosive epidemics may occur during war and famine.[1]

(3) Reservoir. Humans are the reservoirs and are responsible for maintaining the infection between epidemic periods.[1]

(4) Transmission. The body louse, *Pediculus humanus corporis,* is infected by feeding on the blood of a patient with acute typhus fever. Infected lice excrete rickettsiae in their feces and usually defecate at the time of feeding. People are infected by rubbing feces or crushed lice into the bite or into superficial abrasions. Inhalation of infective louse feces in dust may account for some infections.[1]

(5) Symptoms. Symptoms have variable onsets, often are sudden, and are marked by headache, chills, exhaustion, fever, and general pains. A macular eruption appears on the fifth to sixth day, initially on the upper trunk, followed by spreading to the entire body (but usually not to the face, palms, or soles). The case-fatality rate increases with age and varies from 10 to 40 percent in the absence of specific therapy.[1]

(6) Incubation period. From 1 to 2 weeks, commonly 12 days.[1]

(7) Communicability. No direct person-to-person transmission. Patients are infective for lice during the febrile illness and possibly for 2 to 3 days after the temperature returns to normal. Infected lice pass rickettsiae in their feces within 2 to 6 days after the

blood meal; it is infective earlier if crushed. The louse invariably dies within 2 weeks after infection; rickettsiae may remain viable in the dead louse for weeks.[1]

(8) Prevention. Dust clothing and body with an effective insecticide.[1] No vaccine is available.[5] Washing sheets and clothing in hot water will eliminate the louse. Good hygiene will help prevent this condition.

(9) Delivery. The primary threat would be by aerosol release or infected vectors.[5]

4. Viral Agents of Potential Concern

The groups of microorganisms called viruses are all parasites that live in the cells of their selected hosts. Viruses cause about 60 percent of all infectious diseases. Once a virus enters a living cell, it is capable of replicating itself by taking over the metabolic processes of the invaded cell. Cells infected with viruses show one of the following responses: degeneration and death, transformation to a nonfunctioning state, or survival without transformation but with the evidence of the presence of one or more viral components. Diseases of viral origin do not respond to treatment with antibiotics.[2] Viruses have primary military potential as antipersonnel agents.

a. Chikungunya Virus Disease.

(1) Infectious Agent. Chikungunya Virus (*alphavirus*).[1]

(2) Occurrence. This virus is found in Africa, India, southeast Asia, and the Philippine Islands.[1]

(3) Reservoir. The reservoir is unknown for most viruses. An insect (mosquito) reservoir is a possibility.[1]

(4) Transmission. Mosquitoes transmit the virus.[1]

(5) Symptoms. The usual onset of symptoms is characterized by arthalgia (severe pain in a joint, especially one not inflammatory) or arthritis, primarily in the wrist, knee, ankle, and small joints of the extremities, which lasts from days to months. Rashes are common. Polyarthritis is a characteristic feature. Minor hemorrhages have been attributed to this virus.[1]

(6) Incubation period: From 3 to 11 days.[1]

(7) Communicability. No direct person-to-person transmission.[1]

(8) Prevention. Insecticides and proper field sanitation are examples of preventive measures.[1] An experimental vaccine is being tested.[5]

(9) Delivery. The likely method of dissemination is aerosol.[5]

b. Crimean-Congo Hemorrhagic Fever Virus (HFV).

(1) Infectious Agent. Crimean-Congo HFV, genus *Nairovirus*.[1]

(2) Occurrence. Observed in regions of Russia as well as in Albania and Bosnia-Herzogovina, Bulgaria, Iraq, the Arabian Peninsula, Pakistan, western China, tropical Africa, and South Africa.[1]

(3) Reservoir. In nature, hares, birds and ticks in Eurasia and South Africa are believed to be reservoirs; reservoir hosts remain undefined in tropical Africa, but ticks,

insectivores, and rodents may be involved. Domestic animals (sheep, goats, and cattle) may act as amplifying hosts.[1]

(4) Transmission. Transmission is by a bite of infective adult ticks. Immature ticks are believed to acquire infection from the animal hosts and by transovarian transmission. Transmission from patients to medical workers after exposure to blood and secretions has been important in recent outbreaks. Infection is also associated with butchering infected animals.[1]

(5) Symptoms. This is a disease with sudden onset of fever, malaise, weakness, irritability, headache, severe pain in limbs and loins, and marked anorexia. Vomiting, abdominal pain, and diarrhea occur occasionally. Flush on face and chest develops early. There may be some bleeding from gums, nose, lungs, uterus, and intestine, but in large amounts only in serious or fatal cases. Fever is constantly elevated for 5 to 12 days. The reported case fatality rate ranges from 2 to 50 percent.[1]

(6) Incubation period. Range is from 1 to 12 days, usually 1 to 3 days.[1]

(7) Communicability. Moderate person-to-person transmission.[5]

(8) Prevention. Insecticides and proper field sanitation are examples of preventive measures.[1] An experimental vaccine is available.[5]

(9) Delivery. Aerosol is the likely method of dissemination.[5]

c. Dengue Fever.

(1) Infectious Agent. Dengue-1, Dengue-2, Dengue-3, and Dengue-4. They are all *flaviviruses*.[1]

(2) Occurrence. Most countries in the tropics, Africa, Saudi Arabia, and the Americas.[1]

(3) Reservoir. The viruses are maintained in a human mosquito cycle in tropical urban centers; a monkey mosquito cycle serves as a reservoir in southest Asia and west Africa.[1]

(4) Transmission. By the bite of infective mosquitoes.[1]

(5) Symptoms. An acute febrile viral disease that is characterized by sudden onset, fever for 3 to 5 days, intense headache, myalgia, arthralgia, retro-orbital pain, anorexia, GI disturbances, and rash. Minor bleeding may occur. Epidemics are explosive, but fatalities in the absence of dengue hemorrhagic fever are rare.[1]

(6) Incubation period. From 3 to 14 days, commonly 4 to 7 days.[1]

(7) Communicability. Not directly transmitted from person to person. Patients are infective for mosquitoes from shortly before to the end of the febrile period, usually a period of 3 to 5 days. The mosquito becomes infective 8 to 12 days after the viremic blood meal and remains so for life.[1]

(8) Prevention. Use screening, protective clothing, and repellents and try to eliminate breeding grounds.[1] An experimental vaccine is available.[5]

(9) Delivery. The primary threat is delivery by aerosol release.[5]

d. Eastern and Western Equine Encephalitis (EEE and WEE), Japanese Encephalitis (JE).

(1) Infectious Agent. A specific virus causes each disease; EEE and WEE are caused by *alphaviruses* and JE by a *flavivirus*.[1]

(2) Occurrence. The EEE is recognized in eastern and north central US, Canada, Central and South America, and the Caribbean islands; WEE occurs in western and central US, Canada, and parts of South America. The JE occurs in western Pacific islands from Japan to the Philippines.[1]

(3) Reservoir. The true reservoir or means of winter carryover for these viruses is unknown—possibly birds, rodents, bats, reptiles, and/or amphibians; or it survives in mosquito eggs or adult mosquitoes, with the mechanisms probably differing in each virus.[1]

(4) Transmission. Transmission is from the bite of infective mosquitoes.[1]

(5) Symptoms. Symptoms of these diseases are similar, but vary in severity and rate of progress. A large percentage of patients with vector-borne infections are either asymptomatic or present with a nonspecific febrile illness or aseptic meningitis.[12] Severe infections are usually marked by acute onset, headache, high fever, stupor, disorientation, coma, tremors, occasional convulsions, and spastic (but rarely flaccid) paralysis. Case-fatality rates range from 0.3 to 60 percent (JE and EEE among the highest).[1]

(6) Incubation period. Usually 5 to 15 days.[1]

(7) Communicability. Not directly transmitted from person to person. Virus is not usually demonstrable in the blood of humans after onset of the disease. Mosquitoes remain infective for life. Viremia in birds usually lasts 2 to 5 days, but may be prolonged in bats, reptiles, and amphibians, particularly if interrupted by hibernation. Humans and horses are uncommon sources of mosquito infection.[1]

(8) Prevention. Use screening, protective clothing, and repellents, and try to eliminate breeding grounds. Vaccines are available.[1]

(9) Delivery. The primary threat would be delivery from aerosol release.[6]

e. Ebola Viral Hemorrhagic Fever (VHF) and Marburg Viral Disease.

(1) Infectious Agent. The Ebola virus and the Marburg virus are in the *Filoviridea* group.[1]

(2) Occurrence. Confirmed cases of Ebola VHF have been reported in the Democratic Republic of the Congo, Gabon, Sudan, the Ivory Coast, Uganda, and the Republic of the Congo.[13] Marburg disease was found in Germany and Yugoslavia after exposure to green monkeys from Uganda, Zimbabwe, Kenya, and the Democratic Republic of the Congo.[1]

(3) Reservoir. Unknown despite extensive studies.[1]

(4) Transmission. Person-to-person transmission occurs by direct contact with infected blood, secretions, organs, or semen.[1]

(5) Symptoms. These diseases are usually characterized by sudden onset with malaise, fever, myalgia, headache, and pharyngitis, followed by vomiting, diarrhea, rash, and hemorrhagic diathesis. Approximately 25 percent of reported primary cases of Marburg virus infection have been fatal; case-fatality rates of Ebola infections have ranged from 50 to nearly 90 percent.[1]

(6) Incubation period. Ebola—2 to 21 days; Marburg—3 to 9 days.[1]

(7) Communicability. Person-to-person transmission is possible as long as blood and secretions contain the virus.[1]

(8) Prevention. No vaccines are available.[5]

(9) Delivery. The primary threat would be delivery from aerosol release.[5]

f. Far Eastern Tick-Borne Encephalitis (Russian Spring-Summer Encephalitis).

(1) Infectious Agent. A complex within the flaviviruses.[1]

(2) Occurrence. The disease is distributed spottily over much of the former Soviet Union, other parts of eastern and central Europe, Scandinavia, and the United Kingdom (UK). Areas of highest incidence are those where humans have intimate association with large numbers of infected ticks, generally in rural or forested areas, but also in some urban populations.[1]

(3) Reservoir. Ticks, or ticks and mammals in combination, appear to be the true reservoir. Rodents, other mammals, or birds serve as sources of tick infections.[1]

(4) Transmission. By the bite of infective ticks or by consumption of milk from certain infected animals.[1]

(5) Symptoms. This is a viral disease clinically resembling the mosquito-borne encephalitides (EEE, WEE, and JE). Symptoms also include focal epilepsy and flaccid paralysis (particularly of the shoulder girdle).[1]

(6) Incubation period. Usually 7 to 14 days.[1]

(7) Communicability. No direct person-to-person transmission. A tick infected at any stage remains infective for life. Viremia in a variety of vertebrates may last for several days; in humans, up to 7 to 10 days.[1]

(8) Prevention. Preventive measures can include effective personal hygiene, field sanitation, and insect control.[1] A vaccine is available.[6]

(9) Delivery. The likely method of dissemination is aerosol or through milk.[5]

g. Hantaviral Disease (Korean Hemorrhagic Fever [KHF]).

(1) Infectious Agent. Hantaviruses, a genus of the Bunyaviridae family.[1] The prototype virus from this group, Hantaan, is the cause of the KHF.[14]

(2) Occurrence. Worldwide.[1]

(3) Reservoir. Field rodents; man is an accidental host.[1]

(4) Transmission. Aerosol transmission from rodent excreta is presumed (aerosol infectivity has been demonstrated experimentally). Virus is present in urine, feces, and saliva of persistently infected rodents; highest virus concentration is found in the lungs.[1]

(5) Symptoms. An acute viral disease characterized by an abrupt onset of fever, lower back pain, varying degrees of hemorrhagic manifestations, and renal involvement. The case fatality rate is variable, but generally ranges from 5 to 15 percent.[1]

(6) Incubation period. As short as a few days, as long as nearly 2 months, but usually 2 to 4 weeks.[1]

(7) Communicability. Not well-defined. Person-to-person transmission is rare.[1]

(8) Prevention. Follow rodent control procedures. Experimental vaccine is available.[5]

(9) Delivery. The primary threat would be delivery from aerosol release.[5]

h. Junin Hemorrhagic Fever (Argentine Disease) and Machupo Hemorrhagic Fever (Bolivian Disease).

(1) Infectious Agent. Junin virus for the Argentine disease and closely related Machupo virus for the Bolivian disease.[1]

(2) Occurrence. Junin virus in Argentina and Machupo virus in northeastern Bolivia.[1]

(3) Reservoir. Rodents.[1]

(4) Transmission. Airborne transmission may occur via dust contaminated with infected rodent excreta; both saliva and excreta of infected rodents contain the virus. Abraded skin may also be a portal of entry for infection.[1]

(5) Symptoms. Acute febrile viral illnesses. The onset is gradual with malaise, headache, retro-orbital pain, sustained fever, and sweats, followed by prostration. Case fatality rates range from 15 to 30 percent or more.[1]

(6) Incubation period: Usually 7 to 16 days.[1]

(7) Communicability. Not often directly transmitted from person to person.[1]

(8) Prevention. Follow rodent control procedures. A live-attenuated vaccine is currently in field trials (unlicensed in the US).[1]

(9) Delivery. The primary threat would be delivery from an aerosol release.

i. Lassa Fever.

(1) Infectious Agent. Lassa virus.[1]

(2) Occurrence. Africa.[1]

(3) Reservoir. Wild rodents.[1]

(4) Transmission. Primarily through aerosol or direct contact with excreta of infected rodents deposited on surfaces such as floors, beds, or in food and water. Person-to-person and laboratory infections occur by direct contact with blood, pharyngeal secretions, or urine, or by sexual contact.[1]

(5) Symptoms. Lassa fever is an acute viral illness of 1 to 4 weeks duration. Onset is gradual with malaise, fever, headache, sore throat, cough, nausea, vomiting, diarrhea, myalgia, and chest and abdominal pain. Fever is persistent or intermittently spiking. In severe cases, hemorrhage, seizures, and edema of the face and neck are frequent. The Case-fatality rate is about 15 percent among hospitalized cases.[1]

(6) Incubation period. Commonly 6 to 21 days.[1]

(7) Communicability. Person-to-person transmission spread may occur during the acute febrile phase when the virus is present in the throat. The virus may be excreted in the urine of patients for 3 to 9 weeks from onset of illness.[1]

(8) Prevention. Follow specific rodent control procedures.[1] No vaccine is available.[5]

(9) Delivery. The primary threat would be delivery from aerosol release.[5]

j. Lymphocytic Choriomeningitis.

(1) Infectious Agent. Lymphocytic choriomeningitis virus, an arenavirus.[1]

(2) Occurrence. Not uncommon in Europe and the Americas.[1]

(3) Reservoir. The infected house mouse is the natural reservoir.[1]

(4) Transmission. Virus is excreted in urine, saliva, and feces of infected animals—usually mice. Transmission to humans is probably through oral or respiratory contact with virus-contaminated excreta, food, or dust or by contamination of skin lesions or cuts.[1]

(5) Symptoms. This disease is a viral infection of animals, especially mice, transmissible to humans, with a marked diversity of clinical manifestations. At times, there may be influenza-like symptoms, with myalgia and headache. The acute course is usually short, very rarely fatal.[1]

(6) Incubation period. Probably 8 to 13 days, 15 to 21 days until meningeal symptoms appear.[1]

(7) Communicability. Transmission from person to person has not been demonstrated and is unlikely. Infected female mice transmit infection to the offspring, which become asymptomatic persistent viral shedders.[1]

(8) Prevention. Field hygiene and personal hygiene are effective individual measures.[1]

(9) Delivery. The primary threat would be delivery from an aerosol release.

k. Monkeypox.

(1) Infectious Agent. *Monkeypox virus;* it belongs to the genus *orthopoxvirus,* which includes the smallpox virus (variola), the virus used in the smallpox vaccine (vaccinia), and the cowpox virus.[15]

(2) Occurrence. The rain forest countries of central and western Africa and the US.[16]

(3) Reservoir. Studies suggest several species of squirrels and Gambian rats in Africa[16] and pet prairie dogs in the US may be animal reservoirs.[15] Other animals may be possible reservoirs.[16] Rats, mice, and rabbits can get monkeypox.[13]

(4) Transmission. Limited data on the transmission of monkeypox virus are available from studies conducted in Africa. Person-to-person transmission is believed to occur primarily through direct contact and also by respiratory droplet spread. Airborne transmission cannot be excluded, especially in patients presenting with cough.[17]

(5) Symptoms. In humans, the symptoms are similar to those of smallpox, but usually milder. The illness begins with fever, headache, muscle aches, backache, swollen lymph nodes, a general feeling of discomfort, and exhaustion. A papular rash develops, often first on the face. The lesions usually develop through several stages before crusting and falling off. In Africa, the reported case-fatality rate is 10 percent.[15]

(6) Incubation period. About 12 days.[15]

(7) Communicability. The Centers for Disease Control and Prevention (CDC) suggests there is a relatively low risk of person-to-person transmission.[17]

(8) Prevention. Because the monkeypox virus is related to the virus that causes smallpox, the smallpox vaccine can protect people from getting monkeypox as well as smallpox.[18]

(9) Delivery. The likely method of dissemination is an aerosol release.

l. Rift Valley Fever (RVF).

(1) Infectious Agent. RVF Virus (RVF is a phlebovirus).[1]

(2) Occurrence. This virus is found in Africa.[1]

(3) Reservoir. Unknown.[1]

(4) Transmission. Mosquitoes transmit the virus. Many human infections of RVF are associated with handling infective material of animal tissues during autopsy and butchering. Mechanical transmission by flies and transmission by aerosols or contact with highly infective blood may contribute to the explosive nature of RVF outbreaks.[1]

(5) Symptoms. The usual onset is fever, headache, malaise, arthralgia, myalgia, occasional nausea and vomiting, and generally some conjunctivitis and photophobia. Encephalitis, hemorrhage, or retinitis may develop.[1]

(6) Incubation period. Usually 3 to 12 days.[1]

(7) Communicability. No direct person-to-person transmission. Infected mosquitoes probably transmit the virus throughout life. Epidemics may involve thousands of people.[1]

(8) Prevention. Insecticides, field sanitation, and personal hygiene are examples of preventive measures. Precautions in the care and handling of infected animals and their products are important.[1] A vaccine is available.[5]

(9) Delivery. The likely method of dissemination is aerosol or through infected vectors.[5]

m. Smallpox. In response to concerns that variola stocks may by needed for counterterrorism research in the event that clandestine stocks held by other countries fall into terrorist hands, the World Health Assembly (WHA), in May 1999, authorized that the virus be held at laboratories in the US and Russia until no later than 2002. The World Health Organization (WHO) reaffirmed that destruction of all the remaining virus stocks is still the organization's ultimate goal and will appoint a group of experts to consider what research needs to be carried out before the virus can be destroyed. The WHO will also set up an inspection schedule for the two laboratories where the official stocks are kept to make sure that they are secure and that research can be carried out safely.[1]

(1) Infectious Agent. Variola virus, a species of *Orthopoxvirus*.[1]

(2) Occurrence. Formerly, smallpox was a worldwide disease. The last naturally acquired case of smallpox occurred in October 1977 in Somalia; global eradication was certified two years later by the WHO and sanctioned by the WHA in May 1980.[1]

(3) Reservoir. Humans were the only natural reservoir of variola. All known variola virus stocks are held under security at the CDC, Atlanta, Georgia, or the State Research Center (SRC) of Virology and Biotechnology, Koltsovo, Novosibirsk Region, Russian Federation (RF).[3]

(4) Transmission. Usually by respiratory droplet transmission, following close face-to-face contact. Smallpox was also transmitted by direct contact with skin lesions or drainage, or with contaminated objects. Although uncommon, airborne transmission (long-distance) also occurred.[3]

(5) Symptoms. Smallpox was a systemic viral disease that generally presented with a characteristic skin eruption. Onset was sudden, with fever, malaise, headache, prostration, severe backache, and occasional abdominal pain and vomiting—a clinical picture that resembled influenza. After 2 to 4 days, the fever began to fall and a deep-seated rash developed in which individual lesions containing infectious virus progressed through successive stages of macules, papules, vesicles, pustules, and crusted scabs, which fell off after 3 to 4 weeks. The lesions were first evident on the face and extremities and subsequently on the trunk. Two epidemiologic types of smallpox were recognized during the twentieth century: variola minor (alastrim), which had a case-fatality rate of less than 1 percent, and variola major (ordinary) with a fatality rate among unvaccinated populations of 20 to 40 percent or more. Fatalities normally occurred in 5 to 7 days, occasionally as late as the second week.[1]

(6) Incubation period. From 7 to 19 days; commonly 10 to 14 days to onset of illness and 2 to 4 days more to onset of rash.[1]

(7) Communicability. Smallpox is very contagious from the time of development of the earliest lesions to disappearance of all scabs—about 3 weeks. The patient is most contagious during the pre-eruptive period by aerosol droplets from lesions.[1]

(8) Prevention. There are no routine immunizations of US forces for smallpox. When the threat indicates, senior leadership may direct vaccination of personnel with vaccinia.[3]

(9) Delivery. The primary threat is delivery by aerosol release.[3]

n. Venezuelan Equine Encephalitis.

(1) Infectious Agent. VEE virus (*alphavirus*).[3]

(2) Occurrence. Present in northern South America, Trinidad, and Central America.[1]

(3) Reservoir. During outbreaks, VEE virus is transmitted in a cycle involving horses—which serve as the major source of the virus—to mosquitoes, which in turn infect humans. Humans also serve as hosts in a human-to-mosquito-to-human transmission cycle.[1]

(4) Transmission. By the bite of an infected mosquito; laboratory infections by aerosol are common.[1]

(5) Symptoms. Clinical manifestations of this viral infection are influenza-like, with an abrupt onset of severe headache, chills, fever, myalgia, retro-orbital pain, nausea, and vomiting. Most infections are relatively mild, with symptoms lasting from 3 to 5 days. There may be CNS involvement ranging from somnolence to disorientation, convulsions, paralysis, coma, and death.[1]

(6) Incubation period. Usually 2 to 6 days, can be as short as 1 day.[1]

(7) Communicability. Low risk of person-to-person transmission.[5] Infected humans and horses are infectious for mosquitoes for up to 72 hours; infected mosquitoes probably transmit the virus throughout life.[1]

(8) Prevention. Use general mosquito control procedures.[1] Vaccine is available.[5]

(9) Delivery. The primary threat is by aerosol release or through infected vectors.[5]

o. Yellow Fever.

(1) Infectious Agent. Yellow Fever Virus of the genus *Flavivirus*.[1]

(2) Occurrence. Sylvatic type: tropical regions of Africa and Latin America. Urban type: North, Central, and South America, and Trinidad.[1]

(3) Reservoir. Sylvatic: vertebrates other than humans—mainly monkeys, possibly marsupials, and mosquitoes. Urban: humans and mosquitoes.[1]

(4) Transmission. Sylvatic: from infective monkey to mosquito to man. Urban: from the bite of an infective mosquito.[1]

(5) Symptoms. Typical attacks are characterized by a sudden onset of fever, chills, headache, backache, generalized muscle pain, prostration, nausea, and vomiting. Jaundice is moderate early in the disease and is intensified later.[1]

(6) Incubation period. From 3 to 6 days.[1]

(7) Communicability. No person-to-person transmission.[5] Blood of patients is infective for mosquitoes shortly before onset of fever and for the first 3 to 5 days of illness. The disease is highly communicable where many susceptible people and abundant vector mosquitoes co-exist. The incubation period in mosquitoes is commonly 9 to 12 days at the usual tropical temperatures. Once infected, mosquitoes remain so for life.[1]

(8) Prevention. A vaccine is available.[5]

(9) Delivery. The likely method of dissemination is aerosol.[5]

5. Toxins of Potential Concern

Toxins are poisonous byproducts of living organisms. They are very stable and produce severe illness when ingested, inhaled, or introduced into the body by any other means. Some toxins are susceptible to heat, while others are heat-stable. Their effects on the human body range from minor illness to death.[3]

a. Botulinum Toxins.

(1) Toxin Origin. Botulinum toxins are a group of seven toxins produced by *Clostridium (C.) botulinum*. The spores are ubiquitous; they germinate to give rise to

vegetative bacteria that produce toxins during anaerobic incubation. Industrial-scale fermentation can produce large quantities of toxins for use as a BW agent. There are three forms of naturally occurring botulism—food-borne, infantile, and wound. Botulinum toxin is the most potent neurotoxin known.[3]

(2) Occurrence. Worldwide.[1]

(3) Reservoir. The reservoirs are soil, animals, and fish. The organisms can be recovered from honey and other agriculture products. High-risk foods are primarily improperly canned foods and dried meat or fish.[3]

(4) Transmission. Transmission is from consumption of contaminated food or contamination of a wound by ground-in soil or gravel.[1]

(5) Symptoms. The autonomic features of botulism are typical anticholinergic signs and symptoms: dry mouth, ileus, constipation, and urinary retention. Nausea and vomiting can occur. Dilated pupils occur in approximately 50 percent of cases. The motor complications of botulism feature a descending paralysis leading to blurred vision and, eventually, respiratory failure may occur. Symptoms usually begin 12 to 36 hours following intoxication; time can vary according to the amount of toxin absorbed and could be reduced to hours following a BW attack.[3] The case-fatality rate in the US for food-borne botulism is 5 to 10 percent.[1]

(6) Incubation period. The incubation period for food-borne botulism is usually 24 to 36 hours; for wound botulism it is usually 3 or more days.[3]

(7) Communicability. No person-to-person transmission.[6]

(8) Prevention. The vaccine currently available is used under Investigational New Drug (IND) status.[19]

(9) Delivery. The primary threats are by sabotage of food and water supply and by aerosol release.[5]

b. *Clostridium Perfringens* (C. perfringens) Toxins.

(1) Toxin Origin. *C. perfringens* is a common anaerobic bacillus that produces at least 12 toxins.[3]

(2) Occurrence. Worldwide.[3]

(3) Reservoir. The reservoirs are soil and the GI tract of healthy people and animals.[3]

(4) Transmission. Transmission is by the ingestion of food that was contaminated by soil or feces and then held under conditions that permit multiplication of the organism. Almost all outbreaks are associated with inadequately heated or reheated meats. Spores survive cooking and then germinate and multiply in storage at ambient temperatures, slow cooling, or inadequate rewarming.[1] Gas gangrene results from wound contamination with soil containing spores.[3]

(5) Symptoms. Gas gangrene features pathological death of skeletal muscles and overlying soft tissue and constitutes a surgical emergency.[3] Food poisoning is an intestinal disorder characterized by sudden onset of colic followed by diarrhea; nausea is common, but vomiting and fever are usually absent. Generally, it is a mild disease of short duration, one day or less, and rarely fatal in healthy persons.[1]

(6) Incubation period. The incubation period is 1 to 6 hours.[3]

(7) Communicability. No person-to-person transmission.[5]

(8) Prevention. Educate food handlers.[1] No vaccine is available.[5]

(9) Delivery. The primary threat is delivery of *C. perfringens* alpha toxin as an aerosol to the respiratory tract. This would result in pulmonary disease, vastly different from the naturally occurring diseases associated with *C. perfringens*. The toxin may also be delivered in combination with other toxins to produce a variety of effects.[3]

c. Conotoxins.

(1) Toxin Origin. Conotoxins are a neurotoxin from the mollusks of the genus *Conus*, commonly known as cone snails. There are potentially over 50,000 different conotoxins present in the venoms of species in the genus Conus.[20]

(2) Occurrence. These cone snails live in intertidal regions of the Indian and Pacific Oceans, east to Hawaii, north to Japan, and south to New Zealand and Australia.[20]

(3) Reservoir. Cone snails.[20]

(4) Transmission. Conotoxins are injected by the sting of a cone snail.[17]

(5) Symptoms. The sting produces a small, deep, triangular puncture wound. Pain results immediately, ranging from mild to severe, followed by numbness and swelling at the point of envenomation. Nausea, vomiting, difficulty swallowing, malaise, and weakness may occur. Paralysis of the respiratory muscles might cause the airway to collapse. The heart may beat irregularly. Neurological effects may include impaired coordination, decreased visual acuity, altered level of consciousness, diminished or absent reflexes, and paralysis. The untreated case-fatality rate is 70 percent.[20]

(6) Incubation period. Symptoms begin immediately.[20]

(7) Communicability. No person-to-person transmission.

(8) Prevention. No specific antivenom exists.[20]

(9) Delivery. The likely dissemination is by respirable aerosols or contamination of food or water supply.[19]

d. Diarrhea Caused by Enterohemorrhagic (EHEC) Strains (Shiga toxin-producing *E. coli* [STEC], *E. coli* O157:H7, Verotoxin-producing *E. coli* [VTEC]).

(1) Toxin Origin. The main EHEC serotype in North America is *E. coli* O157:H7. This category of diarrheogenic *E. coli* produces cytotoxins. Previously, these toxins were called Verotoxins 1 and 2 or Shiga-like toxins I and II.[1]

(2) Occurrence. These infections are now recognized to be an important problem in North America, Europe, South Africa, Japan, South America, and Australia.[1]

(3) Reservoir. Cattle are the most important reservoir; however, humans may also serve as a reservoir for person-to-person transmission. There is increasing evidence in North America that deer may also serve as a reservoir.[1]

(4) Transmission. Transmission occurs mainly by ingestion of contaminated food. It is most often due to inadequately cooked beef (especially ground beef), raw milk,

and fruit or vegetables contaminated with ruminant feces. Waterborne transmission has been documented. Transmission also occurs directly from person to person.[1]

(5) Symptoms. The diarrhea may range from mild and nonbloody to stools that are virtually all blood but no fecal leukocytes. Approximately 2 to 7 percent of subjects develop hemolytic uremic syndrome.[1]

(6) Incubation period. Ranges from 2 to 8 days, with a median of 3 to 4 days.[1]

(7) Communicability. Person-to-person transmission is possible. The duration of excretion of the pathogen is typically a week or less in adults. Prolonged carriage is uncommon.[1]

(8) Prevention. Field sanitation and personal hygiene are useful tools to help prevent exposure.[1]

(9) Delivery. The likely method of dissemination is by respirable aerosols or contamination of food or water supply.[19]

e. Microcystin.

(1) Toxin Origin. Common species of the blue-green algae that produce microcystins include Anabaena, Nostoc, Oscillatoria, Hapalosiphon, and Microcystic. Their structural class is cyanobacteria. Freshwater cyanobacteria accumulate in surface water supplies and appear as blue-green "scums."[21] For microcystin, unlike most toxins, the toxicity is the same, no matter what the route of exposure.[22] Microcystin is extremely stable and resistant to chemical hydrolysis or oxidation and remain potent even after boiling.[23]

(2) Occurrence. Worldwide.[23]

(3) Reservoir. Blue-green algae.[22]

(4) Transmission. Drinking or swimming in contaminated water. Very limited available information suggests that inhalation in aerosols may be an equally important route of exposure.[23]

(5) Symptoms. Microcystin cause gastroenteritis.[20] It is a membrane-damaging toxin. Unless uptake of the toxin by the liver is blocked, irreversible damage to the organ occurs within 15 to 60 minutes after exposure to a lethal dose. When this happens, the tissue damage to the liver is so severe that therapy may have little or no value.[22]

(6) Incubation period. Data not available.

(7) Communicability. In natural waters and in the dark, microcystins may persist for months or years.[23]

(8) Prevention. Data not available.

(9) Delivery. The likely method of dissemination is by respirable aerosol or contamination of food and water supply.

f. Ricin.

(1) Toxin Origin. Ricin is a potent cytotoxin derived from the beans of the castor plant (*Ricinus communis*). Large quantities of ricin are easily and inexpensively produced. Ricin toxins are potent inhibitors of DNA replication and protein synthesis.[3]

(2) Occurrence. N/A.

(3) Reservoir. Castor beans.[3]

(4) Transmission. Transmission has been by inhalation of organism during industrial operations and ingestion of castor bean meal.[3]

(5) Symptoms. The clinical signs, symptoms, and pathological manifestations of ricin toxicity vary with the dose and the route of exposure. Inhalation results in respiratory distress and airway and pulmonary lesions; ingestion causes GI signs and GI hemorrhage with necrosis of liver, spleen, and kidneys; and intramuscular intoxication causes severe localized pain, muscle and regional lymph node necrosis, and moderate involvement of the visceral organs.[24]

(6) Incubation period. From 18 to 24 hours.[3]

(7) Communicability. No person-to-person transmission.[3]

(8) Prevention. Candidate vaccines are under development.[3]

(9) Delivery. The primary threat is delivery by aerosol release. A large quantity is required to cover a significant area on a battlefield; however, it can be used for small-scale operations. The agent may also be delivered through contamination of food and water supplies.[3]

g. Saxitoxin.[3]

(1) Toxin Origin. Saxitoxin is the parent compound of a group of related neurotoxins produced by marine dinoflagellates of the genus *Gonyaulax*.

(2) Occurrence. N/A.

(3) Reservoir. Shellfish.

(4) Transmission. Saxitoxin is transmitted to humans by ingesting bivalve mollusks, which accumulate dinoflagellates during filter feeding.

(5) Symptoms. Paralytic shellfish poisoning (PSP) is a severe, life-threatening neuromuscular condition. Saxitoxin is rapidly absorbed from the GI tract following ingestion of contaminated shellfish. Symptoms begin as early as 10 minutes to several hours after ingestion, depending on the ingested dose and host factors. Initial symptoms include numbness and tingling of the lips, tongue, and fingertips followed by numbness of the neck and extremities and motor incoordination. Other symptoms may include light-headedness, dizziness, weakness, confusion, memory loss, and headache. Flaccid paralysis and respiratory failure are life-threatening complications and occur within 2 to 12 hours after ingestion. Supportive care is essential.

(6) Incubation period. Minutes to hours.

(7) Communicability. No person-to-person transmission.

(8) Prevention. No vaccine is available.

(9) Delivery. The primary threat is delivery by aerosol release. Saxitoxin may also be delivered by projectiles or by contamination of food and water.

h. Staphylococcal Enterotoxin B.

(1) Toxin Origin. Staphylococcal enterotoxin B (SEB) is one of numerous exotoxins produced by *Staphylococcus (S.) aureus*. The SEB toxin is heat-stable and is the second most common source of outbreaks of food poisoning.[3]

(2) Occurrence. Worldwide and relatively frequent.[1]

(3) Reservoir. The reservoirs of *S. aureus* are humans and contaminated milk and milk products. The SEB is usually produced in foods contaminated with *S. aureus*.[3]

(4) Transmission. Ingestion of food, milk, or milk products containing preformed toxin.[3]

(5) Symptoms. Symptoms include the acute onset of fever, nausea, vomiting, and diarrhea within hours of intoxication. Illness due to inhalation will result in respiratory tract disease not encountered in the endemic disease. Symptoms may also occur in the GI tract due to inadvertent swallowing of SEB delivered via aerosol and deposited in the upper aero-digestive tract.[3]

(6) Incubation period. Variable, 4 to 10 hours for GI illness.[3]

(7) Communicability. No person-to-person transmission.[3]

(8) Prevention. Follow strict food hygiene and sanitation controls.[1] Protecting food and water supplies from contamination and avoiding potentially contaminated food and water will protect individuals from the effects of ingested toxins. No vaccine is available.[3]

(9) Delivery. The primary threat is SEB aerosol release. The SEB may also be employed by sabotage contamination of food and/or water supplies.[5]

i. Tetrodotoxin (Puffer Fish Poisoning).

(1) Toxin Origin. The causative toxin is tetrodotoxin, a heat-stable, nonprotein neurotoxin[1] that even in small amounts can cause rapid and violent death in humans.[17]

(2) Occurrence. Majority of cases occur in Japan where puffer fish is a traditional delicacy.[20]

(3) Reservoir. A variety of marine species—notably the puffer fish, California newt,[20] porcupine fish, ocean sunfish, and species of newts and salamanders.[1]

(4) Transmission. Poisoning results from consuming tetrodotoxin.[20]

(5) Symptoms. Puffer fish poisoning is characterized by onset of paresthesias, dizziness, GI symptoms, and ataxia, which often progresses rapidly to paralysis and death within several hours after eating. The case-fatality rate approaches 60 percent.[1]

(6) Incubation period. Symptoms can appear 20 minutes to 3 hours after toxin is introduced. Death can occur 4 to 6 hours after introduction.[20]

(7) Communicability. No person-to-person transmission.[20]

(8) Prevention. There is no specific antidote available.[20] Avoid ingesting any of the fish or amphibians that produce tetrodotoxin.[1]

(9) Delivery. The likely method of dissemination is by respirable aerosols or contamination of food or water supply.[22]

j. Trichothecene Mycotoxins.

(1) Toxin Origin. Trichothecene (T-2) mycotoxins are a diverse group of over 40 compounds produced by molds of the genus *Fusarium*. These toxins inhibit protein and DNA synthesis and mitochondrial respiration and alter cell membrane structure and function.[3]

(2) Occurrence. Worldwide. Naturally occurring mycotoxicosis occurs in livestock following ingestion of grains contaminated with molds. When maintained as either crystalline powders or liquid solutions, these compounds are stable when exposed to air, light, or both. They are not inactivated by autoclaving but require heating at 900 degrees F for 10 minutes or 500 degrees F for 30 minutes for complete inactivation. A 3 to 5 percent solution of bleach is an effective inactivation agent for them.[25]

(3) Reservoir. Moldy grain.[3]

(4) Transmission. Ingestion of moldy grains.[3]

(5) Symptoms. Mycotoxins are highly cytotoxic and have effects similar to vesicants, especially mustard agents. Delivery to the skin may cause a burning skin pain, redness, tenderness, blistering, and progression to skin necrosis with eschar formation and sloughing. Respiratory exposure may result in nasal itching with pain, rhinorrhea, sneezing, wheezing, and cough.[3]

(6) Incubation period. The incubation period is minutes after exposure.[3]

(7) Communicability. No person-to-person transmission.[3]

(8) Prevention. As a pre-exposure prophylaxis, the use of topical antivesicant cream or ointment may provide limited protection of skin surfaces. Food and water contaminated with mycotoxins must not be consumed.[3] Washing the exposed skin areas with soap and water will decrease absorption through the skin.[25]

(9) Delivery. The toxin may be delivered by aerosol release or through the contamination of food and water supplies. These toxins are the agents allegedly delivered via aerosol during the "Yellow Rain" attacks in Afghanistan and Southeast Asia during 1970s and 1980s. The T-2 mycotoxins are the only potential BW agents that can harm and be absorbed through the intact skin.[3]

NOTES

[1]James Chin (ed.), *Control of Communicable Diseases Manual,* 17th ed., United Book Press, Inc., Baltimore, MD, 2000.

[2]TM 3-216/AFM 355-6, *Technical Aspects of Biological Defense,* 12 January 1971.

[3]FM 8-284/NAVMED P-5042/AFMAN (I) 44-156/MCRP 4-11.1C, *Treatment of Biological Warfare Agent Casualties,* 17 July 2000.

[4]BG Russ Zajtchuk, et al. (eds), *Textbook of Military Medicine: Medical Aspects of Chemical and Biological Warfare,* Office of the Surgeon General, 1997, Chap. 22, "Anthrax."

[5] FM 8-9/NAVMED P-5059/AFJMAN 44-151, *NATO Handbook on the Medical Aspects of NBC Defense Operations AMEDP-6(B),* 1 February 1996.

[6]CDC, Division of Bacterial and Mycotic Diseases, "Disease Information: Glanders (Technical Information), 7 March 2003, <http://www.cdc.gov/ncidod/dbmd/diseaseinfo/glanders_t.htm>, 11 August 2003.

[7]CDC, Division of Bacterial and Mycotic Diseases, "Disease Information: Glanders (General Information), 20 June 2001, <http://www.cdc.gov/ncidod/dbmd/diseaseinfo/glanders_g.htm>, 29 August 2003.

[8]BG Russ Zajtchuk, et al. (eds), *Textbook of Military Medicine: Medical Aspects of Chemical and Biological Warfare,* Office of the Surgeon General, 1997, Chap. 23, "Plague."

[9]US HHS, CDC, "Compendium of Measures to Control *Chlamydia psittaci* Infection Among Humans (Psittacosis) and Pet Birds (Avian Chlamydiosis), 1998," *Morbidity and Mortality Weekly Report,* Vol. 47, N0. RR-10, 10 July 1998.

[10]BG Russ Zajtchuk, et al. (eds), *Textbook of Military Medicine: Medical Aspects of Chemical and Biological Warfare,* Office of the Surgeon General, 1997, Chapter 24, "Tularemia."

[11]BG Russ Zajtchuk, et al. (eds), *Textbook of Military Medicine: Medical Aspects of Chemical and Biological Warfare,* Office of the Surgeon General, 1997, Chapter 26, "Q Fever."

[12]BG Russ Zajtchuk, et al. (eds), *Textbook of Military Medicine: Medical Aspects of Chemical and Biological Warfare,* Office of the Surgeon General, 1997, Chapter 28, "Viral Encephalitides."

[13]CDC, Special Pathogens Branch, "Ebola Hemorrhagic Fever," 6 August 2003, http://www.cdc.gov/ncidod/dvrd/spb/mnpages/dispages/ebola.htm, 2 September 2003.

[14]BG Russ Zajtchuk, et al. (eds), *Textbook of Military Medicine: Medical Aspects of Chemical and Biological Warfare,* Office of the Surgeon General, 1997, Chapter 29, "Viral Hemorrhagic Fevers."

[15]CDC, Monkeypox, "Fact Sheet: Basic Information About Monkeypox," 12 June 2003, http://www.cdc.gov/ncidod/monkeypox/factsheet.htm, 2 September 2003.

[16]Yvan J.F. Hutin (CDC), et. al., "Research: Outbreak of Human Monkeypox, Democratic Republic of Congo, 1996-1997," *Emerging Infectious Diseases,* Vol. 7, No. 3, May-June 2001.

[17]CDC, Monkeypox, "Updated Interim Infection Control and Exposure Management Guidance in the Health-Care and Community Setting for Patients with possible Monkeypox

Virus Infection," 18 July 2003, http://www.cdc.gov/ncidod/monkeypox/infectioncontrol.htm, 2 September 2003.

[18]CDC, Monkeypox, "Fact Sheet: Smallpox Vaccine and Monkeypox," 9 July 2003, http://www.cdc.gov/ncidod/monkeypox/smallpoxvaccine_mpox.htm, 2 September 2003.

[19]BG Russ Zajtchuk, et al. (eds), *Textbook of Military Medicine: Medical Aspects of Chemical and Biological Warfare,* Office of the Surgeon General, 1997, Chapter 33, "Botulinum Toxins."

[20]CDC, HHS, "Regulatory Impact Analysis: 42 CFR Part 73: Select Biological Agents and Toxins, Interim Final Rule (Draft)," 9 December 2002.

[21]National Cancer Institute (NCI) Nomination Submitted to the National Toxicity Program (NTP), "Blue-Green Algae," September 2000.

[22]BG Russ Zajtchuk, et al. (eds), *Textbook of Military Medicine: Medical Aspects of Chemical and Biological Warfare,* Office of the Surgeon General, 1997, Chap. 30, "Defense Against Toxin Weapons."

[23]Ingrid Chorus and Jamie Bartram ed., *Toxic Cyanobacteria in Water: A guide to their public health consequences, monitoring and management*, Geneva, WHO, 1999.

[24]BG Russ Zajtchuk, et al. (eds), *Textbook of Military Medicine: Medical Aspects of Chemical and Biological Warfare,* Office of the Surgeon General, 1997, Chap. 32, "Ricin Toxin."

[25]BG Russ Zajtchuk, et al. (eds), *Textbook of Military Medicine: Medical Aspects of Chemical and Biological Warfare,* Office of the Surgeon General, 1997, Chap. 34, "Trichothecene Mycotoxins."

[26]Office of the US President, *The Biological and Chemical Warfare Threat,* 1999.

Chapter V

TOXIC INDUSTRIAL CHEMICALS AND THEIR PROPERTIES

1. Background

In April 1915, the Germans released 150 tons of a TIC (chlorine gas) across a 4-mile front in one of the first gas attacks of WWI. It is estimated that up to 5,000 allied soldiers were killed and 20,000 became casualties. In addition to chlorine, other gases were used against the Allies in other periods during this war. Arsine, cyanogen chloride, hydrogen cyanide, phosgene, and hydrogen sulfide were used during WWI. In most cases, these gases were first developed for industrial purposes, but the potential of these chemicals as war gases was recognized. In Bhopal, India, in December 1984, an accidental release of methyl isocyanate from a production facility killed an estimated 2,500 persons and left many thousands injured.[1]

a. Potential Hazard. The US troops—through the deliberate release, accidental release, or as a result of collateral damage—could encounter TIC. Military personnel are trained to operate in an environment in which CW weapons may be used. However, the potential hazards of TIC are a relatively new threat. These forces must be able to safely operate, survive, and sustain operations in those instances where they may be exposed to toxic industrial hazards. Commanders have direct responsibility for protecting their forces against these threats and, therefore, must become aware of and plan for defense against the potential release of TIC in their AO. In future operations, failure to properly plan for the release of TIC may result in significant casualties, disruption of operations, and mission degradation.

b. Characteristics. A number of hazards are associated with a release of TIC. Some key factors are—

(1) Certain chemicals have the ability to bypass or penetrate current military protective equipment and, in many cases, are not detectable by current military equipment.

(2) Exposure can be through several routes, including inhalation, ingestion, or surface contact with the material.

(3) Many of these chemicals cause a variety of immediate and delayed symptoms that make detection, diagnosis, and treatment difficult.

(4) Symptoms are often difficult to trace back to a specific chemical.

(5) Determining the exposure level that constitutes a hazard for the individual is often difficult due to the variances of susceptibility among individuals.

c. Protection. It is important that commanders and troops be aware that the best defense against TIC is to escape the path of the TIC immediately. The military field protective mask can provide limited protection and should only be used to escape the hazard area. Personnel or equipment that have been contaminated with TIC can be decontaminated by washing with large amounts of cold, soapy water. Contaminated clothing should be immediately removed and disposed of in a safe manner.

d. TIC of Concern. The TIC have important physical properties (such as reactivity, flammability, or corrosivity), and these properties cause many chemical injuries and damage annually. For example, the United Nations Environmental Protection Organization (UNEPO) developed a list of accidents (1970-98) involving hazardous substances in which there were at least 25 deaths, 125 injured, 10 thousand evacuated or deprived of water, or $10 million US damages to third parties. See Table V-1 for a listing of the hazardous substances, primarily TIC, that were involved.

Table V-1. Accidents Involving Hazardous Substances

Chemical Class	Frequency %	Ever used as a weapon?
• Fuel gases	15	Yes – Columbia, 2000
• Explosives	12	Yes – Oklahoma City bomb
• Liquid fuels	8	Yes – animal rights and environmental arson
• Petrochemicals	8	Yes – Kutina, Croatia
• Chlorine	8	Yes – Jovan, Croatia
• Oil	7	Yes –Texas, Columbia, Guatemala, Kuwait
• Ammonia	6	Yes – Kutina, Croatia
• Acids & bases	6	Yes – abortion clinics
• Plastics manufacture	3	No
• Pesticides, including precursors	3	Yes – Bhopal (sabotage)
• PCBs (slow onset)	2	No
• Fertilizers	2	Yes – Oklahoma City bomb
• Dusts (e.g., grain elevator)	1	Yes – IEDs
• Radiation	1	Yes – RDDs (Moscow)
• Other Chemicals	13	Yes – arsenic, cyanides, phosgene, metals
• Not specified	5	NA

(1) As shown, fuel gases, explosives, liquid fuel, petrochemicals, and oil account for nearly 50 percent of the incidents.

(2) Consideration of a chemical's many hazardous properties is reinforced by the criteria governing the US chemical manufacturing Risk Management Program (RMP). The RMP, instituted in June 1999, under US federal regulations, requires certain chemical facilities to implement chemical accident prevention and preparedness measures and to submit summary reports to the government every 5 years. The RMP database contains information on 14,828 facilities containing 20,210 chemical processes. Of these processes, 17,529 contain at least one TIC and 8,107 contain at least one flammable chemical. Chemicals such as anhydrous ammonia, chlorine, propane, and flammable mixtures are present in nearly 70 percent of all RMP processes.

2. Other Information Sources

Multiple data sources are available to support the need for data and information to support the military decision-making process (MDMP). These sources include—

a. North American Emergency Response Guidebook (NAERG) 2000 (http://hazmat.dot.gov/gydebook.htm).

(1) The US DOT, Transport Canada, and the Secretariat of Communications and Transportation (SCT) of Mexico developed the *Emergency Response Guide (ERG) 2000* jointly for use by firefighters, police, and other emergency services personnel who may be the first to arrive at the scene of a transportation incident involving a hazardous material. It is primarily a guide to aid first responders in quickly identifying the specific or generic classification of the material(s) involved in the incident, and protecting themselves and the general public during this initial response phase of the incident. The ERG is updated every 3 years to accommodate new products and technology. The ERG 2000 incorporates

dangerous goods lists from the most recent United Nations (UN) recommendations as well as from other international and national regulations.

(2) The DOT goal is to place one ERG 2000 in each emergency service vehicle, nationwide, through distribution to state and local public safety authorities. To date, more than 7 million copies have been distributed without charge to the emergency responder community. Copies are made available free of charge to public emergency responders through the state coordinators (US only). In Canada, contact the Canadian Transport. Emergency Centre (CANUTEC) at 613-992-4624 or e-mail canutec@tc.gc.ca for information. In Mexico, call SCT at 52-5-684-1275.

b. NIOSH Pocket Guide (NPG) to Chemical Hazards (http://www.cdc.gov/niosh/npg/npg.html).

(1) The NPG is intended as a source of general industrial hygiene information for workers, employers, and occupational health professionals. The NPG presents key information and data in abbreviated tabular form for 677 chemicals or substance groupings (e.g., manganese compounds, tellurium compounds, inorganic tin compounds, etc.) that are found in the work environment. The industrial hygiene information found in the pocket guide should help users recognize and control occupational chemical hazards. The chemicals or substances contained in this revision include all substances for which the NIOSH has recommended exposure limits (RELs) and those with permissible exposure limits (PELs) as found in the Occupational Safety and Health Administration (OSHA) General Industry Air Contaminants Standard (29 Code of Federal Regulations [CFR] 1910.1000).

(2) The information in the NPG includes chemical structures and formulas, identification codes, synonyms, exposure limits, chemical and physical properties, incompatibilities and reactivities, measurement methods, respirator selections, signs and symptoms of exposure, and procedures for emergency treatment.

c. Registry of Toxic Effects of Chemical Substances (RTECS) (http://www.cdc.gov/niosh/rtecs.html).

(1) The RTECS provides toxicological information with citations on over 140,000 chemical substances. These detailed profiles include toxicological data and reviews, international workplace exposure limits, references to US standards and regulations, analytical methods, and exposure and hazard survey data.

(2) The data are compiled into substance records for ease of use, and updated data are fully integrated.

d. International Chemical Safety Cards (ICSCs) (http://www.cdc.gov/niosh/ipcs/icstart.html).

(1) The International Program on Chemical Safety (IPCS) is a joint activity of three cooperating International Organizations: the United Nations Environment Program (UNEP), the International Labor Office (ILO), and the WHO. The main objective of the IPCS is to carry out and disseminate evaluations of the hazards posed by chemicals to human health and the environment.

(2) An ICSC summarizes essential health and safety information on TIC. The ICSCs are not legally binding documents, but consist of a series of standard phrases, mainly summarizing health and safety information collected, verified, and peer-reviewed by

internationally recognized experts, taking into account advice from manufacturers and poison control centers.

e. Chemical Hazard Response Information System (CHRIS) (http://www.chrismanual.com).

(1) The CHRIS database is a comprehensive source of emergency response information for those involved in the transport of hazardous materials (HAZMAT). The database is based upon the text of the United States Coast Guard (USCG) CHRIS manual. At printing, this site does not yet have an electronic version of CHRIS available.

(2) Records for more than 1,300 HAZMAT are provided in English. Each record for a HAZMAT contains key identification data, such as synonyms, Chemical Abstracts Service (CAS) numbers, hazard labels, and observable characteristics. In addition, information useful for emergency response situations—such as health, fire and reactivity hazards, first aid, water pollution, shipping and hazard classifications, and physical and chemical properties—is included in each record.

f. Hazardous Substances Data Bank (HSDB) (http: www.nlm.nih.gov/pubs/factsheets/hsdbfs.html).

(1) The HSDB is a nonbibliographic data bank created and maintained by the National Library of Medicine (NLM) in the US.

(2) The data bank provides extensive information on identification, manufacturing, use, chemical and physical properties, safety and handling, human and nonhuman toxicity, pharmacology, environmental fate and exposure, regulations, and analytical determinations of chemical substances.

(3) HSDB information is organized into chemical records, with records for over 4,500 chemical substances.

g. Extremely Hazardous Substance (EHS) Chemical Profiles and Emergency First Aid Guides. (http://www.epa.gov/swercepp/ehs/ehslist.html).

(1) This guide contains information on over 300 EHSs currently listed as part of Section 302 of the *Emergency Planning and Community Right-to-Know Act.*

(2) Each chemical profile includes physical/chemical properties, health hazards, fire and explosion hazards, reactivity data, precautions for safe handling and use, and protective equipment of emergency situations.

(3) The first aid guide provides signs and symptoms of poisoning and emergency treatment for first responders. The chemical profiles and first aid guides may be accessed from either the CAS number or alphabetical list of EHSs.

3. Reach-Back Capability[2]

a. General. Technical reach-back is the ability to contact technical subject matter experts (SMEs) when a technical issue exceeds the on-scene SME's capability. Reach-back should be conducted using established unit protocols. Many of the reach-back resources have other primary missions and are not specifically resourced for reach-back. See FM 3-11.21 for additional resources.

b. National Response Center (NRC). The NRC mans the hotline service and serves as an emergency resource for first responders to request technical assistance during an incident. The intended users include trained emergency personnel (such as emergency operators) and first responders (such as firefighters, police, and emergency medical technicians [EMTs]) who arrive at the scene of a CB terrorist incident. Other potential users may include the state emergency operations centers (EOCs) and hospitals that may treat victims of agent exposure.

(1) The USCG operates the NRC, and trained operators staff the hotline 7 days a week, 24 hours a day. NRC duty officers take reports of actual or potential domestic terrorism and link emergency calls with applicable SMEs (such as United States Army Soldier and Biological Chemical Command [SBCCOM] and United States Army Medical Research Institute of Chemical Defense [USAMRICD]) for technical assistance and with the FBI for federal response actions. The NRC also provides reports and notifications to other federal agencies as necessary. Specialty areas include the following:

(a) Detection equipment.

(b) Personal protective equipment (PPE).

(c) Decontamination systems and methods.

(d) Physical properties of CB agents.

(e) Toxicology information.

(f) Medical symptoms from exposure to CB agents.

(g) Treatment of exposure to CB agents.

(h) Hazard prediction models.

(i) Federal response assets.

(j) Applicable laws and regulations.

(2) The CB hotline is a joint effort of the USCG, Federal Bureau of Investigation (FBI), Federal Emergency Management Agency (FEMA), EPA, Department of Health and Human Services (DHHS), and the DOD. The NRC is the entry point for the CB hotline. The NRC receives basic incident information and links the caller to the DOD and FBI CB and terrorism experts. These and other federal agencies can be accessed within a few minutes to provide technical assistance during a potential CB incident. If the situation warrants, a federal response action may be initiated.

(3) Use the local established policies and procedures for requesting federal assistance before contacting the CB hot line. State and local officials can access the hotline in emergency circumstances by calling 1-800-424-8802.

NOTES

[1]A.K. Steumpfle, et al., *Final Report of International Task Force-25: Hazard From Toxic Industrial Chemicals,* March 18, 1996.

[2]FM 3-11.22, *Weapons of Mass Destruction Civil Support Team Tactics, Techniques, and Procedures*, 6 June 2003.

Appendix A
TABLE OF EQUIVALENTS

Table A-1 Table of Equivalents

- Length -

1 kilometer (km)	1,000 meters (m)
	0.5397 nautical miles (nm)
	0.6215 statute miles (mi)
1 meter (m)	1.0936 yards (yd)
	3.2808 feet (ft)
	39.3701 inches (in)

- Area -

1 hectare (ha)	10,000 square meters (m^2)
	0.0039 square miles (mi^2)
	0.01 square kilometers (km^2)
	2.4710 acres (ac)
	100 ares (a)

- Volume -

1 liter (l)	0.001 cubic meters (m^3)
	0.2642 gallons (gal)

- Mass -

1 kilogram (kg)	1,000 grams (g)
	2.205 pounds (lb)
	0.001 tons (t)

- Pressure -

1 torr	1 milimeter Mecury (mm Hg)
1 atmosphere (atm)	760 torr

Table of A-2 Table of Commonly Used Prefixes

Multiplier	Symbol	Prefix
10^{24}	Y	yotta
10^{21}	Z	zetta
10^{18}	E	exa
10^{15}	P	peta
10^{12}	T	tera
10^{9}	G	giga
10^{6}	M	mega
10^{3}	k	kilo
10^{2}	h	hecto
10^{1}	da	deca
10^{0}	***(unit)***	***(unit)***
10^{-1}	d	deci
10^{-2}	c	centi
10^{-3}	m	milli
10^{-6}	μ	micro
10^{-9}	n	nano
10^{-12}	p	pico
10^{-15}	f	femto
10^{-18}	a	atto
10^{-21}	z	zepto
10^{-24}	y	yocto

Appendix B
TEMPERATURE CONVERSIONS

To convert degrees Fahrenheit to degrees Centigrade:

$$°C = (°F - 32) / 1.8$$

To convert degress Centigrade to degrees Fahrenheit:

$$°F = 1.8 \times °C + 32$$

Table B-1 Temperature Conversions

	Centigrade (°C)	Fahrenheit (°F)
Water Boils	100	212
	40	104
	35	95
	30	86
	25	77
Room Temperature	20	68
	15	59
HD Freezes	14.45	58.01
	10	50
	5	41
HT Freezes	1.3	34.34
Water Freezes	0	32
	-5	23
	-10	14
	-15	5
	-20	-4
	-25	-13
	-30	-22
HL Freezes	-42	-43.6

Appendix C
PERIODIC TABLE OF THE ELEMENTS

Table C-1 Periodic Table of the Elements

1a	2a	3b	4b	5b	6b	7b	8			1b	2b	3a	4a	5a	6a	7a	Noble Gases 0	Orbit
1 H																	2 He	K
3 Li	4 Be											5 B	6 C	7 N	8 O	9 F	10 Ne	K-L
11 Na	12 Mg	Transition Elements										13 Al	14 Si	15 P	16 S	17 Cl	18 Ar	K-L-M
19 K	20 Ca	21 Sc	22 Ti	23 V	24 Cr	25 Mn	26 Fe	27 Co	28 Ni	29 Cu	30 Zn	31 Ga	32 Ge	33 As	34 Se	35 Br	36 Kr	L-M-N
37 Rb	38 Sr	39 Y	40 Zr	41 Nb	42 Mo	43 Tc	44 Ru	45 Rh	46 Pd	47 Ag	48 Cd	49 In	50 Sn	51 Sb	52 Te	53 I	54 Xe	M-N-O
55 Cs	56 Ba	57 La	72 Hf	73 Ta	74 W	75 Re	76 Os	77 Ir	78 Pt	79 Au	80 Hg	81 Tl	82 Pb	83 Bi	84 Po	85 At	86 Rn	N-O-P
87 Fr	88 Ra	89 Ac	104 Rf	105 Db	106 Sg	107 Bh	108 Hs	109 Mt								Halides		O P Q
							Group 8											

Lanthanides	58 Ce	59 Pr	60 Nd	61 Pm	62 Sm	63 Eu	64 Gd	65 Tb	66 Dy	67 Ho	68 Er	69 Tm	70 Yb	71 Lu	N O P
Actinides	90 Th	91 Pa	92 U	93 Np	94 Pu	95 Am	96 Cm	97 Bk	98 Cf	99 Es	100 Fm	101 Md	102 No	103 Lr	O P Q

Table of C-2 Chemical Elements and Symbols

Element	Symbol	Atomic Weight*	Element	Symbol	Atomic Weight*
Actinium	Ac	(227)	Mendelevium	Md	(258)
Aluminum	Al	26.98	Mercury	Hg	200.59
Americium	Am	(243)	Molybdenum	Mo	95.94
Antimony	Sb	121.75	Neodymium	Nd	144.24
Argon	Ar	39.95	Neon	Ne	20.17
Arsenic	As	74.92	Neptunium	Np	237.05
Astatine	At	(210)	Nickel	Ni	58.71
Barium	Ba	137.33	Niobium	Nb	92.91
Berkelium	Bk	(247)	Nitrogen	N	14.01
Beryllium	Be	9.01	Nobelium	No	(259)
Bismuth	Bi	208.98	Osmium	Os	190.20
Bohrium	Bh	(262)	Oxygen	O	16.00
Boron	B	10.81	Palladium	Pd	106.40
Bromine	Br	79.90	Phosphorus	P	30.97
Cadmium	Cd	112.41	Platinum	Pt	195.09
Calcium	Ca	40.08	Plutonium	Pu	(244)
Californium	Cf	(251)	Polonium	Po	(209)
Carbon	C	12.01	Potassium	K	39.10
Cerium	Ce	140.12	Praseodymium	Pr	140.91
Cesium	Cs	132.91	Promethium	Pm	(145)
Chlorine	Cl	35.45	Protactinium	Pa	231.04
Chromium	Cr	52.00	Radium	Ra	226.03
Cobalt	Co	58.93	Radon	Rn	(222)
Copper	Cu	63.55	Rhenium	Re	186.20
Curium	Cm	(247)	Rhodium	Rh	102.91
Dubnium	Db	(262)	Rubidium	Rb	85.47
Dysprosium	Dy	162.50	Ruthenium	Ru	101.07
Einsteinium	Es	(254)	Rutherfordium	Rf	(261)
Erbium	Er	167.26	Samarium	Sm	150.40
Europium	Eu	151.96	Scandium	Sc	44.96
Fermium	Fm	(257)	Seaborgium	Sg	(266)
Fluorine	F	19.00	Selenium	Se	78.96
Gadolinium	Gd	157.25	Silicon	Si	28.09
Gallium	Ga	69.74	Silver	Ag	107.87
Germanium	Ge	72.59	Sodium	Na	22.99
Gold	Au	196.97	Strontium	Sr	87.62
Hafnium	Hf	178.49	Sulfur	S	32.06
Hassium	Hs	(265)	Tantalum	Ta	180.95
Helium	He	4.00	Technetium	Tc	98.91
Holmium	Ho	164.93	Tellurium	Te	127.60
Hydrogen	H	1.00	Terbium	Tb	158.93
Indium	In	114.82	Thallium	Tl	204.37
Iodine	I	126.90	Thorium	Th	232.04
Iridium	Ir	192.22	Thulium	Tm	168.93
Iron	Fe	55.85	Tin	Sn	118.69
Krypton	Kr	83.80	Titanium	Ti	47.90
Lanthanum	La	138.91	Tungsten	W	183.85
Lawrencium	Lr	(260)	Uranium	U	238.03
Lead	Pb	207.20	Vanadium	V	50.94
Lithium	Li	6.94	Xenon	Xe	131.30
Lutetium	Lu	174.97	Ytterbium	Yb	173.04
Magnesium	Mg	24.31	Yttrium	Y	88.91
Manganese	Mn	54.94	Zinc	Zn	65.38
Meitnerium	Mt	(266)	Zirconium	Zr	91.22

*Values in parantheses are for the most stable isotope of element.

Appendix D

CHEMICAL WEAPONS CONVENTION SCHEDULE 1, 2 AND 3 CHEMICALS

Table D-1. CWC Schedule 1 Chemicals[1]

Substance	CAS Registry Number
Toxic Chemicals	
O-Alkyl (≤ C_{10}, including cycloalkyl) alkyl (Me, Et, n-Pr or I-Pr)-phosphonofluoridates	
e.g. Sarin: O-Isopropyl methylphosphonofluoridate	107-44-8
e.g. Soman: O-Pinacolyl methylphosphonofluoridate	96-64-0
O-Alkyl (≤ C_{10}, including cycloalkyl) N,N-dialkyl (Me, Et, n-Pr or i-Pr)-phosphoramidocyanidates	
e.g. Tabun: O-Ethyl N, N-dimethyl phosphoramidocyanidate	77-81-6
O-Alkyl (H or ≤ C_{10}, including cycloalkyl) S-2-dialkyl (Me, Et, n-Pr, or i-Pr)-aminoethyl alkyl (Me, Et, n-Pr, or i-Pr)-phosphonothiolates and corresponding alkylated or protonated salts	
e.g. VX: O-ethyl S-2-diisopropylaminoethyl methyl phosphonothiolate	50782-69-9
Sulphur mustards: 2-Chlorethychloromethylsulfide	2625-76-5
Mustard gas: Bis(2-chloroethyl)sulfide	505-60-2
Bis(2-chloroethylthio)methane	63869-13-6
Sesquimustard: 1,2-Bis(2-chloroethylthio)ethane	3563-36-8
1,3-Bis(2-chloroethylthio)-n-propane	63905-10-2
1,4-Bis(2-chloroethylthio)-n-butane	142868-93-7
1,5-Bis(2-chloroethylthio)-n-pentane	142868-94-8
Bis(2-chloroethylthiomethyl)ether	63918-90-1
O-Mustard: Bis(2-chloroethylthioethyl)ether	63918-89-8
Lewisites	
Lewisite 1: 2-Chlorovinyldichloroarsine	541-25-3
Lewisite 2: Bis(2-chlorovinyl)chloroarsine	40334-69-8
Lewisite 3: Tris(2-chlorovinyl)arsine	40334-70-1
Nitrogen mustards	
HN1: Bis(2-chloroethyl)ethylamine	538-07-8
HN2: Bis(2-chloroethyl)methylamine	51-75-2
HN3: Tris(2-chloroethyl)amine	555-77-1
Saxitoxin	35523-89-8
Ricin	9009-86-3

Table D-1. CWC Schedule 1 Chemicals[1] (Continued)

Precursors	
Alkyl (Me, Et, n-Pr, or i-Pr) phosphonyldifluorides	
e.g. DF: Methylphosphonyldifluoride	676-99-3
O-Alkyl (H or $\leq C_{10}$, including cycloalkyl) O-2-dialkyl (Me, Et, n-Pr or i-Pr)-aminoethyl alkyl (Me, Et, n-Pr or i-Pr) phosphonites and corresponding alkylated or protonated salts	
e.g. QL: O-Ethyl O-2-diisopropylaminoethyl methylphosphonite	57856-11-8
Chlorosarin: O-Isopropyl methylphosphonochloridate	1445-76-7
Chlorosoman: O-Pinacolyl methylphosphonochloridate	7040-57-5

Table D-2. CWC Schedule 2 Chemicals[1]

Substance	CAS Registry Number
Toxic Chemicals	
Amiton: O, O-Diethyl S-[2-(diethylamino) ethyl] phosphorothiolate and corresponding alkylated or protonated salts	78-53-5
PFIB: 1,1,3,3,3-Pentafluoro-2-(trifluoromethyl)-1-propene	382-21-8
BZ: 3-Quinuclidinyl benzilate*	6581-06-2
Precursors	
Chemicals, except for those listed in Schedule 1, containing a phosphorous atom to which is bonded one methyl, ethyl, or propyl (normal or iso) group but not further carbon atoms	
e.g. Methylphosphonyl dichloride	676-97-1
e.g. Dimethyl methylphosphonate	756-79-6
Exemption: Fonofos: O-Ethyl S-phenyl ethylphosphonothiolothionate	944-22-9
N,N-Dialkyl (Me, Et, n-Pr or i-Pr) phosphoramidic dihalides	
Diakyl (Me, Et, n-Pr or i-Pr) N,N-dialkyl (Me, Et, n-Pr or i-Pr)-phosphoramidates	
Arsenic trichloride	7784-34-1
2,2-Diphenyl-2-hydroxyacetic acid	76-93-7
Quinuclidine-3-ol	1619-34-7
N,N-Dialkyl (Me, Et, n-Pr or i-Pr) aminoethyl-2-chlorides and corresponding protonated salts	
N,N-Dialkyl (Me, Et, n-Pr or i-Pr) aminoethane-2-ols and corresponding protonated salts	
Exemptions: N,N-Dimethylaminoethanol and corresponding protonated salts	108-01-0
N,N-Diethylaminoethanol and corresponding protonated salts	100-37-8
N,N-Dialkyl (Me, Et, n-Pr or i-Pr) aminoethane-2-thiols and corresponding protonated salts	
Thiodiglycol: Bis(2-hydroxyethyl)sulfide	111-48-8
Pinacolyl alcohol: 3,3-Dimethylbutane-2-ol	464-07-3
NOTES	
*Subject to special thresholds for declaration and verification, as specified in Part VII of the Verification Annex of the CWC.	

Table D-3. CWC Schedule 3 Chemicals[1]

Substance	CAS Registry Number
Toxic Chemicals	
Phosgene: Carbonyl dichloride	75-44-5
Cyanogen chloride	506-77-4
Hydrogen cyanide	74-90-8
Chloropicrin: Trichloronitromethane	76-06-2
Precursors	
Phosphorus oxychloride	10025-87-3
Phosphorus trichloride	7719-12-2
Phosphorus pentachloride	10026-13-8
Trimethyl phosphite	121-45-9
Triethyl phosphite	122-52-1
Dimethyl phosphite	868-85-9
Diethyl phosphite	762-04-9
Sulfur monochloride	10025-67-9
Sulfur dichloride	10545-99-0
Thionyl chloride	7719-09-7
Ethyldiethanolamine	139-87-7
Methyldiethanolamine	105-59-9
Triethanolamine	102-71-6

NOTES

[1]Office of the US President, *The Biological and Chemical Warfare Threat*, 1999.

Appendix E

CHEMICAL WARFARE AGENT PRECURSOR CHEMICALS: USES AND EQUIVALENTS

Table E-1. CW Agent Precursor Chemicals: Uses and Equivalents[1]

Precursor Chemical/CAS Registry Number	Civil Uses	CW Agent Production	Units of Agent per Unit of Precursor[a]
1. Thiodiglygol 111-48-8	Organic synthesis	Sulfur mustard (HD)	1.3
	Carrier for dyes in textile industry		
	Lubricant additives	Sesqui mustard (Q)	1.79
	Manufacturing plastics		
2. Phosphorus oxychloride 10025-87-3	Organic synthesis	Tabun (GA)	1.05
	Plasticizers		
	Gasoline additives		
	Hydraulic fluids		
	Insecticides		
	Dopant for semiconductors grade silicon		
	Flame retardants		
3. Dimethyl methylphosphonate (DMMP) 756-79-6	Flame retardant	Sarin (GB)	1.12
		Soman (GD)	
		Cyclosarin (GF)	1.45
4. Methylphosphonyl difluoride 676-99-3	Organic synthesis	Sarin (GB)	1.40
	Specific uses not identified	Soman (GD)	1.82
		Cyclosarin (GF)	1.80
5. Methylphosphonyl dichloride 676-97-1	Organic synthesis	Sarin (GB)	1.05
	Specific uses not identified	Soman (GD)	1.36
		Cyclosarin (GF)	1.35
6. Dimethyl phosphite 868-85-9	Organic synthesis	Sarin (GB)	1.27
	Lubricant additive	Soman (GD)	1.65
		Cyclosarin (GF)	1.65
7. Phosphorus trichloride 7719-12-2	Organic synthesis	Amiton (VG)	1.95
	Insecticides	Tabun (GA)	1.18
	Gasoline additives	Sarin (GB)	1.02
		Salt process	(0.34) [b]
	Plasticizers	Rearrangement process	1.02/(0.68) [b]
	Surfactants	Soman (GD)	1.32
		Salt process	(0.44) [b]
	Dyestuffs	Rearrangement process	1.32/(0.88) [b]
		Cyclosarin (GF)	1.31
		Salt process	(0.44) [b]
		Rearrangement process	1.31/(0.87) [b]

Table E-1. CW Agent Precursor Chemicals: Uses and Equivalents (Continued)

Precursor Chemical/CAS Registry Number	Civil Uses	CW Agent Production	Units of Agent per Unit of Precursor[a]
8. Trimethyl phosphite 121-45-9	Organic synthesis	Used to make dimethyl methylphosphonate (DMMP)—molecular rearrangement	See dimethyl methylphosphonate
9. Thionyl chloride[b] 7719-09-7 Could serve as chlorinating agent in all of these processes—other chlorinating agents could be substituted	Organic synthesis	Sarin (GB)	1.18
		Soman (GD)	1.53
		Cyclosarin (GF)	1.51
		Sulfur mustard (HD)	1.34
	Chlorinating agent	Sesqui mustard (Q)	1.84
	Catalyst	Nitrogen mustard (HN-1)	0.714
	Pesticides	Nitrogen mustard (HN-2)	0.655
	Engineering plastics	Nitrogen mustard (HN-3)	1.145
10. 3-Hydroxy-1-methylpiperidine 3554-74-3	Specific uses not identified; probably used in pharmaceutical industry	Nonidentified; could be used in the synthesis of psychoactive compounds such as BZ	
11. N,N-diisopropyl-(beta)-aminoethyl chloride 96-79-7	Organic Synthesis	VX	1.64
		VS	1.72
12. N,N-diisopropyl-aminoethanethiol 5842-07-9	Organic Synthesis	VX	1.66
		VS	1.75
13. 3-Quinuclidinol 1619-34-7	Hypotensive agent	BZ	2.65
	Probably used in synthesis of pharmaceuticals		
14. Potassium fluoride 7789-23-3	Fluorination of organic compounds	Sarin (GB)	2.41
	Cleaning and disinfecting brewery, dairy, and other food processing equipment	Soman (GD)	3.14
	Glass and porcelain manufacturing	Cyclosarin (GF)	3.10
15. 2-Chloroethanol 107-07-3	Organic synthesis	Sulfur mustard (HD)	0.99
	Manufacturing of ethylene oxide and ethylene-glycol	Sesqui mustard (Q) Nitrogen mustard (HN-1)	0.99 1.06
	Insecticides		
	Solvent		
16. Dimethylamine 124-40-3	Organic synthesis	Tabun (GA)	3.61
	Pharmaceuticals		
	Detergents		
	Pesticides		
	Gasoline additive		
	Missile fuels		
	Vulcanization of rubber		
17. Diethyl ethylphosphonate 78-38-6	Heavy metal extraction	Ethyl sarin (GE)	0.93
	Gasoline additive		
	Antifoam agent		
	Plasticizer		

Table E-1. CW Agent Precursor Chemicals: Uses and Equivalents (Continued)

Precursor Chemical/CAS Registry Number	Civil Uses	CW Agent Production	Units of Agent per Unit of Precursor[a]
18. Diethyl N,N-dimethyl phosphoramidate 2404-03-7	Organic synthesis	Tabun (GA)	0.90
	Specific uses not identified.		
19. Diethyl phosphite 762-04-9	Organic synthesis	Amiton (VG)	Catalyst
	Paint solvent	Sarin (GB)	1.02
	Lubricant additive	Soman (GD)	1.32
		Cyclosarin (GF)	1.30
20. Dimethylamine HCl 506-59-2	Organic synthesis	Tabun (GA)	1.99
	Pharmaceuticals		
	Surfactants		
	Pesticides		
	Gasoline additives		
21. Ethylphosphonous dichloride 1498-40-4	Organic synthesis	VE	1.93
	Specific uses not identified but could be used in manufacturing of flame retardants, gas additives, pesticides, surfactants, etc.	VS	2.14
		Ethyl sarin (GE)	1.18
22. Ethylphosphonyl dichloride 1066-50-8	Organic synthesis	Ethyl sarin (GE)	2.10
	Specific uses not identified; see ethylphosphonous dichloride		
23. Ethylphosphonyl difluoride 753-98-0	Organic synthesis	Ethyl sarin (GE)	2.70
	Specific uses not identified; see ethylphosphonous dichloride		
24. Hydrogen fluoride 7664-39-3	Fluorinating agent in chemical reactions	Sarin (GB)	7.0
	Catalyst in alkylation and polymerization reactions	Soman (GD)	9.11
	Additives to liquid rocket fuels	Ethyl sarin (GE)	7.7
	Uranium refining	Cyclosarin (GF)	9.01
25. Methyl benzilate 76-89-1	Organic synthesis	BZ	1.39
	Tranquilizers		
26. Methylphosphonous dichloride 676-83-5	Organic synthesis	VX	2.28
27. N,N-diisopropyl-(beta)-aminoethanol 96-80-0	Organic synthesis	VX	1.84
	Specific uses not identified		
28. Pinacolyl alcohol 464-07-3	Specific uses not identified	Soman (GD)	1.79
29. O-ethyl,2-diisopropyl aminoethyl methyl-phosphonate (QL), 57856-11-8	Specific uses not identified	VX	1.14
30. Triethyl phosphite 122-52-1	Organic synthesis	Amiton (VG)	1.62
	Plasticizers		
	Lubricant additives		

Table E-1. CW Agent Precursor Chemicals: Uses and Equivalents (Continued)

Precursor Chemical/CAS Registry Number	Civil Uses	CW Agent Production	Units of Agent per Unit of Precursor[a]
31. Arsenic trichloride 7784-34-1	Organic synthesis	Arsine (SA)	0.43
	Pharmaceuticals	Lewisite (L)	1.14
	Insecticides		
	Ceramics	Adamsite (DM)	1.53
		Diphenylchloroarsine (DA)	1.45
32. Benzilic acid 76-93-7	Organic synthesis	BZ	1.48
33. Diethyl methylphosphonite 15715-41-0	Organic synthesis	VX	1.97
34. Dimethyl ethylphosphonate 6163-75-3	Organic synthesis	Ethyl sarin (GE)	1.12
35. Ethylphosphonous difluoride 430-78-4	Organic synthesis	VE	2.58
		Ethyl sarin (GE)	1.57
36. Methylphosphonous difluoride 753-59-3	Organic synthesis	VX	3.18
		VM	2.84
		Sarin (GB)	1.67
		Soman (GD)	2.17
		Cyclosarin (GF)	2.15
37. 3-Quinuclidone 1619-34-7	Same as 3-Quinuclidinol3-quinuclidinol	BZ	2.65
38. Phosphorous pentachloride 10026-13-8	Organic synthesis	Tabun (GA)	0.78
	Pesticides		
	Plastics		
39. Pinacolone 75-97-8	Specific uses not identified	Soman (GD)	1.82
40. Potassium cyanide 151-50-8	Extraction of gold and silver from ores	Tabun (GA)	1.25
	Pesticide		
	Fumigant	Hydrogen cyanide (AC)	0.41
	Electroplating		
41. Potassium bifluoride 7789-29-9	Fluorine production	Sarin (GB)	1.79
	Catalyst in alkylation	Soman (GD)	2.33
	Treatment of coal to reduce slag formation	Cyclosarin (GF)	2.31
	Fluid in silver solder		
42. Ammonium bifluoride 1341-49-7	Ceramics	Sarin (GB)	2.46
	Disinfectant for food equipment	Soman (GD)	3.20
	Electroplating	Cyclosarin (GF)	3.16
	Etching glass		
43. Sodium fluoride 7681-49-4	Pesticide	Sarin (GB)	3.33
	Disinfectant	Soman (GD)	4.34
	Dental prophylaxis	Cyclosarin (GF)	4.29
	Glass and steel manufacturing		

Table E-1. CW Agent Precursor Chemicals: Uses and Equivalents (Continued)

Precursor Chemical/CAS Registry Number	Civil Uses	CW Agent Production	Units of Agent per Unit of Precursor[a]
44. Sodium bifluoride 1333-83-1	Antiseptic	Sarin (GB)	2.26
	Neutralizer in laundry operations	Soman (GD)	2.94
	Tin plate production	Cyclosarin (GF)	2.91
45. Sodium cyanide 143-33-9	Extraction from gold and silver ores	Tabun (GA)	1.65
	Fumigant	Hydrogen cyanide (AC)	0.55
	Manufacturing dyes and pigments	Cyanogen chloride (CK)	1.25
	Core hardening of metals		
	Nylon production		
46. Triethanolamine 102-71-6	Organic synthesis	Nitrogen mustard (HN-3)	1.37
	Detergents		
	Cosmetics		
	Corrosion inhibitor		
	Plasticizer		
	Rubber accelerator		
47. Phosphorus pentasulfide 1314-80-3	Organic synthesis	Amiton (VG)	1.21
	Insecticide	VX	1.20
	Mitocides		
	Lubricant oil additives		
	Pyrotechnics		
48. Diisopropylamine 108-18-9	Organic synthesis	VX	3.65
	Specific uses not identified		
49. Diethylaminoethanol 100-37-8	Organic synthesis	Amiton (VG)	2.30
	Anticorrosion compositions	VM	2.05
	Pharmaceuticals		
	Textile softeners		
50. Sodium sulfide 1313-82-2	Paper manufacturing	Sulfur mustard (HD)	2.04
	Rubber manufacturing		
	Metal refining		
	Dye manufacturing		
51. Sulfur monochloride sulfur chloride 10025-67-9	Organic synthesis	Sulfur mustard (HD)	1.18
	Pharmaceuticals		
	Sulfur dyes		
	Insecticides		
	Rubber vulcanization		
	Polymerization catalyst		
	Hardening of soft woods		
	Extraction of gold from ores		
52. Sulfur dichloride 10545-99-0	Organic synthesis	Sulfur mustard (HD)	1.54
	Rubber vulcanizing		
	Insecticides		
	Vulcanizing oils		
	Chlorinating agents		

Table E-1. CW Agent Precursor Chemicals: Uses and Equivalents (Continued)

Precursor Chemical/CAS Registry Number	Civil Uses	CW Agent Production	Units of Agent per Unit of Precursor[a]
53. Triethanolamine hydrochloride 637-39-8	Organic synthesis	Nitrogen mustard (HN)	1.10
	Insecticides		
	Surface active agents		
	Waxes and polishes		
	Textile specialties		
	Lubricants		
	Toiletries		
	Cement additive		
	Petroleum demulsifier		
	Synthetic resin		
54. N,N-diisopropyl-2-aminoethyl chloride hydrochloride 4261-68-1	Organic synthesis	VX	1.34

Notes:
[a]Assumes quantitative reaction yields
[b]Figures in parentheses are based on the use of PCl_3 as a chlorine donator in the reaction.

NOTES

[1]Office of the US President, *The Biological and Chemical Warfare Threat*, 1999.

Appendix F

CHEMICAL WARFARE AGENTS AND OTHER MILITARY CHEMICAL COMPOUNDS

Table F-1. Symbols for CW Agents and Military Chemical Compounds

Agent Class	Agent	Symbol
Choking Agents	Phosgene	CG
	Diphosgene	DP
	Chloropicrin	PS
Nerve Agents	Tabun	GA
	Sarin	GB
	Soman	GD
	Cyclosarin	GF
	Ethyl Sarin	GE
	VX	VX
	Vx	Vx
	VE	VE
	Amiton	VG
	VM	VM
	VS	VS
Blood Agents	Hydrogen Cyanide	AC
	Cyanogen Chloride	CK
	Arsine	SA
Blister Agents	Distilled Mustard	H/HD
	Nitrogen Mustard	HN-1, HN-2, HN-3
	Mustard-T Mixture	HT
	Sesqui Mustard	Q
	Lewisite	L
	Mustard-Lewisite Mixture	HL
	Phenyldichloroarsine	PD
	Ethyldichloroarsine	ED
	Methyldichloroarsine	MD
	Phosgene Oxime	CX
Incapacitating Agent	3-Quinuclidinyl benzilate	BZ
Riot Control Agents	O-Chlorobenzylidene Malononitrile	CS
	Dibenz(b,f)-1:4-oxazepine	CR
	Capsaicin	OC
Respiratory Irritants	Diphenylchloroarsine	DA
	Diphenylcyanoarsine	DC
	Adamsite	DM
	Chlorine	Cl_2

Appendix G

PROPERTIES OF CHEMICAL WARFARE AGENTS AND MILITARY CHEMICAL COMPOUNDS

Tables G-1 through G-6 (pages G-2 through G-18) and Figures G-1 through G-7 (pages G-19 through G-25) provide a ready source of data for key CW agents and military chemical compounds. The data were compiled from the detailed data presented for each agent described in Chapters II and III.

Table G-1. Physical Properties of Choking, Nerve, and Blood Agents

Agents	Chemical Agent; Formula; Symbol	Molecular Weight	State at 20°C	Odor	Boiling Point (°C)	FP/MP (°C)	Liquid Density (g/mL)	Vapor Density (Air=1)	Vapor Pressure (torr)
Choking Agents	**Phosgene; $COCl_2$; CG**	98.92	Colorless gas that is readily liquefied	Musty hay or rotting fruit	7.8	-128 (MP)	1.360 (25°C); 1.402 (7.8°C); 1.420 (0°C)	3.4	1400 (25°C); 760 (7.8°C); 560 (0°C)
	Diphosgene; $C_2Cl_4O_2$; DP	197.83	Colorless oily liquid	Musty hay	127	-57 (MP)	Munitions grade: 1.656 (20°C); 1.687 (0°C)	6.8	4.41 (20°C); 0.914 (0°C)
Nerve Agents	**Tabun; $C_5H_{11}N_2O_2P$; GA**	162.13	Colorless to brown liquid	Faintly fruity; none when pure	248	-50 (FP)	1.0756 (25°C); 1.0999 (0°C)	5.6	0.0570 (25°C); 0.00475 (0°C)
	Sarin; $C_4H_{10}FO_2P$; GB	140.09	Colorless liquid	None when pure	150	-56 (FP)	Pure: 1.0887 (25°C); 1.1182 (0°C)	4.8	248 (25°C); 0.410 (0°C)
	Soman; $C_7H_{16}FO_2P$; GD	182.17	Colorless liquid when pure	Fruity; camphor when impure	198	-42 (MP)	1.0222 (25°C); 1.0456 (0°C)	6.3	0.401 (25°C); 0.0496 (0°C)
	Cyclosarin; $C_7H_{14}FO_2P$; GF	180.16	Colorless liquid	None when pure	228	-30 to -50 (FP); -12 (MP)	1.1276 (25°C); 1.1525 (0°C)	6.2	0.0927 (25°C); 0.00978 (0°C)
	VX; $C_{11}H_{26}NO_2PS$	267.37	Colorless liquid when pure	Odorless when pure	292	Below -51 (FP); -39 to –60 (FP)	1.0083 (25°C); 1.0209 (0°C)	9.2	0.000878 (25°C); 0.0000422 (0°C)
	"V-Sub X"; $C_7H_{18}NO_2PS$; Vx	211.26	Liquid	Odorless	256	Data not available	1.060 (25°C); 1.0820 (0°C)	7.3	0.00673 (25°C); 0.00057 (0°C)
Blood Agents	**Hydrogen cyanide; HCN; AC**	27.03	Colorless liquid	Bitter almonds; peach kernels	25.5	-13.3 (MP)	0.6797 (25°C); 0.7162 (0°C)	0.93	760 (25.5°C); 746 (25.0°C); 265 (0°C)
	Cyanogen chloride; CNCl; CK	61.47	Colorless gas	Lacrimatory and irritating	12.8	-6.9 (FP)	1.202 (10°C); 1.222 (0°C)	2.1	760 (12.8°C); 680 (10°C); 48 (0°C)
	Arsine; AsH_3; SA	77.95	Colorless gas	Disagreeable, garlic-like	-62.2	-116 (MP)	1.667 (-75°C); 1.734 (-100°C)	2.7	400 (-75°C); 86.9 (-100°C)

Table G-1. Physical Properties of Choking, Nerve, and Blood Agents (Continued)

Agents		Volatility (mg/m^3)	Latent Heat of Vaporization (kcal/mol)	Viscosity (cP)	Viscosity of Vapor (cP)	Surface Tension (dynes/cm)	Flash Point
Choking Agents	CG	7,460,000 (25°C); 4,290,000 (7.8°C); 3,530,000 (0°C)	5.92 (25°C); 5.95 (7.8°C); 5.96 (0°C)	Data Not Available in Samuel, et al. reference	Data Not Available in Samuel, et al. reference	Data Not Available in Samuel, et al. reference	Nonflammable
	DP	47,700 (20°C); 10,600 (0°C)	12.2 (20°C); 12.8 (0°C)	Data Not Available in Samuel, et al. reference	Data Not Available in Samuel, et al. reference	Data Not Available in Samuel, et al. reference	None
Nerve Agents	GA	497 (25°C); 45.2 (0°C)	15.5 (25°C); 16.7 (0°C)	2.277 (25.0°C), 4.320 (0°C)	0.00620 (25.0°C), 0.00560 (0°C)	32.5 (25.0°C), 35.0 (0°C)	78°C
	GB	18,700 (25°C); 3,370 (0°C)	11.6 (25°C); 11.7 (0°C)	1.397 (25.0°C), 2.583 (0°C)	0.00719 (25.0°C), 0.00651 (0°C)	25.9 (25.0°C), 28.8 (0°C)	Nonflammable
	GD	3,930 (25°C); 531 (0°C)	13.2 (25°C); 13.8 (0°C)	3.167(25.0°C), 6.789 (0°C)	0.00590 (25.0°C), 0.00533 (0°C)	24.5 (26.5°C)	121°C
	GF	898 (25°C); 103 (0°C)	14.3 (25°C); 14.8 (0°C)	5.431 (25.0°C), 14.762 (0°C)	0.00615 (25.0°C), 0.00555 (0°C)	32.3 (25.5°C)	94°C
	VX	12.6 (25°C); 0.662 (0°C)	19.2 (25°C); 20.1 (0°C)	10.041 (25.0°C), 37.532 (0°C)	0.00513 (25.0°C), 0.00463 (0°C)	31.3 (25.0°C), 34.7 (0°C)	127°C
	Vx	76.4 (25°C); 7.02 (0°C)	16.0 (25°C); 16.1 (0°C)	5.628 (25.0°C), 15.335 (0°C)	0.00556 (25.0°C), 0.00502 (0°C)	31.4 (25.0°C), 33.7 (0°C)	Data not available
Blood Agents	AC	1,100,000 (25.5°C); 1,080,000 (25.0°C); 420,000 (0°C)	6.72 (25.5°C); 6.72 (25.0°C); 6.71 (0°C)	Data Not Available in Samuel, et al. reference	Data Not Available in Samuel, et al. reference	Data Not Available in Samuel, et al. reference	-18°C; can ignite when explosively disseminated
	CK	2,620,000 (12.8°C); 2,370,000 (10°C); 1,620,000 (0°C)	6.40 (12.8°C); 6.41 (10°C); 6.44 (0°C)	Data Not Available in Samuel, et al. reference	Data Not Available in Samuel, et al. reference	Data Not Available in Samuel, et al. reference	Nonflammable
	SA	2,550,000 (-75°C); 627,000 (-100°C)	4.17	Data Not Available in Samuel, et al. reference	Data Not Available in Samuel, et al. reference	Data Not Available in Samuel, et al. reference	Flammable, forms explosive mixtures with air

Table G-1. Physical Properties of Choking, Nerve, and Blood Agents (Continued)

Agents		Decomposition Temperature	Solubility	Rate of Hydrolysis	Hydrolysis Products
Choking Agents	CG	Complete @ 800°C	Limited in water; miscible with common organic solvents, petroleum, and lubricating oil	$t_{½}$ = 0.25 sec (13°C); does not react quickly with water vapor but does with liquid water	$HCl_{(aq)}$[1], CO_2
	DP	300 to 350°C	In water, 44.6 g DP/L (20°C); soluble in common organic solvents	Slow at ambient temperature and fairly rapid at 100°C	HCl, CO_2
Nerve Agents	GA	150°C (3 to 3¼ hrs)	In water, 7.2 g GA/100 g (20°C); 9.8 g GA/100 g (0°C); readily soluble in common organic solvents	$t_{1/2}$ = 8.5 hrs (20°C, pH 7); slow in water, fairly rapid with strong acids and alkalis with self-buffering (pH 4 to 5)	HCN and other products
	GB	150°C (2½ hrs)	Completely miscible with water and common organic solvents	At 20°C, $t_{1/2}$ = 27 min (pH 1); $t_{1/2}$ = 3½ hrs (pH 2); $t_{1/2}$ = 80 hrs (pH 7); $t_{1/2}$ = 5.4 min (pH 10); $t_{1/2}$ = 0.6 min (pH 11)	HF, IMPA, MFPA, MPA, IPA
	GD	> 150°C; stabilized: 130°C (200 hrs); unstabilized: 130°C (4 hrs)	In water, 2.1 g GD/100 g solution (20°C); 3.4 g GD/100 g (0°C); very soluble in organic solvents	0.003M[2] GD at 25°C, $t_{1/2}$ = 3 hrs (pH 2); $t_{1/2}$ = 45 hrs (pH 6.65); $t_{1/2}$ = 60 hrs (pH 10); < 5 min in 5% NaOH[3] solution	PMPA, HF
	GF	150°C (2 hrs)	In water, 3.7 g GF/100 g (20°C); 5.1 g GF/100 g (0°C)	0.003M GF (distilled water), $t_{1/2}$ = 42 hrs (25°C)	HF and another product
	VX	$t_{1/2}$ = 502 days (71°C); $t_{1/2}$ = 41 days (100°C); $t_{1/2}$ = 34.5 hrs (150°C); $t_{1/2}$ = 36 sec (295°C)	Water solubility = 5% (21.5°C); miscible with water below 9.4°C; soluble in common organic solvents	At 22°C; $t_{½}$ = 1.8 min (1.25M NaOH); $t_{½}$ = 10.8 min (0.25M NaOH); $t_{½}$ = 31 min (0.10M NaOH); $t_{1/2}$ = 60 hrs (pure water)	EMPA, EA 2192, and other products
	Vx	Data not available	Soluble in organic solvents; slightly soluble in water	Rate coefficient is 0.0078 hrs^{-1}	Ethanol and other toxic products
Blood Agents	AC	Stabilized: >65.5°C; forms explosive polymer on standing	Miscible with water and common organic solvents including alcohol and ether	Slow under acidic conditions; rapid with traces of base or basic salts	NH_3 and other products
	CK	≈ 149°C	Liquefied CK in water, 71.4 g/L (20°C); soluble in common organic solvents, sulfur mustard, and HCN	With tap water, $t_{½}$ = 180 hrs (ambient temperature, pH 7)	HCl, CNOH
	SA	300°C	In water, 0.028 g/100 g (20°C); soluble in alkalis, halogen alkanes, hydrocarbons, and benzene	Rapid in light; slow in absence of light and air (15.5°C, pH 7); 32% hydrolyzed in 5 hrs, 66% in 24 hrs	Arsenic (toxic)

Table G-1. Physical Properties of Choking, Nerve, and Blood Agents (Continued)

Agents		Stability In Storage	Action on Metals	Skin and Eye Toxicity	Inhalation Toxicity
Choking Agents	CG	If dry, stable in steel for 1 year at ambient temperatures; stability decreases as temperature increases	None when dry; acidic and corrosive when moist	Initial effects resemble those of tear gas	Causes pulmonary edema
	DP	Unstable; converts to CG	Metals act as catalyzers in conversion to CG; attacks rubber, cork, and cement	Lacrimator	Causes pulmonary edema
Nerve Agents	GA	Stabilized GA: stable in steel for several years (ambient temperatures); stability decreases as temperature increases, 6 months (50°C); 3 months (65°C)	Corrosion rate of steel on crude GA (5-20% chlorobenene) is 0.000034 in/month (65°C)	Very high toxicity; greater through eyes than through skin; liquid penetrates skin	Primarily inhalation hazard
	GB	Stabilized GB: Stable in steel for 5-10 years (ambient temperatures); At elevated temperatures up to 71°C storage life decreases slightly	At 71°C; slightly corrosive on steel, brass, copper, inconel, K-monel, and lead; slight to severe corrosion on aluminum	Very high toxicity; greater through eyes than through skin; liquid penetrates skin	Most toxic route of exposure
	GD	Relatively stable in glass 5½ months (ambient temperature); stabilized GD: 6 months (71°C) in glass, steel, and aluminum containers	Corrosion rate on steel is 0.00001 in/month (65°C)	Very high toxicity; greater through eyes than through skin; liquid penetrates skin	Most toxic route of exposure
	GF	Stabilized GF (71°C): 6 months in glass, and 1 year in steel and aluminum containers	Corrosion rate on steel is 0.000053 in/month (65°C)	Very high toxicity; greater through eyes than through skin; liquid penetrates skin	Most toxic route of exposure
	VX	Stable at ambient temperature; unstabilized VX (95% purity) decomposes 5% a month (71°C); purified VX is stable in glass and steel.	Negligible on brass, steel, and aluminum, slight corrosion with copper	Extremely toxic by skin and eye absorption	Extremely potent
	Vx	Data not available	Data not available	Extremely toxic by skin and eye absorption	Extremely potent
Blood Agents	AC	Pure: unstable on storage; stabilized AC: may be stored in metal containers for long periods (65°C)	Corrodes iron, cast iron, chromium steel, and lead	None	Can cause death within minutes
	CK	Stable in glass for long periods and elevated temperatures; stable in steel for 1 year at ambient temperatures, only 9 weeks (60°C); stabilized CK: long periods (up to 100°C)	None if dry; unstabilized CK slowly polymerizes in steel and other common metals (elevated temperatures)	Irritation to eyes similar to RCAs	Can cause death within minutes
	SA	Unstable in most metal containers; metals catalyze decomposition on exposure to light, moist arsine decomposes quickly	Corrosive to most metals	Exposure to liquid causes frostbite	Acute toxicity is high

Table G-1. Physical Properties of Choking, Nerve, and Blood Agents (Continued)

Agents		Rate of Action	Means of Detection	Protection Required	Decontamination	Use
Choking Agents	CG	Immediate to 3 hours depending on concentration	M18A2, MM1	Protective mask	Not required in field except in very cold climates	Delayed-action casualty agent
	DP	Immediate to 3 hours	MM1	Protective mask	Not required in field except in very cold climates	Quick-acting casualty agent
Nerve Agents	GA	Rapid	M256A1, M8A1, M18A2, M21, M9, M22, CAPDS, CAM/ICAM, M272, IPDS, MM1, M90, AN/KAS-1, M8, M18A3	MOPP4	SDK, IEDK, STB	Quick-acting casualty agent
	GB	Rapid	M256A1, M8A1, M18A2, M21, M9, M22, CAPDS, CAM/ICAM, M272, IPDS, MM1, M90, AN/KAS-1, M8, M18A3	MOPP4	SDK, IEDK, STB	Quick-acting casualty agent
	GD	Rapid	M256A1, M8A1, M18A2, M21, M9, M22, CAPDS, CAM/ICAM, M272, IPDS, MM1, M90, AN/KAS-1, M8, M18A3	MOPP4	SDK, IEDK, STB	Quick-acting casualty agent
	GF	Rapid	M256A1, M8A1, M18A2, M21, M9, M22, CAPDS, CAM/ICAM, M272, IPDS, MM1, M90, AN/KAS-1, M8, M18A3	MOPP4	SDK, IEDK, STB	Quick-acting casualty agent
	VX	Rapid	M256A1, M8A1, M18A2, M21, M9, M22, CAPDS, CAM/ICAM, M272, IPDS, MM1, M90, AN/KAS-1, M8, M18A3	MOPP4	SDK, IEDK, STB, HTH, household bleach	Quick-acting casualty agent
	Vx	Rapid	M8, M9, M18A2, MM1, IPDS, M8A1, CAM/ICAM	MOPP4	SDK, IEDK, STB, HTH, household bleach	Quick-acting casualty agent
Blood Agents	AC	Rapid	M256A1, M272, M18A2, M18A3, MM1	Protective mask	Not required under field conditions	Quick-acting casualty agent
	CK	Rapid	M256A1, M272, M18A2, M18A3, MM1	Protective mask	Not required under field conditions	Quick-acting casualty agent
	SA	1-24 hours (dependent on concentration and exposure duration)	MM1	Protective mask	Not required under field conditions	Delayed-action casualty agent

NOTES:
[1]**Hydrochloric acid, $HCl_{(aq)}$**
[2]**M = molarity; molarity is a term for concentration and means moles per liter.**

Table G-2. Physical Properties of Blister Agents and Incapacitating Agents

Agents	Chemical Agent; Formula; Symbol	Molecular Weight	State at 20°C	Odor	Boiling Point (°C)	FP/MP (°C)	Liquid Density (g/mL)	Vapor Density (Air=1)	Vapor Pressure (torr)
Blister Agents	**Distilled mustard; $C_4H_8Cl_2S$; HD**	159.07	Pale yellow to dark brown oily liquid; colorless when pure	Garlic-like or horseradish	218	14.45 (FP)	1.2685 (25°C); Solid Density: 1.372 (0°C), 1.333 (10°C)	5.5	0.106 (25°C)
	Nitrogen mustard; $C_6H_{13}Cl_2N$; HN-1	170.08	Dark oily liquid; colorless when pure	Faint, fishy or soapy	192	-34.2 (MP)	1.086 (25°C); 1.110 (0°C)	5.9	0.244 (25°C); .0332 (0°C)
	Nitrogen mustard; $C_5H_{11}Cl_2N$; HN-2	156.05	Colorless liquid when pure	Fishy or soapy	177	-70 (FP)	1.118 (25°C); 1.1425 (0°C)	5.4	0.416 (25°C); 0.0570 (°C)
	Nitrogen mustard; $C_6H_{12}Cl_3N$; HN-3	204.53	Oily dark liquid, colorless when pure	Geranium-like; none if pure	257	-3.74 (MP)	1.2352 (25°C); 1.2596 (0°C)	7.1	0.011 (25°C) 0.00092 (0°C)
	Mustard-T Mixture; 60% HD, 40% T; HT	188.96 (average)	Pale yellow to brown liquid	Garlic-like; less pronounced than mustard	None Constant	1.3 (MP)	1.263 (20°C)	6.5	0.077 (25°C)
	Lewisite; $C_2H_2AsCl_3$; L	207.32	Brown liquid; colorless when pure	Geranium-like; none if pure	196	-44.7 to -1.8 (FP)	1.8793 (25°C); 1.9210 (0°C)	7.1	34.6 (25°C); 0.0271 (0°C)
	Mustard-lewisite; 37% HD, 63% L; HL	186.39 (average)	Liquid	Garlic-like (HD)	200	Pure: -25.4 (FP) Munitions: -42 (FP)	1.6383 (20°C)	6.4	0.363 (25°C); 0.0493 (0°C)
	Phenyldichloroarsine $C_6H_5AsCl_2$; PD	222.93	Colorless to yellow liquid	None	233	-22.5 (FP)	1.645 (25°C); 1.677 (0°C)	7.7	0.022 (25°C); 0.0021 (0°C)
	Ethyldichloroarsine; $C_2H_5AsCl_2$; ED	174.89	Colorless liquid	Fruity, biting, and irritating	156	< -65 (MP)	1.742 (14°C)	6.0	2.29 (21.5°C)
	Methyldichloroarsine CH_3AsCl_2; MD	160.86	Colorless liquid	Extremely irritating; none when pure	132.6	-54.8 (FP)	1.839 (20°C); 1.875 (0°C)	5.5	7.593 (20°C); 2.063 (0°C)
	Phosgene Oxime $CHCl_2NO$; CX	113.93	Colorless, crystalline, deliquescent solid when pure	Unpleasant and irritating; low concentrations resembles new-mown hay	129;	39 (MP)	Data not available	3.9	24.3 (50°C)
Incapacitating Agents	**BZ; $C_{12}H_{23}NO_3$**	337.42	White crystalline solid	None	412	167.5 (MP)	Solid Density: (g/cm^3) Bulk: 0.51 Crystal: 1.33	11.6	1.4 x 10^{-10} (25°C); 4.74 x 10^{-13} (0°C)

Table G-2. Physical Properties of Blister Agents and Incapacitating Agents (Continued)

Agents		Volatility (mg/m^3)	Latent Heat of Vaporization (kcal/mol)	Viscosity (cP)	Viscosity of Vapor (cP)	Surface Tension (dynes/cm)	Flash Point
Blister Agents	HD	906 (25°C)	15.0 (25°C)	3.951 (25.0°C), 7.746 (0°C)	0.00597 (25.0°C), 0.00600 (0°C)	42.5 (25.0°C), 45.9 (0°C)	105°C
	HN-1	2,230 (25°C); 331 (0°C)	13.0 (25°C); 12.9 (0°C)	Data Not Available in Samuel, et al. reference	Data Not Available in Samuel, et al. reference	Data Not Available in Samuel, et al. reference	Data not available; flashing on static detonation
	HN-2	3,490 (25°C); 522 (0°C)	12.9 (25°C); 12.8 (0°C)	Data Not Available in Samuel, et al. reference	Data Not Available in Samuel, et al. reference	Data Not Available in Samuel, et al. reference	Data not available
	HN-3	120 (25°C); 11 (0°C)	15.8 (25°C); 16.0 (0°C)	0.073 (25.0°C), 0.177 (0°C)	0.00597 (25.0°C), 0.00538 (0°C)	40.9 (25.0°C), 44.1 (0°C)	Data not available
	HT	783 (25°C)	Data not available	Data Not Available in Samuel, et al. reference	Data Not Available in Samuel, et al. reference	Data Not Available in Samuel, et al. reference	109 to 115°C
	L	3,860 (25C); 330 (0°C)	15.5 (25°C) 17.5 (°C)	2.053 (25.0°C), 3.521 (0°C)	0.00853 (25.0°C), 0.00770 (0°C)	41.1 (25.0°C), 44.2 (0°C)	Nonflammable
	HL	3,640 (25°C); 539 (0°C)	12.8 (25°C); 13.1 (0°C)	Data Not Available in Samuel, et al. reference	Data Not Available in Samuel, et al. reference	Data Not Available in Samuel, et al. reference	Data not available; See HD
	PD	264 (25°C); 23 (0°C)	15.1	Data Not Available in Samuel, et al. reference	Data Not Available in Samuel, et al. reference	Data Not Available in Samuel, et al. reference	Data not available
	ED	21,900 (20°C)	9.18	Data Not Available in Samuel, et al. reference	Data Not Available in Samuel, et al. reference	Data Not Available in Samuel, et al. reference	Data not available
	MD	66,800 (20°C); 19,500 (0°C)	10.5 (20°C); 10.2 (0°C)	Data Not Available in Samuel, et al. reference	Data Not Available in Samuel, et al. reference	Data Not Available in Samuel, et al. reference	Data not available
	CX	137,000 (50°C)	11.2 (50°C)	Data Not Available in Samuel, et al. reference	Data Not Available in Samuel, et al. reference	Data Not Available in Samuel, et al. reference	Data not available
Incapacitating Agents	BZ	0.00000260 (25°) 0.0000000090 (0°C)	21.2	Data Not Available in Samuel, et al. reference	Data Not Available in Samuel, et al. reference	Data Not Available in Samuel, et al. reference	Munitions: 220°C Pure: 246°C

Table G-2. Physical Properties of Blister Agents and Incapacitating Agents (Continued)

Agents		Decomposition Temperature	Solubility	Rate of Hydrolysis	Hydrolysis Products
Blister Agents	HD	180°C	In distilled water, 0.92 g HD/100 g (22°C); freely soluble in fats and oils, gasoline, kerosene, most organic solvents and CW agents	$t_{1/2}$ = 5 min (25°C); $t_{1/2}$ = 60 min (25°C) (salt water); on or under water only if dissolved.	HCl, thiodiglycol, sulfonium ion
	HN-1	For HN-1 · HCl: 12.7% at 149°C and >99% at 426°C	In water, 4g HN-1/L (ambient temperature); miscible with common organic solvents	$t_{1/2}$ = 1.3 min (25°C) in aqueous solution	$HCl_{(aq)}$, EtDEA, and other products
	HN-2	Below boiling point; polymerizes with heat (can cause explosion)	In water, 13 g HN-2/L (ambient temperature, aqueous solution); miscible with common organic solvents	$t_{1/2}$ = 4 min (25°C); slow except where alkali is present; dimerizes fairly rapidly in water	HCl and other products
	HN-3	>150°C; remains stable when explosively disseminated	In water, 0.08 g HN-3/L (ambient temperature); miscible with common organic solvents	Very slow, not complete after several days unless alkali is present	HCl, TEA, and other products
	HT	165 to 180°C	Slightly soluble in water; soluble in most organic solvents	By prolonged boiling with water or treatment with caustic alkalis	$HCl_{(aq)}$, thiodiglycol, sulfonium
	L	5% at 149°C > 99.99% at 493°C	Immediately with water to form Lewisite oxide, which dissolves very slowly; soluble in common organic solvents, oils and CW agents	Rapid	CVAA, lewisite oxide, $HCl_{(aq)}$
	HL	> 100°C	Probably soluble in most organic solvents, slightly soluble in water	$t_{1/2}$ = 5 min (25°C); see HD and L	See HD and L
	PD	Stable to normal boiling point	Immediately with water; miscible with alcohol, benzene, ether, acetone, kerosene, petroleum, and olive oil	Very rapid	$HCl_{(aq)}$, phenylarsine oxide (toxic)
	ED	Stable to boiling point	Immediately with water; soluble in ethyl chloride, alcohol, ether, benzene, acetone, kerosene, and cyclohexane	Very rapid	$HCl_{(aq)}$, ethylarsine oxide (toxic)
	MD	Stable to boiling point	Immediately with water; soluble in common organic solvents (ambient temperatures)	Very rapid; < 2 min (25°C) in dilute solution	HCl, methylarsenic oxide
	CX	< 129°C	Very soluble in both water and common organic solvents	Slow in water; 5% within 6 days (ambient temperature); acids slow down hydrolysis; basic solutions react violently	CO_2, HCl, hydroxylmine
Incapacitating Agents	BZ	Stable to melting point; prolonged heating at ≈170°C (purity and temperature dependent)	In water, 1.18 g/mL; slightly soluble in water; soluble in dilute acids and common organic solvents such as alcohol and chloroform; insoluble in aqueous alkali	$t_{1/2}$ = 6.7 hrs (25°C, pH 9.8); $t_{1/2}$ = 1.8 min (25°C, pH 13); $t_{1/2}$ = 3-4 wks (25°C, pH 7 and moist air); $t_{1/2}$ = 12 min (37°C, pH 12); $t_{1/2}$ = 1.4 hrs (50°C, pH 8.5); $t_{1/2}$ = 9.5 hrs (100°C, pH 0);	3-Quinuclidinol and benzylic acid

Table G-2. Physical Properties of Blister Agents and Incapacitating Agents (Continued)

Agents		Stability In Storage	Action on Metals	Skin and Eye Toxicity	Inhalation Toxicity
Blister Agents	**HD**	Stable in steel for over 50 years (slight degradation)	Very little when pure; for munitions grade, the corrosion rate on steel is 0.0001 in/month (65°C)	Eyes very susceptible to low concentrations; skin absorption requires higher concentrations	Most toxic route of exposure
	HN-1	Polymerizes in steel; slight at ambient temperatures but increases at temperatures > 50°C	Corrosion on steel is $1x10^{-5}$ to $5x10^{-5}$ in/month (65°C)	Eyes very susceptible to low concentrations; skin absorption requires higher concentrations	Most toxic route of exposure
	HN-2	Unstable	None on steel and brass	Eyes very susceptible to low concentrations; skin absorption requires higher concentrations	Most toxic route of exposure
	HN-3	Relatively stable in steel, if dry	If dry, no attack on iron; corrodes steel at a rate of $1x10^{-5}$ to $5x10^{-5}$ in/month (65°C)	Eyes very susceptible to low concentrations; skin absorption requires higher concentrations	Most toxic route of exposure
	HT	Pressure develops in steel	Very little when pure; Canadian HT corrodes steel at a rate of 0.00007 in/month (65°C)	Eyes very susceptible to low concentrations; skin absorption requires higher concentrations	Most toxic route of exposure
	L	Fairly stable in steel and glass; decomposes considerably upon detonation; alkalis decompose L at ambient temperatures	None if dry; corrosive penetration on steel is $1x10^{-5}$ to $5x10^{-5}$ in/month (65°C); extremely corrosive towards aluminum and aluminum alloys	Extremely irritating to the eyes and produces copious tearing; immediate burning sensation on skin	Most toxic route of exposure
	HL	Stable in lacquered steel for 3 months (65°C); 6 months (50°C); >1 yr (ambient temperatures); uncoated: less stable (>50°C); stable in glass @ 65°C	If dry, little or none	Equal to L in vesication action; both H and L are irritating to the eyes.	Most toxic route of exposure
	PD	Data not available	No serious affects on mild steel and cast iron	Fairly potent vesicant	Toxic lung irritant
	ED	Stable	None on steel when pure; noncorrosive toward iron (<50°C) when ED is dry; attacks brass (50°C) and is destructive to rubber and plastics.	Fairly potent vesicant and lacrimator	Primarily a lung irritant
	MD	Stable in steel for 15 weeks (60°C); 1 year (ambient temperatures)	Satisfactory with steel for 1 yr (ambient temperature); crude pitting within 15 weeks (60°C)	Blistering action less than that of HD and L; eye and skin irritant	Toxic lung injurant and respiratory irritant
	CX	Pure unstabilized CX decomposes at ambient temperatures; indefinitely (-20°C); unstable in presence of impurities such as metals	Metals especially iron causes rapid decomposition of CX; iron chloride may cause explosive decomposition; attacks rubber, especially upon heating	Causes pain, irritation and severe tissue damage on skin. Causes pain, conjunctivitis and inflammation of the cornea of the eye.	Can cause pulmonary edema
Incapacitating Agents	BZ	Stable in aluminum and stainless steel for at least 2 years (71°C)	Slight pitting of aluminum and stainless steel occurs after 2 yrs (71°C)	Can cause blurred vision and dilation of pupils	Primary route of exposure

Table G-2. Physical Properties of Blister Agents and Incapacitating Agents (Continued)

Agents		Rate of Action	Means of Detection	Protection	Decontamination	Use
Blister Agents	HD	Delayed – hrs to days	M8, M9, M256A1, M90, M21, M22, CAM/ICAM, M272, M18A2, M18A3, MM1	MOPP4	SDK, IEDK, HTH, household bleach,	Delayed-action casualty agent
	HN-1	Delayed – 12 hrs or longer	M8, M9, M256A1, CAM/ICAM, MM1	MOPP4	HTH, household bleach	Delayed-action casualty agent
	HN-2	Delayed - 12 hrs or longer	M8, M9, M256A1, CAM/ICAM, MM1	MOPP4	HTH, household bleach	Delayed-action casualty agent
	HN-3	Delayed - 12 hrs or longer	M8, M9, M256A1, CAM/ICAM, MM1	MOPP4	HTH, household bleach	Delayed-action casualty agent
	HT	No data available	M8, M9, M256A1, M18A2, M18A3, CAM/ICAM, MM1	MOPP4	HTH, household bleach	Delayed-action casualty agent
	L	Rapid	M9, M256A1, M21, M22, M272, MM1	MOPP4	STB, HTH, household bleach	Quick-acting casualty agent
	HL	Prompt stinging; blistering delayed about 12 hrs	CAM/ICAM, M256A1, M18A3, MM1	MOPP4	HTH, household bleach	Delayed-action casualty agent
	PD	Immediate eye effect; skin effects delayed @ 1 hr	M18A2, M18A3, MM1	MOPP4	Household bleach	Delayed-action casualty agent
	ED	Immediate irritation; delayed blistering	M18A2, M18A3, MM1	MOPP4	Household bleach	Delayed-action casualty agent
	MD	Immediate irritation; delayed blistering	M18A1, M18A3, MM1	MOPP4	Household bleach	Delayed-action casualty agent
	CX	Almost instantaneous	M256A1, M18A2, M18A3, MM1	MOPP4	SDK or large amounts of water; household bleach	Quick-acting casualty agent
Incapacitating Agents	BZ	Delayed action – 1 to 4 hrs, depending on exposure	MM1	MOPP4	Soap and water; SDK if soap and water are not available	Delayed-action incapacitating agent

Table G-3. Physical Properties of RCAs and Respiratory Irritants

Agents	Chemical Agent; Formula; Symbol	Molecular Weight	State at 20°C	Odor	Boiling Point (°C)	FP/MP (°C)	Liquid Density (g/mL)	Vapor Density (Air=1)
Riot Control Agents	O-Chlorobenylidene malononitrile; $C_{10}H_5ClN_2$; CS	188.6	White crystalline solid	Pepper-like	310 to 315	95 to 96 (MP)	Solid Density Bulk: 0.24-0.26; Crystal: 1.04	6.5
	Dibenz-(b,f)-1,4-oxazepine; $C_{13}H_9NO$; CR	195.22	Pale yellow crystalline solid	Pepper-like	Data not available	73 (MP)	Data not available	6.7
	Capsaicin $C_{18}H_{27}N0_3$ OC	305.42	Colorless	Pungent, irritating	340.4	65 (FP)	Data not available	10.5
Respiratory Irritants	Adamsite; $C_{12}H_9AsClN$; DM	277.58	Light yellow to green crystals	None; irritating	410	195 (MP);	Solid Density 1.648 (20°C); 1.672 (0°C)	9.6
	Diphenylchloroarsine $C_{12}H_{10}AsCl$ DA	264.59	Colorless crystalline solid when pure	None	383	37.3; 39 to 44 (MP)	1.3875 (50°C)	9.1
	Diphenylcyanoarsine $C_{13}H_{10}AsN$; DC	255.15	Colorless crystalline solid	Garlic and bitter almonds	341	31.2 (FP)	1.3338 (35°C)	8.8
	Chlorine; Cl_2	70.91	Greenish-yellow diatomic gas	Disagreeable and suffocating	-34.7	-101.6 (FP)	Liquefied: 1.393 (25°C); 1.468 (0°C)	2.4

Table G-3. Physical Properties of RCAs and Respiratory Irritants (Continued)

Agents		Vapor Pressure (torr)	Volatility (mg/m^3)	Latent Heat of Vaporization (kcal/mol)	Flash Point
Riot Control Agents	**CS**	0.00034 (20°C)	0.71 (25°C)	Data not available	Data not available
	CR	Data not available	Data not available	Data not available	Data not available
	OC	0.00000015 (65°C)	0.0022 (65°C)	33.4	Depends on delivery system
Respiratory Irritants	**DM**	Negligible @ ambient temperatures	Negligible @ ambient temperatures	14.2 (410°C)	Nonflammable
	DA	0.0179 (50°C)	236 (50°C)	15.1	350°C
	DC	0.00072 (35°C)	9.56 (35°C)	17.1	Low
	Cl_2	5,750 (25°C); 2,730 (0°C)	21,900,000 (25°C); 11,400,000 (0°C)	4.86 (25°C); 4.80 (0°C)	Nonflammable

Table G-3. Physical Properties of RCAs and Respiratory Irritants (Continued)

Agents		Decomposition Temperature	Solubility	Rate of Hydrolysis	Hydrolysis Products
Riot Control Agents	CS	Data not available	Insoluble in water, moderate in alcohol, and good in acetone, chloroform, methylene dichloride, ethylacetate and benzene	Data not available	Data not available
	CR	Data not available	Sparingly soluble in water	Not hydrolyzed in aqueous solutions	Data not available
	OC	> 150	in water, 0.090 g/L @ 37°C; soluble in alcohol, ether, chloroform, carbon disulfide, conc. HCl, aromatic solvents, hydrocarbons, ketones and aqueous alkali.	Data not available	Alkaline hydrolysis yields vanilylamine and isomeric decenoic acid
Respiratory Irritants	DM	Slight at 195°C; 0.02% per min @ 200°C; 0.15% per min @ 250°C	In water, 0.044 g/L @ 37°C; slightly soluble in benzene exylene, carbon tetrachloride, acetone, alcohols, tetrachoethane	Solid DM - slowly when covered with water; finely divided – rapidly; acidic solutions prevent hydrolysis	Diphenylaminearsenious oxide, $HCl_{(aq)}$
	DA	300 to 350	In water, 0.078 g/L @ 37°C; soluble in acetone, ether, ethanol, benzene, carbon tetrachloride, ethylene chloride, chloroform, and dichloroethylene	Slow in bulk but rapid when finely divided	Diphenylarsenious oxide and HCl
	DC	> 240	In water, 0.021 g/L @ 37°C; soluble in chloroform and other organic solvents	Very slow	HCN and Diphenylarsenious oxide
	Cl_2	> 600°C	In water, 0.63 g/100 g water @25°C, solubility in carbon tetrachloride is 3.5% @ ambient temperatures	Slow	HCl and HOCl

Table G-3. Physical Properties of RCAs and Respiratory Irritants (Continued)

Agents		Stability In Storage	Action on Metals	Skin and Eye Toxicity	Inhalation Toxicity
Riot Control Agents	CS	Combustible material; may burn but does not ignite readily. Containers may explode when heater; incompatible with strong oxiders.	Contact with metals may evolve flammable hydrogen gas.	Burning and irritation to eyes; primary skin irritant	Causes sensation of choking
	CR	Stable in organic solutions	Data not available	Irritant; however, CR does not induce inflammatory cell infiltration, vesication, or contact sensitization	Causes almost no effects in the lower airways and lungs
	OC	Data not available	Data not available	Powerful irritant and lacrimator	Causes bronchoconstriction and edema
Respiratory Irritants	DM	f pure, stable for at least 1 year @ room temperature and in aluminum and stainless steel 2 yrs @ 71°C; munitions grade, stable for 6 months	After 3 months causes extensive corrosion on aluminum and stainless steel @ 71°C; corrodes iron, bronze, and brass	Causes eye irritation and burning	Primary action on upper respiratory tract
	DA	Stable when pure; stable in steel shells for almost 4 months at 60°C and for 1 year at room temperature	None when dry	Irritant	Irritant
	DC	Stable at all ordinary temperatures	None on metals	Irritant	Irritant
	Cl_2	Stable when dry	None if dry; vigorous action with metals if moist	Irritant	Can cause pulmonary edema

Table G-3. Physical Properties of RCAs and Respiratory Irritants (Continued)

Agents		Rate of Action	Protection Required	Decontamination	Use
Riot Control Agents	CS	Instantaneous	Protective mask and dry field clothing; personnel handling/loading bulk CS should wear protective clothing, masks, and gloves	Move to fresh air, do not use any form of bleach	Training and RCA
	CR	Instantaneous	Protective mask and clothing secured at neck, wrists, and ankles	Move to fresh air; do not use any form of bleach	RCA
	OC	Almost immediate	Protective mask and field clothing secured at neck, wrists, and ankles	Move to fresh air; do not use any form of bleach	Incapacitating violent or threatening subjects, SF stability and support operations
Respiratory Irritants	DM	Rapid	Protective mask	Move to fresh air	Former RCA and mask breaker
	DA	Rapid	Protective mask	Move to fresh air	Previously used as mask breaker
	DC	Rapid	Protective mask	Move to fresh air	Previously used as mask breaker
	Cl_2	Delayed (pulmonary edema)	Protective mask	Move to fresh air	Not authorized for military use

Table G-4. Toxicity Estimates for CW Agents

ROE		Liquid (mg/70 kg man)		Inhalation/Ocular (mg-min/m^3)			Inhalation (mg/m^3)	Ocular (mg-min/m^3)		Percutaneous Vapor (mg-min/m^3)					
										Lethal (LCt_{50})		Severe (ECt_{50})		Threshold (ECt_{50})	
Endpoints		Lethal (LD_{50})	Severe (ED_{50})	Lethal (LCt_{50})	Severe (ECt_{50})	Mild (ECt_{50})	Odor Detection (EC_{50})	Severe (ECt_{50})	Mild (ECt_{50})	Moderate	Hot	Moderate	Hot	Moderate	Hot
Choking Agents	CG	-	-	1,500 (2-60)	-	-	6 S	-	-	-	-	-	-	-	-
	DP	-	-	1,500P (10-60)	-	-	4 S	-	-	-	-	-	-	-	-
Nerve Agents	GA	1,500	900	70 (2)	50 (2)	0.4 (2)	-	-	-	15,000 (30-360)	7,500P (30-360)	12,000 (30-360)	6,000P (30-360)	2,000 (30-360)	1,000P (30-360)
	GB	1,700	1,000	35 (2)	25 (2)	0.4 (2)	-	-	-	12,000 (30-360)	6,000P (30-360)	8,000 (30-360)	4,000P (30-360)	1,200 (30-360)	600P (30-360)
	GD	350	200	35 (2)	25 (2)	0.2 (2)	-	-	-	3,000 (30-360)	1,500P (30-360)	2,000 (30-360)	1,000P (30-360)	300 (30-360)	150P (30-360)
	GF	350	200	35 (2)	25 (2)	0.2 (2)	-	-	-	3,000 (30-360)	1,500P (30-360)	2,000 (30-360)	1,000P (30-360)	300 (30-360)	150P (30-360)
	VX	5	2	15 (2-360)	10 (2-360)	0.1 (2-360)	-	-	-	150 (30-360)	75P (30-360)	25 (30-360)	12P (30-360)	10 (30-360)	5 (30-360)
	Vx	NR													
Blood Agent	AC	-	-	2860P (2)	NR		34 S	-	-	-	-	-	-	-	-
	CK	-	-	NR	NR		12 S	-	-	-	-	-	-	-	-
	SA	-	-	7,500P (2)	-	-	-	-	-	-	-	-	-	-	-
Blister Agents	HD	1400	600	1,000 (2)	-	-	0.6 – 1 S	75 (2-360)	25 (2-360)	10,000 (30-360)	5,000P (30-360)	500 (30-360)	200 (30-360)	50 (30)	25 (30)
	HN-1	1400P	600P	1,000P (2)	-	-	-	75P (2)	25P (2)	10,000P (30)	5,000P (30)	500P (30)	200P (30)	50P (30)	25P (30)
	HN-2	1400P	600P	1,000P (2)	-	-	-	75P (2)	25P (2)	10,000P (30)	5,000P (30)	500P (30-360)	200P (30-360)	50P (30-360)	25P (30-360)
	HN-3	1400P	600P	1,000P (2-360)	-	-	-	75P (2-360)	25P (2-360)	10,000P (30-360)	5,000P (30-360)	500P (30-360)	200P (30-360)	50P (30-360)	25P (30-360)
	HT	1400P	600P	1,000P (2-360)	-	-	-	75P (2-360)	25P (2-360)	10,000P (30-360)	5,000P (30-360)	500P (30-360)	200P (30-360)	50P (30-360)	25P (30-360)
	L	1400P	600P	1,000P (2-360)	-	-	8 S	75P (2-360)	25P (2-360)	5,000 - 10,000P (30-360)	2,500 - 5,000P (30-360)	500P (30-360)	200P (30-360)	50P (30-360)	25P (30-360)
	HL	1400P	600P	1,000P (2-360)	-	-	2 S	75P (2-360)	25P (2-360)	10,000P (30-360)	5,000P (30-360)	500P (30-360)	200P (30-360)	50P (30-360)	25P (30-360)
	PD	NR	NR	2600P (2-360)	-	16 (1-2)	1 (1)	-	-	N/R	-	200-500P (30-50)	-	-	-
	ED	-	-	NR	5-10 (1)	-	1 (1)	-	-	NR	-	-	-	-	-
	MD	-	-	NR	25P (1)	-	< 1 (1)	-	-	-	-	-	-	-	-
	CX	-	-	3200P (10)	-	3P (1)	1P (10)	-	-	-	-	-	-	-	-
Incapacitating Agents	BZ	-	-	NR	100P (<5)	-	NR	-	-	-	-	-	-	-	-

NOTES: P = Provisional; NR = None Recommended; S = Seconds. The existing estimates are not supported by the available data
Toxicity estimates are followed by exposure duration in parentheses. Exposure duration given in minutes except when specified.

Table G-5. Toxicity Estimates (Exposure Duration) for Military Chemical Compounds

ROE		Inhalation/Ocular			
		Lethal (LCt_{50})	**Severe (ECt_{50})**	**Intolerable (ECt_{50})**	**Threshold (ECt_{50})**
Riot Control Agents	CS	52,000-61,000P (5-90 minutes)	-	7 (1 minute)	-
	CR	-	-	0.15 (1 minute)	0.002 – 0.004 (1 minute)
	OC	None Recommended	-	-	-
Respiratory Irritants	DM	11,000P (2-240 minutes)	-	22 – 150P (1 minutes)	< 1 (1 minute)
	DA	None Recommended	-	12 (2 minutes)	-
	DC	None Recommended	None Recommended	-	-
	Cl_2	None Recommended	None Recommended	-	10 (seconds)

NOTES

P = Provisional
The existing estimates are not supported by the available data.
Toxicity estimates are followed by exposure duration in parenthesis.

Table G-6. Persistency of Selected CW Agents

CW Agents	Persistency[1,2,3]
CG	Nonpersistent
DP	Nonpersistent
GA	Nonpersistent
GB	Nonpersistent
GD	Nonpersistent
GF	Nonpersistent
VX	Persistent
Vx	Persistent
AC	Nonpersistent
CK	Nonpersistent
SA	Nonpersistent
HD	Persistent
HN-1	Persistent
HN-2	Persistent
HN-3	Persistent
HT	Persistent
L	Persistent
HL	Persistent
PD	Persistent
ED	Persistent
MD	Persistent
CX	Nonpersistent
BZ	Nonpersistent

NOTES

[1]Persistent Agent: A chemical agent that, when released, remains able to cause casualties for more than 24 hours to several days or weeks. (JP 1-02)
[2]Nonpersistent Agent: A chemical agent that when released dissipates and/or loses its ability to cause casualties after 10 to 15 minutes. (JP 1-02)
Note: In general, as temperature increases, persistency decreases.

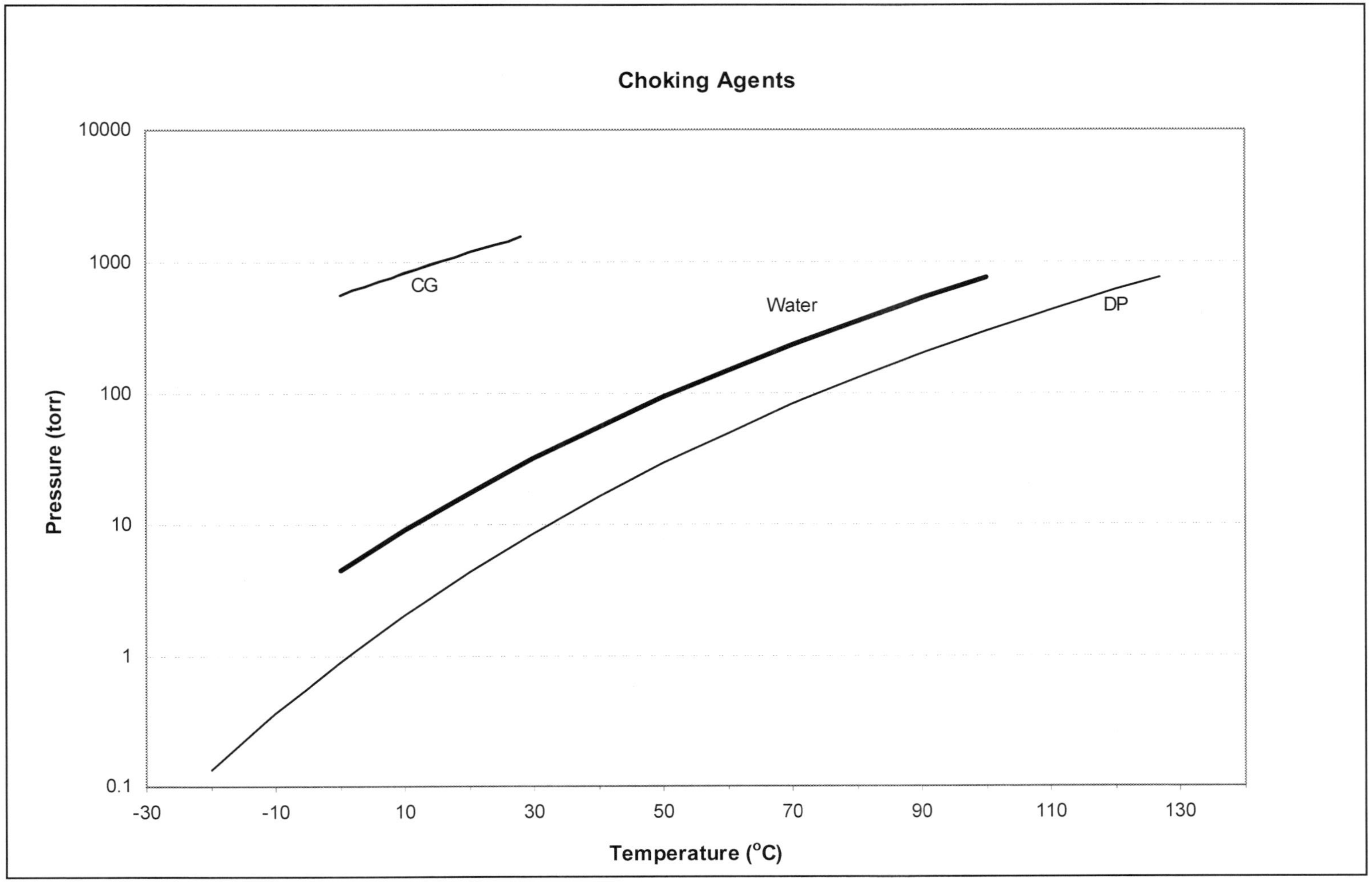

Figure G-1. Vapor Pressure of Choking Agents[1]

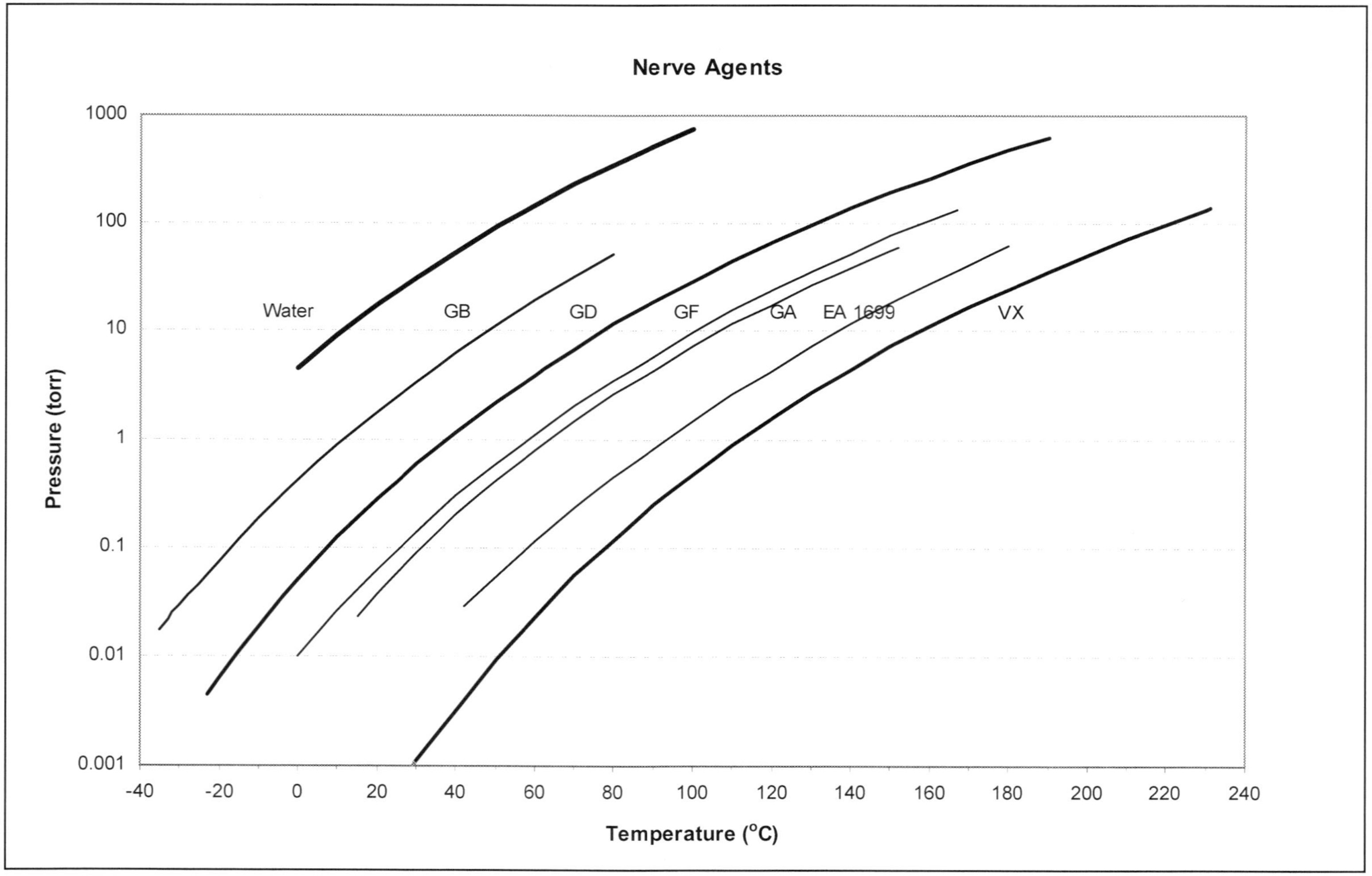

Figure G-2. Vapor Pressure of Nerve Agents[1]

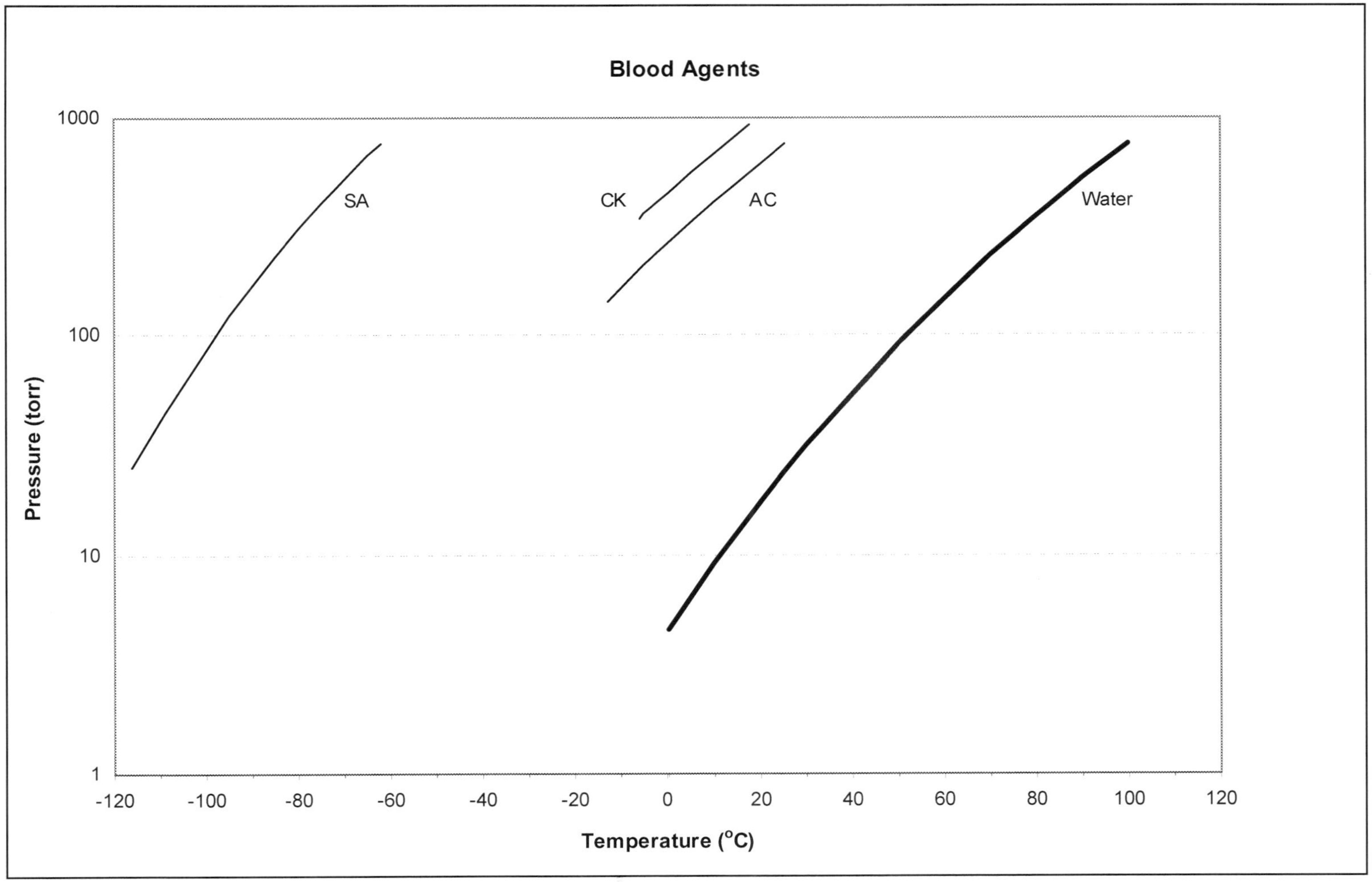

Figure G-3. Vapor Pressure of Blood Agents[1]

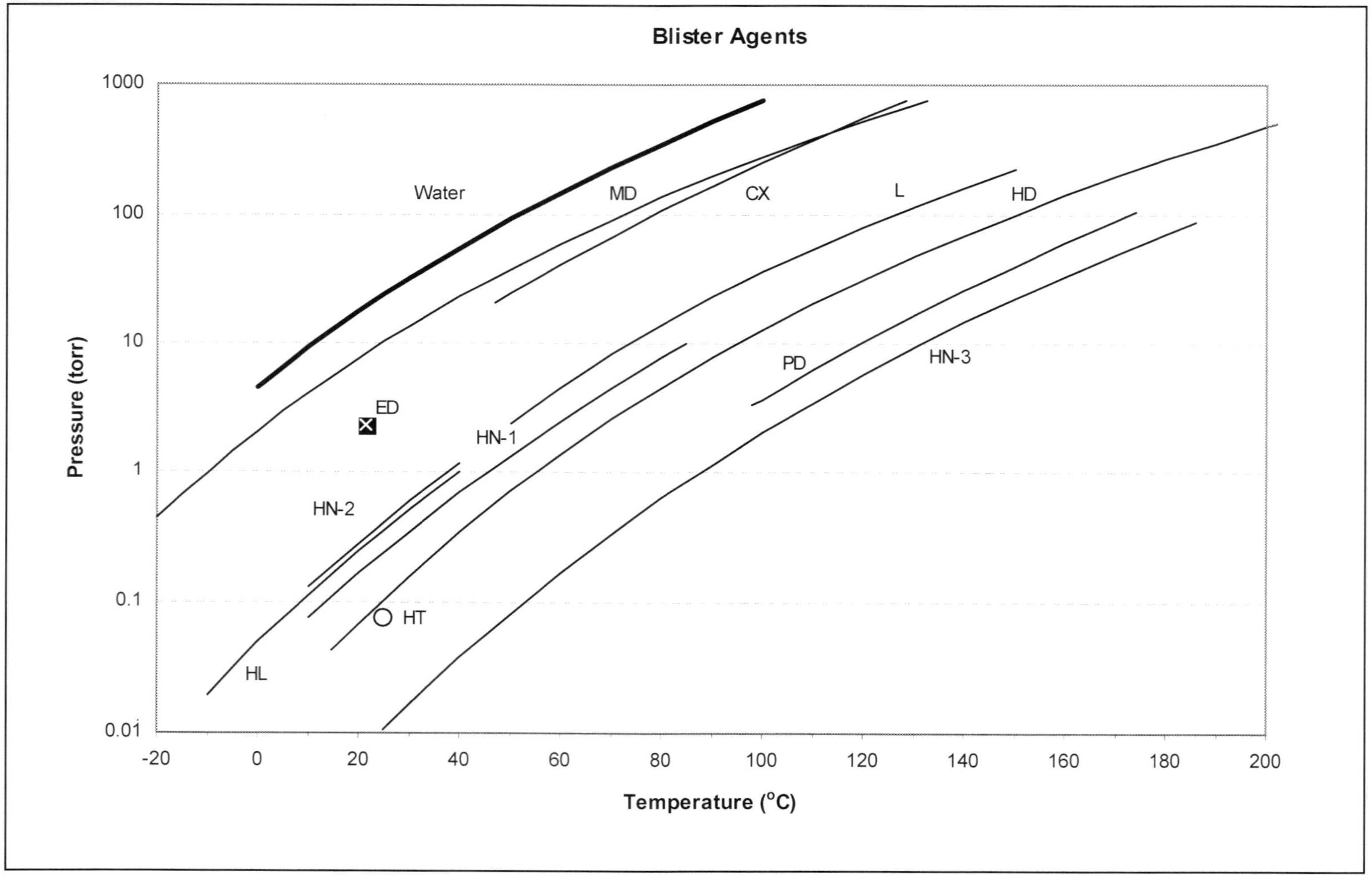

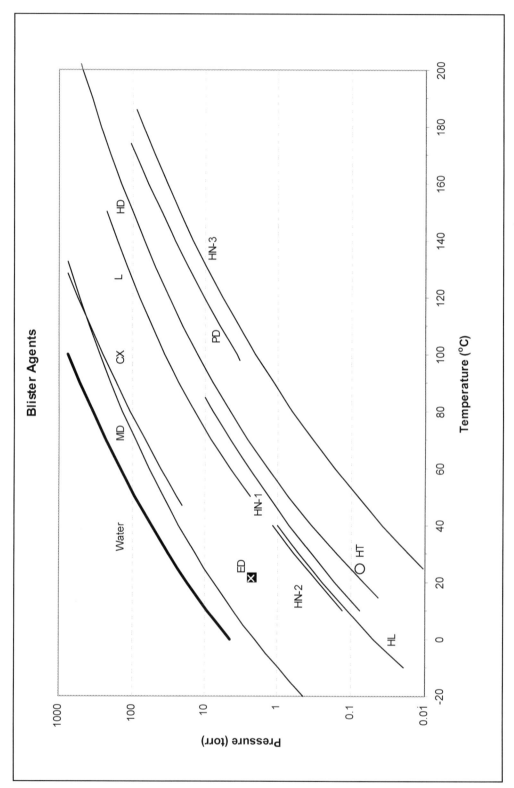

Figure G-4. Vapor Pressure of Blister Agents[1]

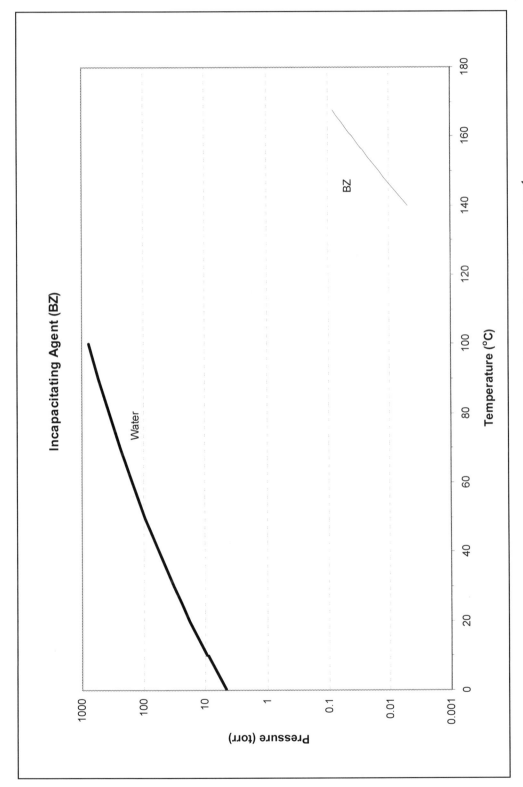

Figure G-5. Vapor Pressure of Incapacitating Agent (BZ)[1]

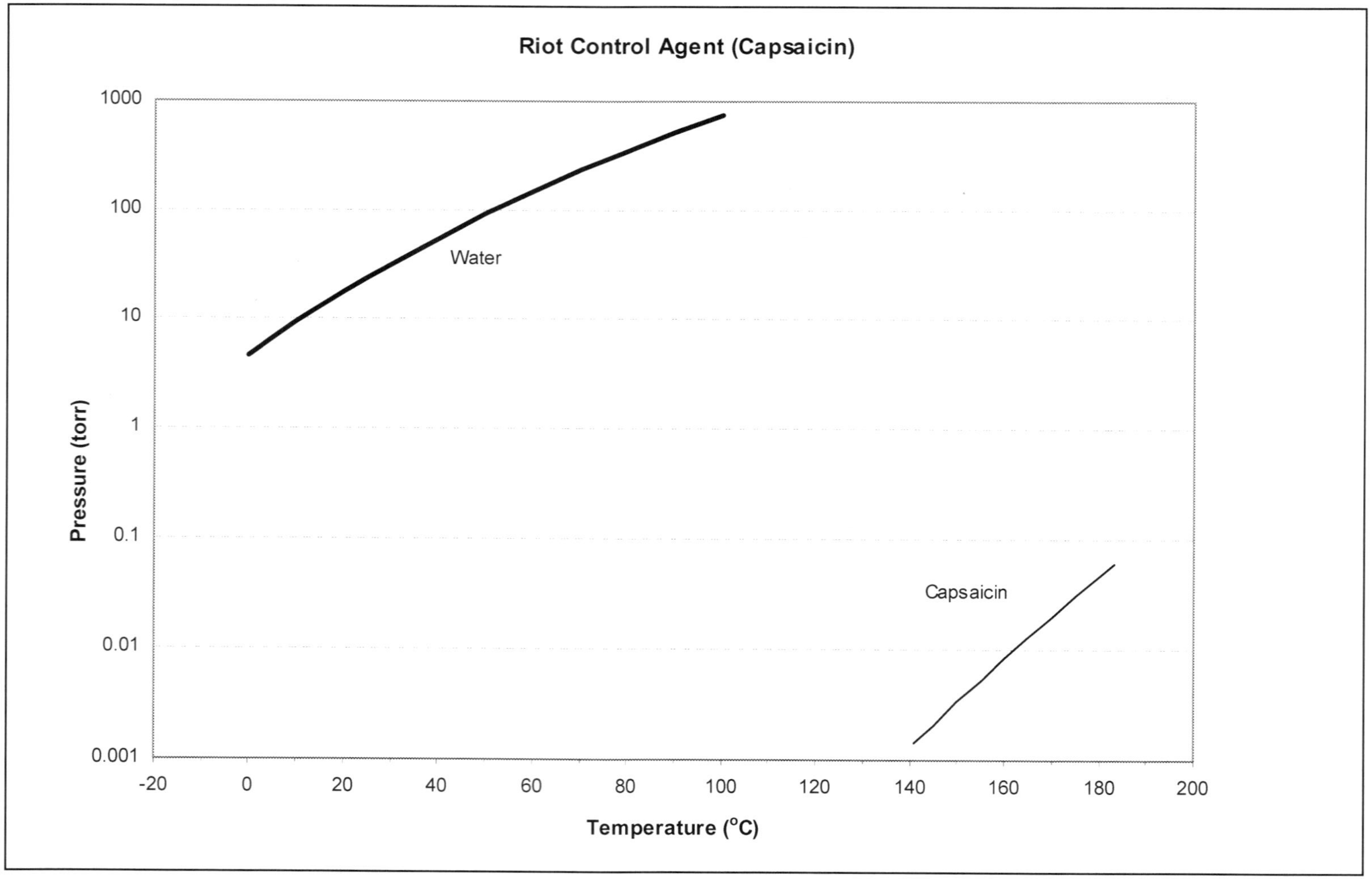

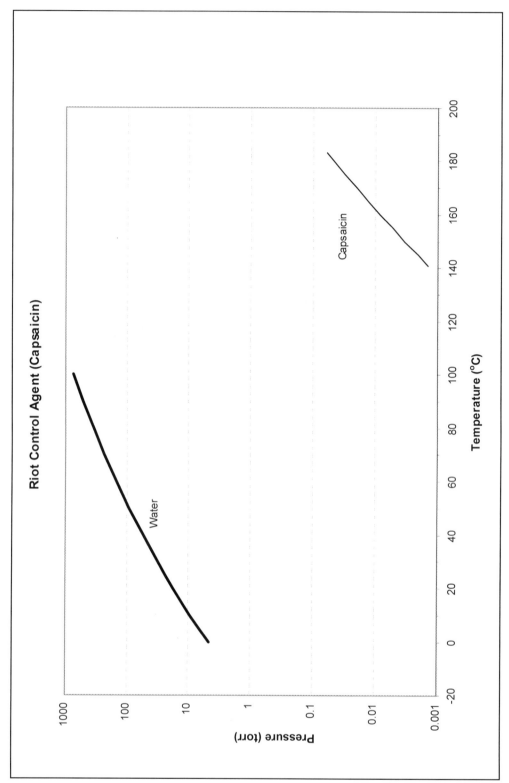

Figure G-6. Vapor Pressure of Riot Control Agent (Capsaicin)[1]

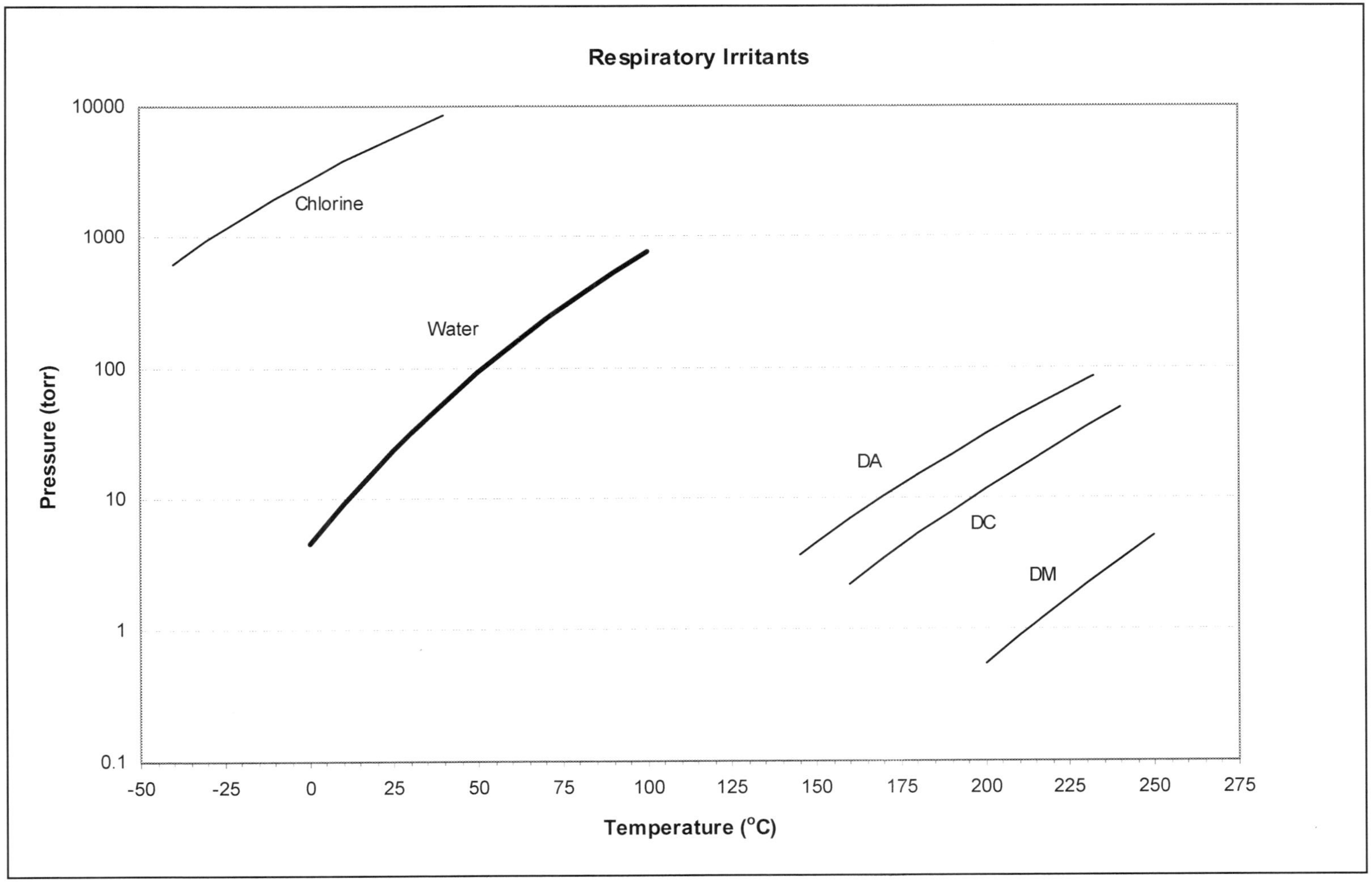

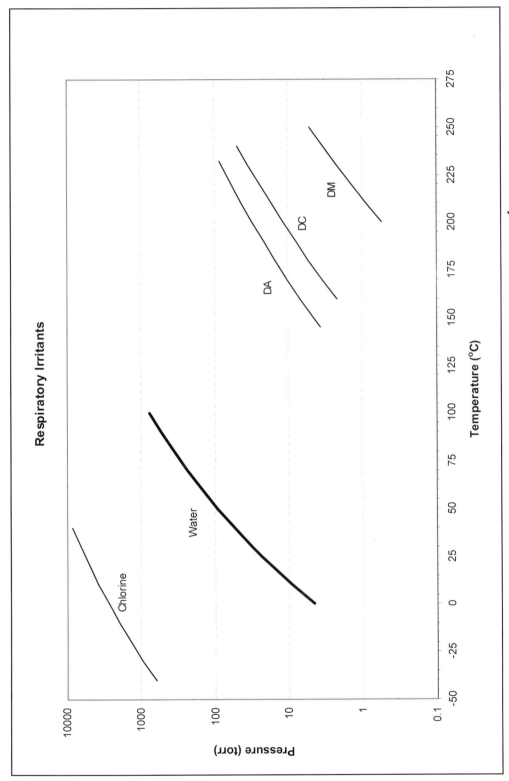

Figure G-7. Vapor Pressure of Respiratory Irritants[1]

NOTES

[1]Patrice L. Abercombie, Physical Property Data Review of Selected Chemical Agents and Related Compounds, September 2003.

Appendix H

TOXICITY PROFILE ESTIMATES

1. Background

This appendix gives the Ct profile estimates and the MV profile estimates (when available) for selected CW agents.

2. Choking Agents

a. CG Profile. Table H-1 provides toxicity estimates for a lethal dose from an inhalation/ocular vapor exposure to CG.

Table H-1. CG Profile Estimates (Lethal Dose, Inhalation/Ocular)[1]

Ct Profile (15L MV)				MV Profile (2-Minute Exposure)	
Exposure Duration (min)	LCt_{50} (mg-min/m³)	Concentration (mg/m³)		MV (L)	LCt_{50} (mg-min/m³)
2	1500	750		10	2250
10	1500	150		15	1500
30	1500	50		30	750
60	1500	25		50	450
Selected Toxicology Information					
Probit Slope: Unknown	TLE: 1		DOC: Moderate		

b. DP Profile. Table H-2 provides toxicity estimates for a lethal dose from an inhalation/ocular vapor exposure to DP.

Table H-2. DP Profile Estimates (Lethal Dose, Inhalation/Ocular)[1]

Ct Profile (15L MV)				MV Profile (2-Minute Exposure)	
Exposure Duration (min)	LCt_{50} (mg-min/m³)	Concentration (mg/m³)		MV (L)	LCt_{50} (mg-min/m³)
2	1500	750		10	2250
10	1500	150		15	1500
30	1500	50		30	750
60	1500	25		50	450
Selected Toxicology Information					
Probit Slope: Unknown	TLE: 1 (Assumed)		DOC: Low		

3. Nerve Agents

a. GA Profile.

(1) Table H-3 provides toxicity estimates for a lethal dose from an inhalation/ocular vapor exposure to GA.

Table H-3. GA Profile Estimates (Lethal Dose, Inhalation/Ocular)[1]

Ct Profile (15L MV)				MV Profile (2-Minute Exposure)	
Exposure Duration (min)	LCt_{50} (mg-min/m³)	Concentration (mg/m³)		MV (L)	LCt_{50} (mg-min/m³)
2	70	35		10	100
10	120	12		15	70
30	170	6.0		30	35
60	220	3.6		50	20
120	270	2.3			
240	345	1.4			
360	395	1.1			
Selected Toxicology Information					
Probit Slope: 12	TLE: 1.5		DOC: Moderate		

(2) Table H-4 provides toxicity estimates for a lethal dose from a percutaneous vapor exposure to GA. Personnel are masked with bare skin.

Table H-4. GA Profile Estimates (Lethal Dose, Percutaneous)[1]

Ct Profiles						
Moderate Temperatures				Hot Temperatures		
Exposure Duration (min)	LCt_{50} (mg-min/m^3)	Concentration (mg/m^3)		Exposure Duration (min)	LCt_{50} (mg-min/m^3)	Concentration (mg/m^3)
2	15,000	7500		2	7500	3750
10	15,000	1500		10	7500	750
30	15,000	500		30	7500	250
60	15,000	250		60	7500	125
120	15,000	125		120	7500	60
240	15,000	60		240	7500	30
360	15,000	40		360	7500	20
Selected Toxicology Information						
Probit Slope: 5	TLE: 1 (Assumed)	DOC: Low		Probit Slope: 5	TLE: 1 (Assumed)	DOC: Low

(3) Table H-5 provides toxicity estimates for severe effects from an inhalation/ocular vapor exposure to GA.

Table H-5. GA Profile Estimates (Severe Effects, Inhalation/Ocular)[1]

Ct Profile (15L MV)				MV Profile (2-Minute Exposure)	
Exposure Duration (min)	ECt_{50} (mg-min/m^3)	Concentration (mg/m^3)		MV (L)	ECt_{50} (mg-min/m^3)
2	50	25		10	75
10	85	8.5		15	50
30	120	4.1		30	25
60	155	2.6		50	15
120	200	1.6			
240	250	1.0			
360	280	0.8			
Selected Toxicology Information					
Probit Slope: 10	TLE: 1.5	DOC: Moderate			

(4) Table H-6 provides toxicity estimates for severe effects from a percutaneous vapor exposure to GA. Personnel are masked with bare skin.

Table H-6. GA Profile Estimates (Severe Effects, Percutaneous)[1]

Ct Profiles						
Moderate Temperatures				Hot Temperatures		
Exposure Duration (min)	ECt_{50} (mg-min/m^3)	Concentration (mg/m^3)		Exposure Duration (min)	ECt_{50} (mg-min/m^3)	Concentration (mg/m^3)
2	12,000	6000		2	6000	3000
10	12,000	1200		10	6000	600
30	12,000	400		30	6000	200
60	12,000	200		60	6000	100
120	12,000	100		120	6000	50
240	12,000	50		240	6000	25
360	12,000	35		360	6000	15
Selected Toxicology Information						
Probit Slope: 5	TLE: 1 (Assumed)	DOC: Low		Probit Slope: 5	TLE: 1 (Assumed)	DOC: Low

(5) Table H-7 provides toxicity estimates for threshold effects from a percutaneous vapor exposure to GA. Personnel are masked with bare skin.

Table H-7. GA Profile Estimates (Threshold Effects, Percutaneous)[1]

Ct Profiles					
Moderate Temperatures			Hot Temperatures		
Exposure Duration (min)	ECt_{50} (mg-min/m^3)	Concentration (mg/m^3)	Exposure Duration (min)	ECt_{50} (mg-min/m^3)	Concentration (mg/m^3)
2	2000	1000	2	1000	500
10	2000	200	10	1000	100
30	2000	65	30	1000	30
60	2000	30	60	1000	15
120	2000	15	120	1000	10
240	2000	10	240	1000	5
360	2000	5	360	1000	2.5
Selected Toxicology Information					
Probit Slope: 5	TLE: 1 (Assumed)	DOC: Moderate	Probit Slope: 5	TLE: 1 (Assumed)	DOC: Moderate

(6) Table H-8 provides toxicity estimates for mild effects from an inhalation/ocular vapor exposure to GA. Mild effects include miosis and rhinorrhea.

Table H-8. GA Profile Estimates (Mild Effects, Inhalation/Ocular)[1]

Ct Profile		
Exposure Duration (min)	ECt_{50} (mg-min/m^3)	Concentration (mg/m^3)
2	0.4	0.16
10	0.6	0.07
30	0.9	0.03
60	1.1	0.02
120	1.4	0.01
240	1.7	0.007
360	2.0	0.005
Selected Toxicology Information		
Probit Slope: 10	TLE: 1.5	DOC: Low

(7) Figure H-1 is a graphical representation of GA vapor Ct profile for dosage from an inhalation/ocular exposure.

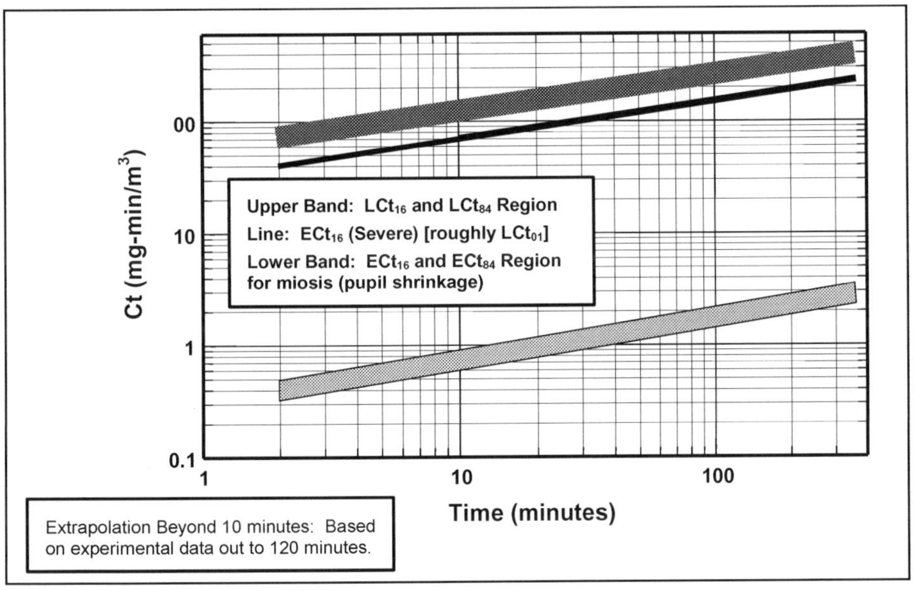

Figure H-1. GA Vapor: Dosage versus Exposure Duration

(8) Figure H-2 is a graphical representation of GA vapor Ct profile for concentration from an inhalation/ocular exposure.

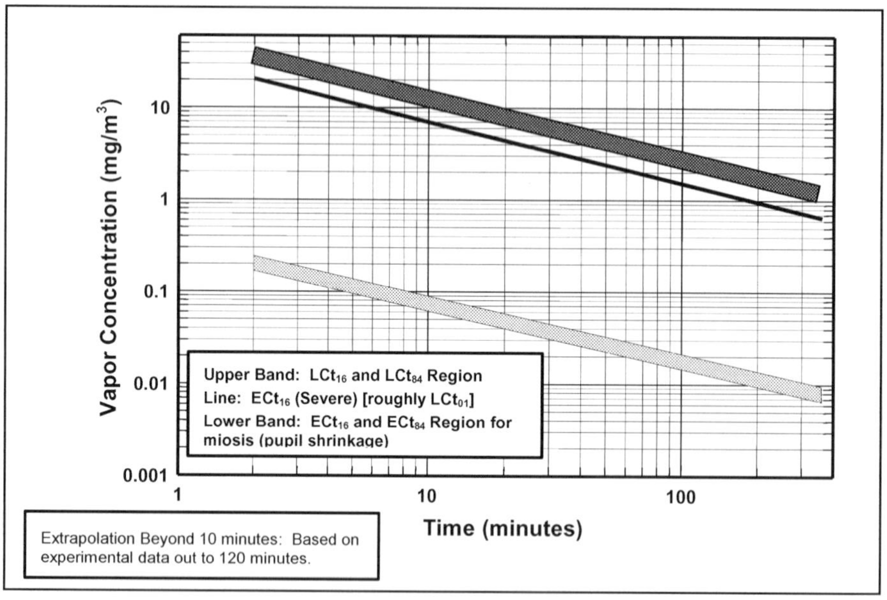

Figure H-2. GA Vapor: Concentration versus Exposure Duration

b. GB Profile.

(1) Table H-9 provides toxicity estimates for a lethal dose from an inhalation/ocular vapor exposure to GB.

Table H-9. GB Profile Estimates (Lethal Dose, Inhalation/Ocular)[1]

Ct Profile (15L MV)			MV Profile (2-Minute Exposure)	
Exposure Duration (min)	**LCt_{50} (mg-min/m^3)**	**Concentration (mg/m^3)**	**MV (L)**	**LCt_{50} (mg-min/m^3)**
2	35	17.5	10	50
10	60	6.0	15	35
30	86	3.0	30	17
60	110	1.8	50	10
120	140	1.1		
240	175	0.7		
360	200	0.6		
Selected Toxicology Information				
Probit Slope: 12	TLE: 1.5	DOC: Moderate		

(2) Table H-10 provides toxicity estimates for a lethal dose from a percutaneous vapor exposure to GB. Personnel are masked with bare skin.

Table H-10. GB Profile Estimates (Lethal Dose, Percutaneous)[1]

Ct Profiles						
Moderate Temperatures				Hot Temperatures		
Exposure Duration (min)	LCt_{50} (mg-min/m^3)	Concentration (mg/m^3)		Exposure Duration (min)	LCt_{50} (mg-min/m^3)	Concentration (mg/m^3)
2	12,000	6000		2	6000	3000
10	12,000	1200		10	6000	600
30	12,000	400		30	6000	200
60	12,000	200		60	6000	100
120	12,000	100		120	6000	50
240	12,000	50		240	6000	25
360	12,000	30		360	6000	15
Selected Toxicology Information						
Probit Slope: 5	TLE: 1 (Assumed)	DOC: Low		Probit Slope: 5	TLE: 1 (Assumed)	DOC: Low

(3) Table H-11 provides toxicity estimates for severe effects from an inhalation/ocular vapor exposure to GB.

Table H-11. GB Profile Estimates (Severe Effects, Inhalation/Ocular)[1]

Ct Profile (15L MV)				MV Profile (2-Minute Exposure)	
Exposure Duration (min)	ECt_{50} (mg-min/m^3)	Concentration (mg/m^3)		MV (L)	ECt_{50} (mg-min/m^3)
2	25	12.5		10	37
10	45	4.3		15	25
30	60	2.1		30	12
60	80	1.3		50	7.5
120	100	0.8			
240	125	0.5			
360	140	0.4			
Selected Toxicology Information					
Probit Slope: 12	TLE: 1.5			DOC: Moderate	

(4) Table H-12 provides toxicity estimates for severe effects from a percutaneous vapor exposure to GB, and personnel are masked with bare skin.

Table H-12. GB Profile Estimates (Severe Effects, Percutaneous)[1]

Ct Profiles						
Moderate Temperatures				Hot Temperatures		
Exposure Duration (min)	ECt_{50} (mg-min/m^3)	Concentration (mg/m^3)		Exposure Duration (min)	ECt_{50} (mg-min/m^3)	Concentration (mg/m^3)
2	8000	4000		2	4000	2000
10	8000	800		10	4000	400
30	8000	250		30	4000	125
60	8000	130		60	4000	65
120	8000	65		120	4000	30
240	8000	30		240	4000	15
360	8000	20		360	4000	10
Selected Toxicology Information						
Probit Slope: 5	TLE: 1 (Assumed)	DOC: Low		Probit Slope: 5	TLE: 1 (Assumed)	DOC: Low

(5) Table H-13 provides toxicity estimates for threshold effects from a percutaneous vapor exposure to GB. Personnel are masked with bare skin.

Table H-13. GB Profile Estimates (Threshold Effects, Percutaneous)[1]

Ct Profiles						
Moderate Temperatures				Hot Temperatures		
Exposure Duration (min)	ECt_{50} (mg-min/m^3)	Concentration (mg/m^3)		Exposure Duration (min)	ECt_{50} (mg-min/m^3)	Concentration (mg/m^3)
2	1200	600		2	600	300
10	1200	120		10	600	60
30	1200	40		30	600	20
60	1200	20		60	600	10
120	1200	10		120	600	5
240	1200	5		240	600	2.5
360	1200	3		360	600	1.5
Selected Toxicology Information						
Probit Slope: 5	TLE: 1 (Assumed)	DOC: Moderate		Probit Slope: 5	TLE: 1 (Assumed)	DOC: Low

(6) Table H-14 provides toxicity estimates for mild effects from an inhalation/ocular vapor exposure to GB.

Table H-14. GB Profile Estimates (Mild Effects, Inhalation/Ocular)[1]

Ct Profile		
Exposure Duration (min)	ECt_{50} (mg-min/m^3)	Concentration (mg/m^3)
2	0.4	0.18
10	0.6	0.06
30	0.9	0.03
60	1.1	0.02
120	1.4	0.01
240	1.7	0.007
360	2.0	0.005
Selected Toxicology Information		
Probit Slope: 10	TLE: 1.5	DOC: Moderate

(7) Figure H-3 is a graphical representation of GB vapor Ct profile for dosage from an inhalation/ocular exposure.

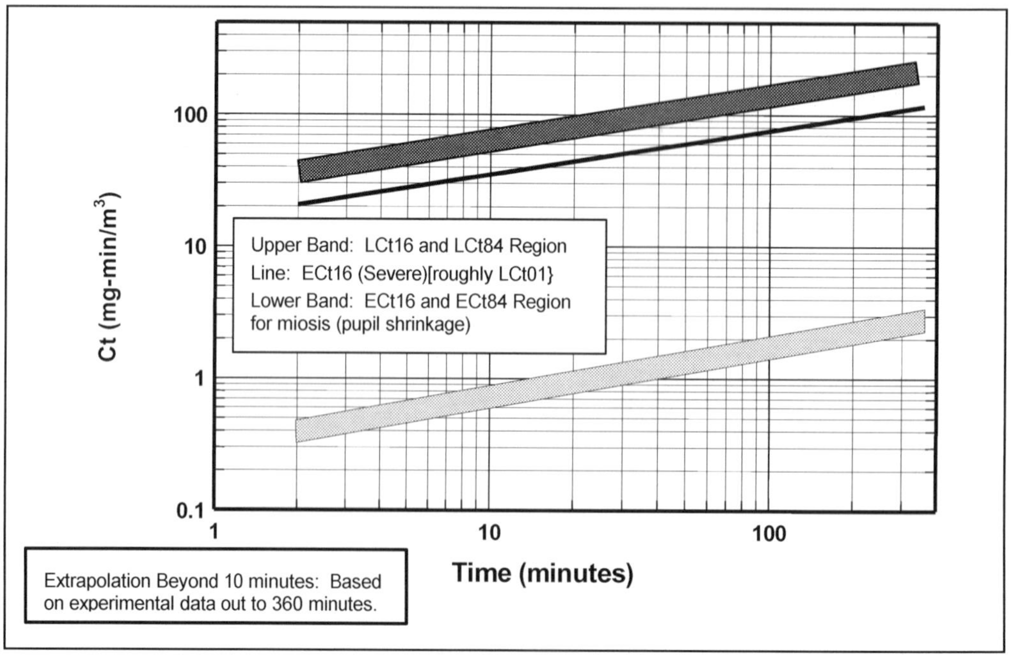

Figure H-3. GB Vapor: Dosage versus Exposure Duration

(8) Figure H-4 is a graphical representation of GB vapor Ct profile for concentration from an inhalation/ocular exposure.

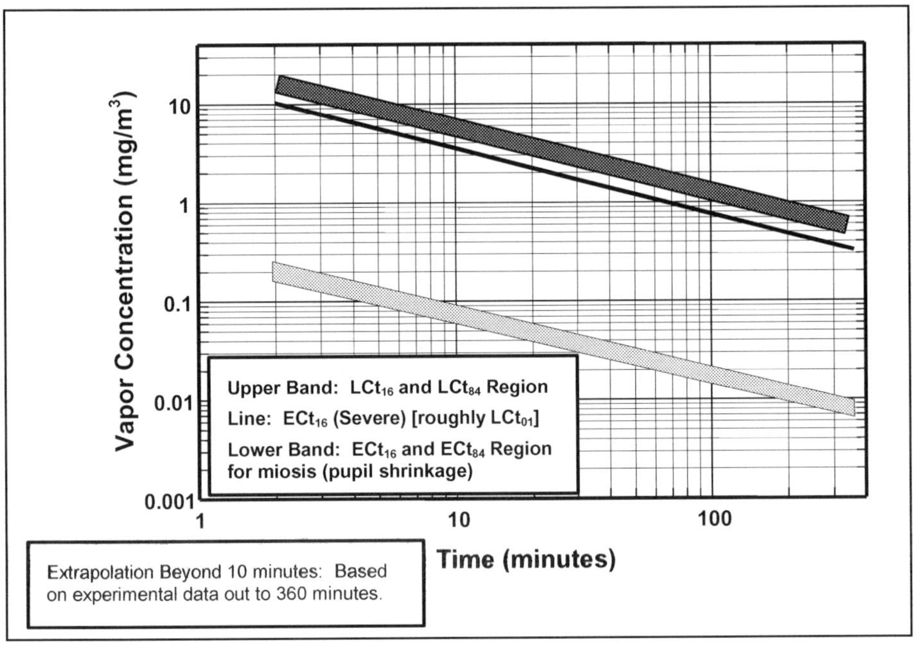

Figure H-4. GB Vapor: Concentration versus Exposure Duration

c. GD Profile.

(1) Table H-15 provides toxicity estimates for a lethal dose from an inhalation/ocular vapor exposure to GD.

Table H-15. GD Profile Estimates (Lethal Dose, Inhalation/Ocular)[1]

Ct Profile (15L MV)			MV Profile (2-Minute Exposure)	
Exposure Duration (min)	LCt_{50} (mg-min/m³)	Concentration (mg/m³)	MV (L)	LCt_{50} (mg-min/m³)
2	35	17.5	10	50
10	50	4.8	15	35
30	60	2.0	30	15
60	70	1.2	50	10
120	80	0.7		
240	90	0.4		
360	100	0.3		
Selected Toxicology Information				
Probit Slope: 12	TLE: 1.25		DOC: Low	

(2) Table H-16 provides toxicity estimates for a lethal dose from a percutaneous vapor exposure to GD. Personnel are masked with bare skin.

Table H-16. GD Profile Estimates (Lethal Dose, Percutaneous)[1]

Ct Profiles						
Moderate Temperatures				Hot Temperatures		
Exposure Duration (min)	LCt_{50} (mg-min/m^3)	Concentration (mg/m^3)		Exposure Duration (min)	LCt_{50} (mg-min/m^3)	Concentration (mg/m^3)
2	3000	1500		2	1500	750
10	3000	300		10	1500	375
30	3000	100		30	1500	150
60	3000	50		60	1500	25
120	3000	25		120	1500	12
240	3000	12		240	1500	5
360	3000	5		360	1500	2
Selected Toxicology Information						
Probit Slope: 6	TLE: 1 (Assumed)	DOC: Low		Probit Slope: 6	TLE: 1 (Assumed)	DOC: Low

(3) Table H-17 provides toxicity estimates for severe effects from an inhalation/ocular vapor exposure to GD.

Table H-17. GD Profile Estimates (Severe Effects, Inhalation/Ocular)[1]

Ct Profile (15L MV)				MV Profile (2-Minute Exposure)	
Exposure Duration (min)	ECt_{50} (mg-min/m^3)	Concentration (mg/m^3)		MV (L)	ECt_{50} (mg-min/m^3)
2	25	12.5		10	35
10	35	3.5		15	25
30	45	1.5		30	10
60	50	0.8		50	7.5
120	55	0.4			
240	65	0.3			
360	70	0.2			
Selected Toxicology Information					
Probit Slope: 12		TLE: 1.25		DOC: Low	

(4) Table H-18 provides toxicity estimates for severe effects from a percutaneous vapor exposure to GD. Personnel are masked with bare skin.

Table H-18. GD Profile Estimates (Severe Effects, Percutaneous)[1]

Ct Profiles						
Moderate Temperatures				Hot Temperatures		
Exposure Duration (min)	ECt_{50} (mg-min/m^3)	Concentration (mg/m^3)		Exposure Duration (min)	ECt_{50} (mg-min/m^3)	Concentration (mg/m^3)
2	2000	1000		2	1000	500
10	2000	200		10	1000	100
30	2000	60		30	1000	30
60	2000	30		60	1000	15
120	2000	15		120	1000	7.5
240	2000	7.5		240	1000	5
360	2000	5		360	1000	2.5
Selected Toxicology Information						
Probit Slope: 6	TLE: 1 (Assumed)	DOC: Low		Probit Slope: 6	TLE: 1 (Assumed)	DOC: Low

(5) Table H-19 provides toxicity estimates for threshold effects from a percutaneous vapor exposure to GD. Personnel are masked with bare skin.

Table H-19. GD Profile Estimates (Threshold Effects, Percutaneous)[1]

Ct Profiles						
Moderate Temperatures				Hot Temperatures		
Exposure Duration (min)	ECt_{50} (mg-min/m³)	Concentration (mg/m³)		Exposure Duration (min)	ECt_{50} (mg-min/m³)	Concentration (mg/m³)
2	300	150		2	150	75
10	300	30		10	150	15
30	300	10		30	150	5
60	300	5		60	150	2.5
120	300	2.5		120	150	1.25
240	300	1.25		240	150	0.5
360	300	0.6		360	150	0.3
Selected Toxicology Information						
Probit Slope: 6	TLE: 1 (Assumed)	DOC: Low		Probit Slope: 6	TLE: 1 (Assumed)	DOC: Low

(6) Table H-20 provides a toxicity estimate given for mild effects from an inhalation/ocular vapor exposure to GD.

Table H-20. GD Profile Estimates (Mild Effects, Inhalation/Ocular)[1]

Ct Profile		
Exposure Duration (min)	ECt_{50} (mg-min/m³)	Concentration (mg/m³)
2	0.2	0.1
10	0.3	0.03
30	0.4	0.01
60	0.5	0.009
120	0.6	0.005
240	0.8	0.003
360	0.9	0.0025
Selected Toxicology Information		
Probit Slope: 10	TLE: 1.4	DOC: Low

(7) Figure H-5 is a graphical representation of Ct profile for GD vapor dosage from an inhalation/ocular exposure.

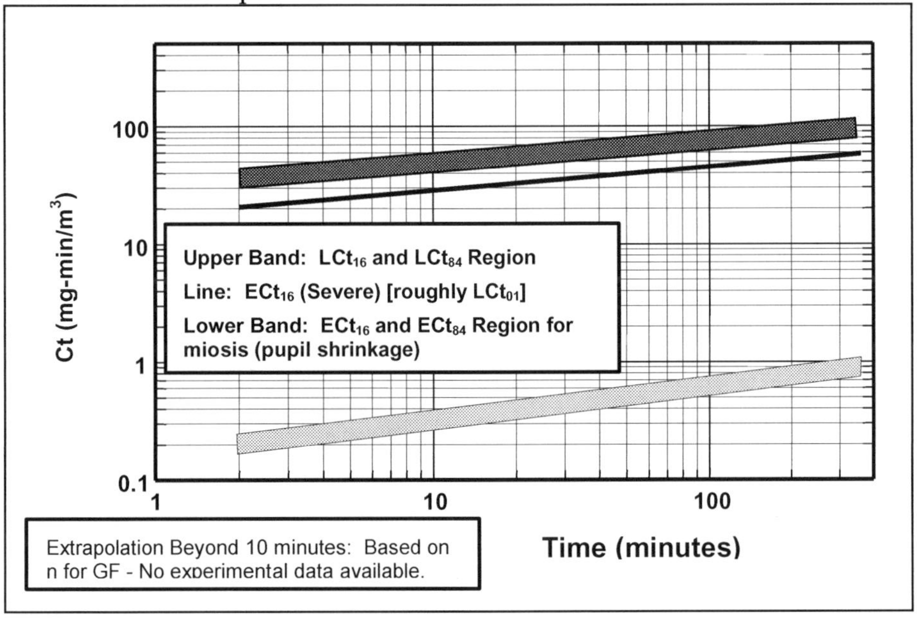

Figure H-5. GD Vapor: Dosage versus Exposure Duration

(8) Figure H-6 is a graphical representation of Ct profile for GD vapor concentration from an inhalation/ocular exposure.

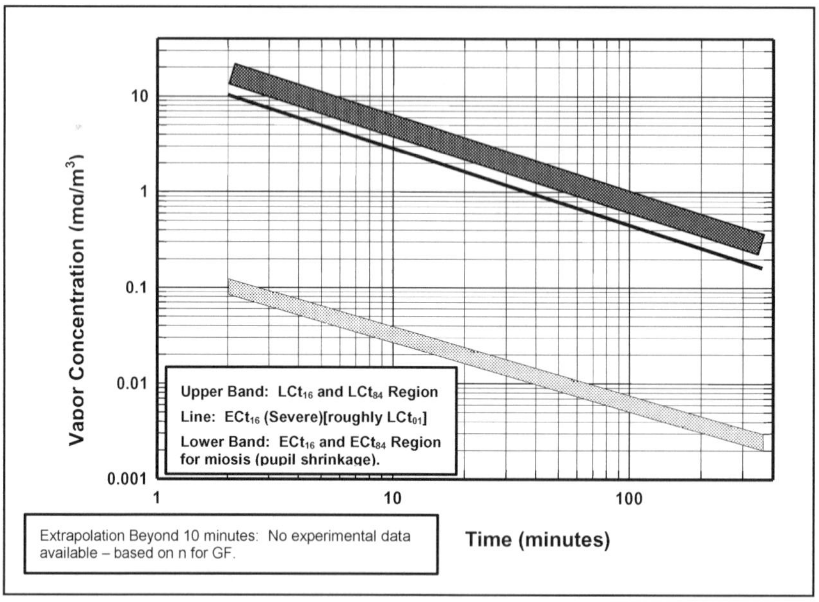

Figure H-6. GD Vapor: Concentration versus Exposure Duration

d. GF Profile.

(1) Table H-21 provides toxicity estimates for a lethal dose from an inhalation/ocular vapor exposure to GF.

Table H-21. GF Profile Estimates (Lethal Dose, Inhalation/Ocular)[1]

Ct Profile (15L MV)				MV Profile (2-Minute Exposure)	
Exposure Duration (min)	LCt_{50} (mg-min/m^3)	Concentration (mg/m^3)		MV (L)	LCt_{50} (mg-min/m^3)
2	35	17.5		10	50
10	48	4.8		15	35
30	60	2.0		30	17
60	69	1.0		50	10
120	79	0.7			
240	91	0.4			
360	99	0.3			
Selected Toxicology Information					
Probit Slope: 12	TLE: 1.25	DOC: Moderate			

(2) Table H-22 provides toxicity estimates for a lethal dose from a percutaneous vapor exposure to GF. Personnel are masked with bare skin.

Table H-22. GF Profile Estimates (Lethal Dose, Percutaneous)[1]

Ct Profiles						
Moderate Temperatures				Hot Temperatures		
Exposure Duration (min)	LCt_{50} (mg-min/m^3)	Concentration (mg/m^3)		Exposure Duration (min)	LCt_{50} (mg-min/m^3)	Concentration (mg/m^3)
2	3000	1500		2	1500	7500
10	3000	300		10	1500	375
30	3000	100		30	1500	150
60	3000	50		60	1500	25
120	3000	25		120	1500	12
240	3000	12		240	1500	5
360	3000	5		360	1500	2
Selected Toxicology Information						
Probit Slope: 5	TLE: 1 (Assumed)	DOC: Low		Probit Slope: 5	TLE: 1 (Assumed)	DOC: Low

(3) Table H-23 provides toxicity estimates for severe effects from an inhalation/ocular vapor exposure to GF.

Table H-23. GF Profile Estimates (Severe Effects, Inhalation/Ocular)[1]

Ct Profile (15L MV)				MV Profile (2-Minute Exposure)	
Exposure Duration (min)	ECt_{50} (mg-min/m³)	Concentration (mg/m³)		MV (L)	ECt_{50} (mg-min/m³)
2	25	12.5		10	35
10	35	3.5		15	25
30	45	1.5		30	10
60	50	0.8		50	7.5
120	55	0.4			
240	65	0.3			
360	70	0.2			
Selected Toxicology Information					
Probit Slope: 12	TLE: 1.25		DOC: Moderate		

(4) Table H-24 provides toxicity estimates for severe effects from a percutaneous vapor exposure to GF. Personnel are masked with bare skin.

Table H-24. GF Profile Estimates (Severe Effects, Percutaneous)[1]

Ct Profiles						
Moderate Temperatures				Hot Temperatures		
Exposure Duration (min)	ECt_{50} (mg-min/m³)	Concentration (mg/m³)		Exposure Duration (min)	ECt_{50} (mg-min/m³)	Concentration (mg/m³)
2	2000	1000		2	1000	500
10	2000	200		10	1000	100
30	2000	60		30	1000	30
60	2000	30		60	1000	15
120	2000	15		120	1000	7.5
240	2000	7.5		240	1000	5
360	2000	5		360	1000	2.5
Selected Toxicology Information						
Probit Slope: 5	TLE: 1 (Assumed)	DOC: Low		Probit Slope: 5	TLE: 1 (Assumed)	DOC: Low

(5) Table H-25 provides toxicity estimates for threshold effects from a percutaneous vapor exposure to GF. Personnel are masked with bare skin.

Table H-25. GF Profile Estimates (Threshold Effects, Percutaneous)[1]

Ct Profiles						
Moderate Temperatures				Hot Temperatures		
Exposure Duration (min)	ECt_{50} (mg-min/m³)	Concentration (mg/m³)		Exposure Duration (min)	ECt_{50} (mg-min/m³)	Concentration (mg/m³)
2	300	150		2	150	75
10	300	30		10	150	15
30	300	10		30	150	5
60	300	5		60	150	2.5
120	300	2.5		120	150	1.25
240	300	1.25		240	150	0.5
360	300	0.6		360	150	0.3
Selected Toxicology Information						
Probit Slope: 6	TLE: 1 (Assumed)	DOC: Low		Probit Slope: 6	TLE: 1 (Assumed)	DOC: Low

(6) Table H-26 provides toxicity estimates for mild effects from an inhalation/ocular vapor exposure to GF.

Table H-26. GF Profile Estimates (Mild Effects, Inhalation/Ocular)[1]

Ct Profile		
Exposure Duration (min)	ECt_{50} (mg-min/m^3)	Concentration (mg/m^3)
2	0.2	0.1
10	0.3	0.03
30	0.4	0.01
60	0.5	0.009
120	0.6	0.005
240	0.8	0.003
360	0.9	0.0025
Selected Toxicology Information		
Probit Slope: 10	TLE: 1.4	DOC: Low

(7) Figure H-7 is a graphical representation Ct profile for GF vapor dosage from an inhalation/ocular exposure.

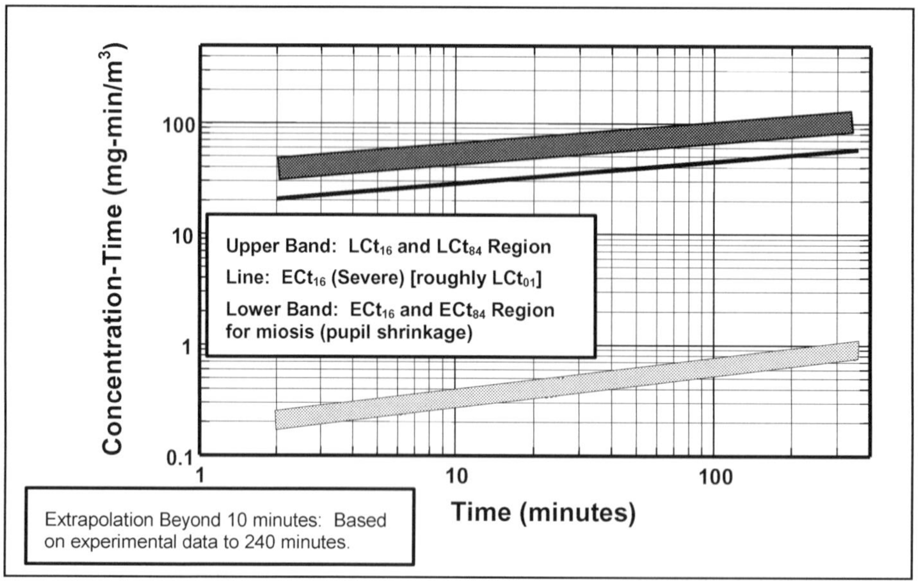

Figure H-7. GF Vapor: Dosage versus Exposure Duration

(8) Figure H-8 is a graphical representation of Ct profile for GF vapor concentration from an inhalation/ocular exposure.

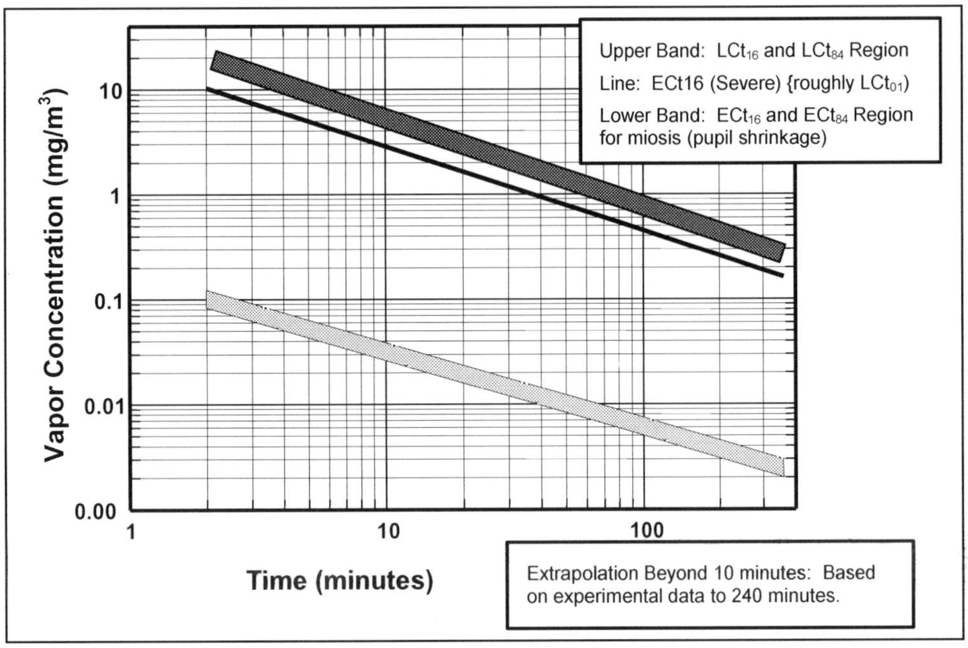

Figure H-8. GF Vapor: Concentration versus Exposure Duration

e. VX Profile.

(1) Table H-27 provides toxicity estimates for a lethal dose from an inhalation/ocular vapor exposure to VX.

Table H-27. VX Profile Estimates (Lethal Dose, Inhalation/Ocular)[1]

Ct Profile (15L MV)			MV Profile (2-Minute Exposure)	
Exposure Duration (min)	LCt_{50} (mg-min/m^3)	Concentration (mg/m^3)	MV (L)	LCt_{50} (mg-min/m^3)
2	15	7.50	10	20
10	15	1.50	15	15
30	15	0.50	30	5
60	15	0.25	50	4
120	15	0.12		
240	15	0.06		
360	15	0.04		
Selected Toxicology Information				
Probit Slope: 6	TLE: 1	DOC: Low (LCt_{50} could be less)		

(2) Table H-28 provides toxicity estimates for a lethal dose from a percutaneous vapor exposure to VX. Personnel are masked with bare skin.

Table H-28. VX Profile Estimates (Lethal Dose, Percutaneous)[1]

Ct Profiles						
Moderate Temperatures				Hot Temperatures		
Exposure Duration (min)	LCt_{50} (mg-min/m^3)	Concentration (mg/m^3)		Exposure Duration (min)	LCt_{50} (mg-min/m^3)	Concentration (mg/m^3)
2	150	75		2	75	38
10	150	15		10	75	7.5
30	150	5		30	75	2.5
60	150	2.5		60	75	1.3
120	150	1.25		120	75	0.6
240	150	0.6		240	75	0.3
360	150	0.4		360	75	0.2
Selected Toxicology Data						
Probit Slope: 6	TLE: 1 (Assumed)	DOC: Low		Probit Slope: 5	TLE: 1 (Assumed)	DOC: Low

(3) Table H-29 provides toxicity estimates for severe effects from an inhalation/ocular vapor exposure to VX.

Table H-29. VX Profile Estimates (Severe Effects, Inhalation/Ocular)[1]

Ct Profile (15L MV)				MV Profile (2-Minute Exposure)	
Exposure Duration (min)	ECt_{50} (mg-min/m^3)	Concentration (mg/m^3)		MV (L)	ECt_{50} (mg-min/m^3)
2	10	5		10	15
10	10	1		15	10
30	10	0.3		30	5
60	10	0.2		50	3
120	10	0.1			
240	10	0.04			
360	10	0.03			
Selected Toxicology Estimates					
Probit Slope: 6		TLE: 1		DOC: Low (ECt_{50} could be less)	

(4) Table H-30 provides toxicity estimates for severe effects from a percutaneous vapor exposure to VX. Personnel are masked with bare skin.

Table H-30. VX Profile Estimates (Severe Effects, Percutaneous)[1]

Ct Profiles						
Moderate Temperatures				Hot Temperatures		
Exposure Duration (min)	ECt_{50} (mg-min/m^3)	Concentration (mg/m^3)		Exposure Duration (min)	ECt_{50} (mg-min/m^3)	Concentration (mg/m^3)
2	25	12.5		2	12	2.5
10	25	2.5		10	12	1.25
30	25	1.25		30	12	0.63
60	25	0.63		60	12	0.30
120	25	0.30		120	12	0.16
240	25	0.16		240	12	0.07
360	25	0.07		360	12	0.04
Selected Toxicology Information						
Probit Slope: 6	TLE: 1 (Assumed)	DOC: Moderate		Probit Slope: 6	TLE: 1 (Assumed)	DOC: Low

(5) Table H-31 provides toxicity estimates for threshold effects from a percutaneous vapor exposure to VX. Personnel are masked with bare skin.

Table H-31. VX Profile Estimates (Threshold Effects, Percutaneous)[1]

Ct Profiles						
Moderate Temperatures				Hot Temperatures		
Exposure Duration (min)	ECt_{50} (mg-min/m³)	Concentration (mg/m³)		Exposure Duration (min)	ECt_{50} (mg-min/m³)	Concentration (mg/m³)
2	10	5		2	5	2.5
10	10	1		10	5	0.5
30	10	0.3		30	5	0.15
60	10	0.15		60	5	0.08
120	10	0.08		120	5	0.04
240	10	0.04		240	5	0.02
360	10	0.03		360	5	0.01
Selected Toxicology Information						
Probit Slope: 6	TLE: 1 (Assumed)	DOC: Moderate		Probit Slope: 6	TLE: 1 (Assumed)	DOC: Low

(6) Table H-32 provides toxicity estimates for mild effects from an inhalation/ocular vapor exposure to VX.

Table H-32. VX Profile Estimates (Mild Effects, Inhalation/Ocular)[1]

Ct Profile		
Exposure Duration (min)	ECt_{50} (mg-min/m³)	Concentration (mg/m³)
2	0.1	0.05
10	0.1	0.01
30	0.1	0.003
60	0.1	0.002
120	0.1	0.001
240	0.1	0.0004
360	0.1	0.0003
Selected Toxicology Information		
Probit Slope: 4	TLE: 1 (Assumed)	DOC: Low

(7) Figure H-9 is a graphical representation of Ct profile for VX vapor dosage from an inhalation/ocular exposure.

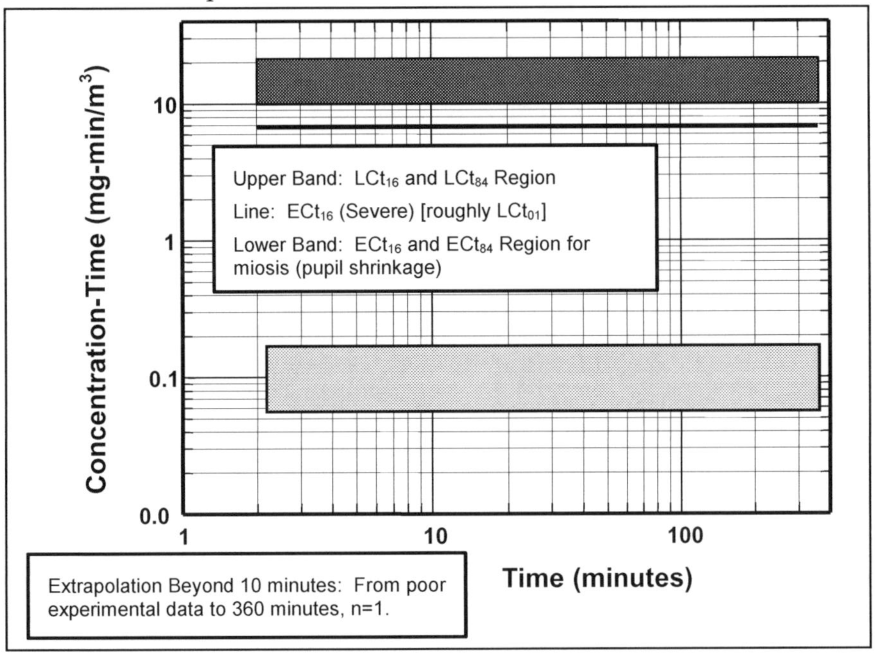

Figure H-9. VX Vapor: Dosage versus Exposure Duration

(8) Figure H-10 is a graphical representation of Ct profile for VX vapor concentration from an inhalation/ocular exposure.

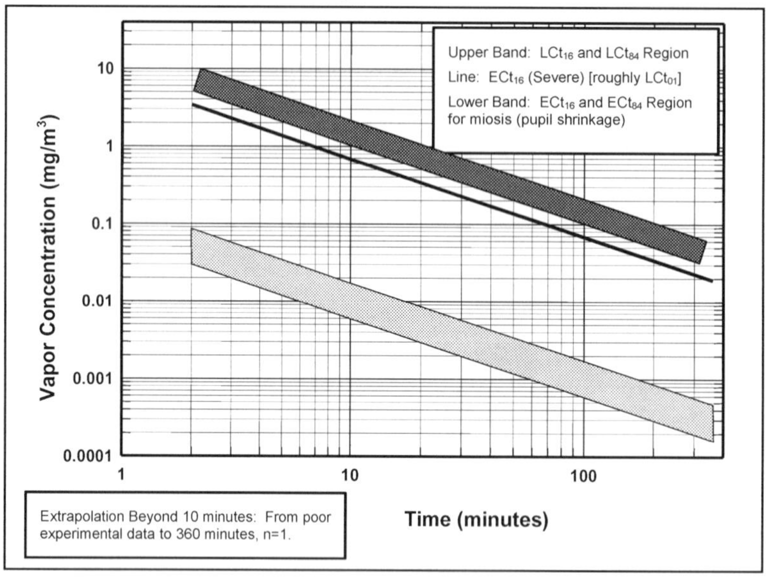

Figure H-10. VX Vapor: Concentration versus Exposure Duration

4. Blood Agents

a. AC Profile. Table H-33 provides provisional toxicity estimates for a lethal dose from an inhalation/ocular vapor exposure to AC.

Table H-33. AC Profile Estimates (Lethal Dose, Inhalation/Ocular)[1]

Ct Profile (15L MV)				MV Profile (2-Minute Exposure)	
Exposure Duration (min)	LCt_{50} (mg-min/m³)	Concentration (mg/m³)		MV (L)	LCt_{50} (mg-min/m³)
2	2860	1430		10	4290
10	6070	607		15	2860
30	20,630	685		30	1430
				50	858
Selected Toxicology Inforamtion					
Probit Slope: 10		TLE: 1.85		DOC: Low	

b. SA Profile. Table H-34 provides provisional toxicity estimates for a lethal dose from an inhalation/ocular vapor exposure to SA.

Table H-34. SA Profile Estimates (Lethal Dose, Inhalation/Ocular)[1]

Ct Profile (15L MV)				MV Profile (2-Minute Exposure)	
Exposure Duration (min)	LCt_{50} (mg-min/m³)	Concentration (mg/m³)		MV (L)	LCt_{50} (mg-min/m³)
1	6000	6000		10	11,250
2	7500	3750		15	7500
5	9500	1900		30	3750
10	11,500	1150		50	2250
20	14,000	700			
30	16,000	530			
60	19,500	325			
120	23,500	195			
240	26,500	120			
Selected Toxicology Information					
Probit Slope: Not Calculated		TLE: 1.4		DOC: Low	

5. Blister Agents

a. HD Profile.

(1) Table H-35 provides toxicity estimates for a lethal dose from an inhalation/ocular vapor exposure to HD.

Table H-35. HD Profile Estimates (Lethal Dose, Inhalation/Ocular)[1]

Ct Profile (15L MV)				MV Profile (2-Minute Exposure)	
Exposure Duration (min)	LCt_{50} (mg-min/m^3)	Concentration (mg/m^3)		MV (L)	LCt_{50} (mg-min/m^3)
2	1000	500		10	1500
10	1710	170		15	1000
30	2466	80		30	500
60	3107	50		50	300
120	3915	30			
240	4932	20			
360	5646	15			
Selected Toxicology Information					
Probit Slope: 6	TLE: 1.5	DOC: Low			

(2) Table H-36 provides toxicity estimates for a lethal dose from a percutaneous vapor exposure to HD. Personnel are masked with clothed skin.

Table H-36. HD Profile Estimates (Lethal Dose, Percutaneous)[1]

Ct Profiles						
Moderate Temperatures				Hot Temperatures		
Exposure Duration (min)	LCt_{50} (mg-min/m^3)	Concentration (mg/m^3)		Exposure Duration (min)	LCt_{50} (mg-min/m^3)	Concentration (mg/m^3)
2	10,000	5000		2	5000	2500
10	10,000	1000		10	5000	500
30	10,000	330		30	5000	165
60	10,000	165		60	5000	85
120	10,000	85		120	5000	40
240	10,000	40		240	5000	20
360	10,000	30		360	5000	15
Selected Toxicology Information						
Probit Slope: 7	TLE: 1 (Assumed)	DOC: Low		Probit Slope: 7	TLE: 1 (Assumed)	DOC: Low
True human LCt_{50} may be lower.						

(3) Table H-37 provides toxicity estimates for severe effects (vesication) from a percutaneous vapor exposure to HD. Personnel are masked with clothed skin.

Table H-37. HD Profile Estimates (Severe Effects, Percutaneous)[1]

Ct Profiles						
Moderate Temperatures				Hot Temperatures		
Exposure Duration (min)	ECt_{50} (mg-min/m^3)	Concentration (mg/m^3)		Exposure Duration (min)	ECt_{50} (mg-min/m^3)	Concentration (mg/m^3)
2	500	250		2	200	100
10	500	50		10	200	20
30	500	15		30	200	5
60	500	10		60	200	3
120	500	5		120	200	1.7
240	500	2		240	200	0.8
360	500	1.4		360	200	0.6
Selected Toxicology Information						
Probit Slope: 3	TLE: 1 (Assumed)	DOC: Moderate		Probit Slope: 3	TLE: 1 (Assumed)	DOC: Moderate
				True human ECt_{50} may be lower.		

(4) Table H-38 provides toxicity estimates for severe effects (eyes) from an ocular vapor exposure to HD.

Table H-38. HD Profile Estimates (Severe Effects, Ocular)[1]

Ct Profile		
Exposure Duration (min)	ECt_{50} (mg-min/m^3)	Concentration (mg/m^3)
2	75	37.5
10	75	7.5
30	75	2.5
60	75	1.25
120	75	0.63
240	75	0.31
360	75	0.21
Selected Toxicology Information		
Probit Slope: 3	TLE: 1	DOC: High

(5) Table H-39 provides toxicity estimates for mild effects from a percutaneous vapor exposure to HD. Personnel are masked with clothed skin.

Table H-39. HD Profile Estimates (Mild Effects, Percutaneous)[1]

Ct Profiles						
Moderate Temperatures				Hot Temperatures		
Exposure Duration (min)	ECt_{50} (mg-min/m^3)	Concentration (mg/m^3)		Exposure Duration (min)	ECt_{50} (mg-min/m^3)	Concentration (mg/m^3)
2	50	25		2	25	12.5
10	50	5		10	25	2.5
30	50	1.7		30	25	0.8
60	50	0.8		60	25	0.4
120	50	0.4		120	25	0.2
240	50	0.2		240	25	0.1
360	50	0.1		360	25	0.07
Selected Toxicology Information						
Probit Slope: 3	TLE: 1	DOC: Moderate		Probit Slope: 3	TLE: 1	DOC: Moderate

(6) Table H-40 provides toxicity estimates for mild effects (eyes) from an ocular vapor exposure to HD.

Table H-40. HD Profile Estimates (Mild Effects, Ocular)[1]

Ct Profile		
Exposure Duration (min)	ECt_{50} (mg-min/m^3)	Concentration (mg/m^3)
2	25	12.5
10	25	2.5
30	25	0.8
60	25	0.4
120	25	0.2
240	25	0.1
360	25	0.07
Selected Toxicology Information		
Probit Slope: 3	TLE: 1	DOC: High

(7) Figure H-11 is a graphical representation of Ct profile for HD vapor dosage from an inhalation/ocular exposure.

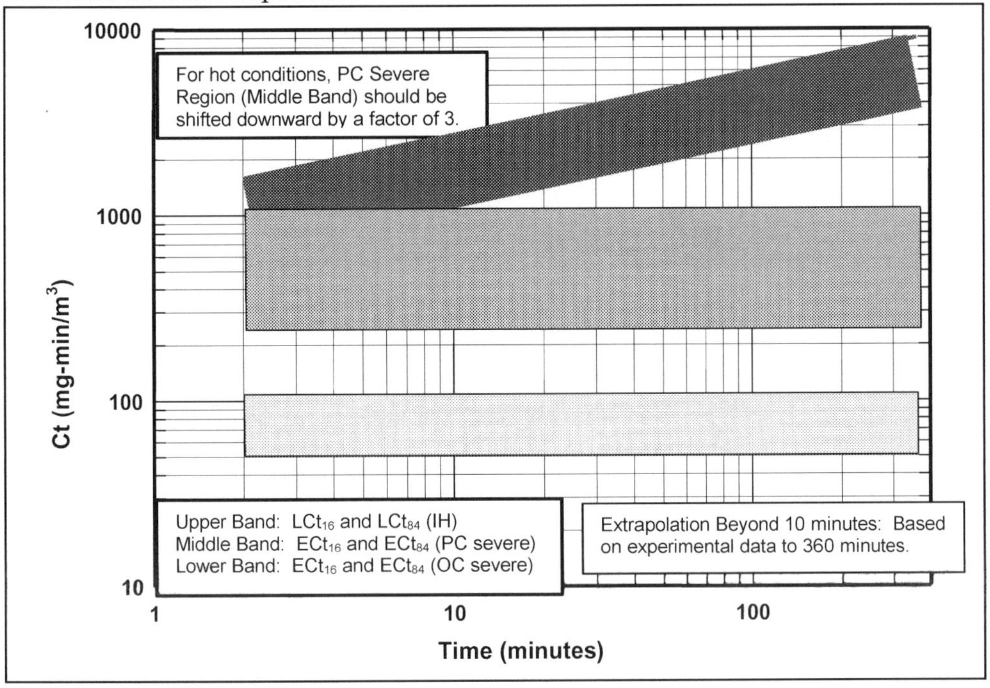

Figure H-11. HD Vapor: Dosage versus Exposure Duration

(8) Figure H-12 is a graphical representation of Ct profile for HD vapor concentration for inhalation/ocular exposure.

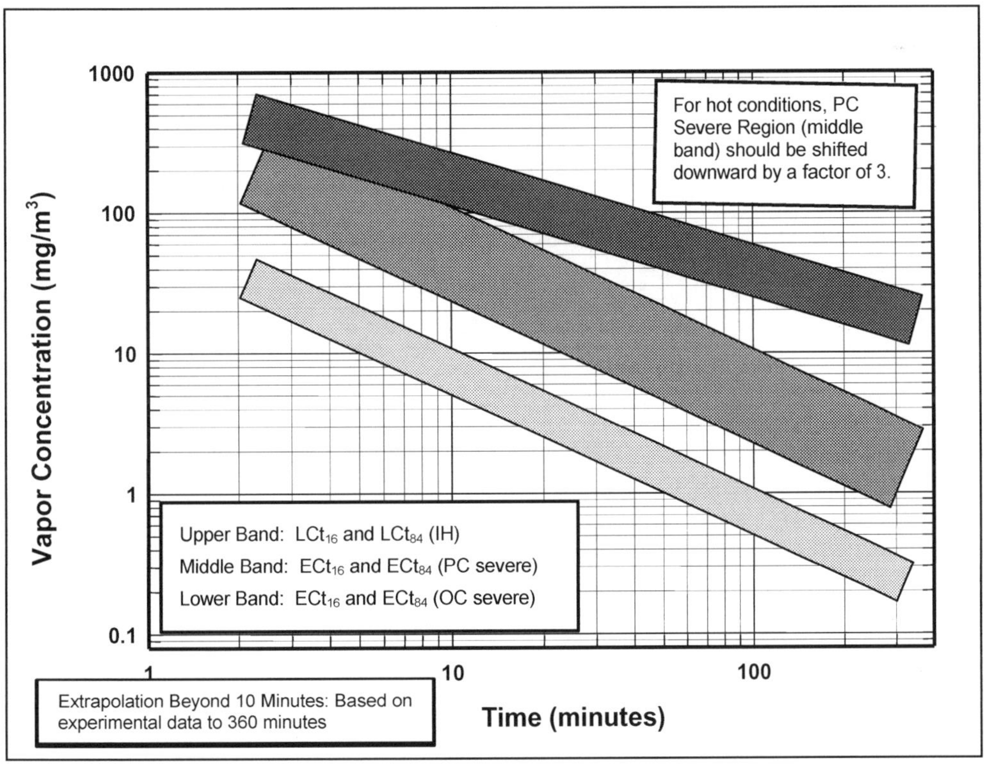

Figure H-12. HD Vapor: Concentration versus Exposure Duration

b. HN-1 Profile.

(1) Table H-41 provides toxicity estimates for a lethal dose from an inhalation/ocular vapor exposure to HN-1.

Table H-41. HN-1 Profile Estimates (Lethal Dose, Inhalation/Ocular)[1]

Ct Profile (15L MV)				MV Profile (2-Minute Exposure)	
Exposure Duration (min)	LCt_{50} (mg-min/m^3)	Concentration (mg/m^3)		MV (L)	LCt_{50} (mg-min/m^3)
2	1000	500		10	1500
10	1000	100		15	1000
30	1000	35		30	500
60	1000	15		50	300
120	1000	10			
240	1000	5			
360	1000	2.5			
Selected Toxicology Information					
Probit Slope: Unknown	TLE: 1 (Assumed)	DOC: Low			

(2) Table H-42 provides toxicity estimates for a lethal dose from a percutaneous vapor exposure to HN-1. Personnel are masked.

Table H-42. HN-1 Profile Estimates (Lethal Dose, Percutaneous)[1]

Ct Profiles						
Moderate Temperatures				Hot Temperatures		
Exposure Duration (min)	LCt_{50} (mg-min/m^3)	Concentration (mg/m^3)		Exposure Duration (min)	LCt_{50} (mg-min/m^3)	Concentration (mg/m^3)
2	10,000	5000		2	5000	2500
10	10,000	1000		10	5000	500
30	10,000	330		30	5000	165
60	10,000	165		60	5000	85
120	10,000	85		120	5000	40
240	10,000	40		240	5000	20
360	10,000	30		360	5000	15
Selected Toxicology Information						
Probit Slope: Unknown	TLE: 1 (Assumed)	DOC: Low		Probit Slope: Unknown	TLE: 1 (Assumed)	DOC: Low

(3) Table H-43 provides toxicity estimates for severe effects (vesication) from a percutaneous vapor exposure to HN-1. Personnel are masked.

Table H-43. HN-1 Profile Estimates (Severe Effects, Percutaneous)[1]

Ct Profiles						
Moderate Temperatures				Hot Temperatures		
Exposure Duration (min)	ECt_{50} (mg-min/m^3)	Concentration (mg/m^3)		Exposure Duration (min)	ECt_{50} (mg-min/m^3)	Concentration (mg/m^3)
2	500	250		2	200	100
10	500	50		10	200	20
30	500	15		30	200	5
60	500	10		60	200	3
120	500	5		120	200	1.7
240	500	2		240	200	0.8
360	500	1.4		360	200	0.6
Selected Toxicology Information						
Probit Slope: Unknown	TLE: 1 (Assumed)	DOC: Low		Probit Slope: Unknown	TLE: 1 (Assumed)	DOC: Low

(4) Table H-44 provides toxicity estimates for severe effects on eyes from an ocular vapor exposure to HN-1.

Table H-44. HN-1 Profile Estimates (Severe Effects, Ocular)[1]

Ct Profile		
Exposure Duration (min)	ECt_{50} (mg-min/m^3)	Concentration (mg/m^3)
2	75	37.5
10	75	7.5
30	75	2.5
60	75	1.25
120	75	0.63
240	75	0.31
360	75	0.21
Selected Toxicology Information		
Probit Slope: Unknown	TLE: 1 (Assumed)	DOC: Low

(5) Table H-45 provides toxicity estimates for mild effects from a percutaneous vapor exposure to HN-1. Personnel are masked.

Table H-45. HN-1 Profile Estimates (Mild Effects, Percutaneous)[1]

Ct Profiles						
Moderate Temperatures				Hot Temperatures		
Exposure Duration (min)	ECt_{50} (mg-min/m^3)	Concentration (mg/m^3)		Exposure Duration (min)	ECt_{50} (mg-min/m^3)	Concentration (mg/m^3)
2	50	25		2	25	12.5
10	50	5		10	25	2.5
30	50	1.7		30	25	0.8
60	50	0.8		60	25	0.4
120	50	0.4		120	25	0.2
240	50	0.2		240	25	0.1
360	50	0.1		360	25	0.07
Selected Toxicology Information						
Probit Slope: Unknown	TLE: 1 (Assumed)	DOC: Low		Probit Slope: Unknown	TLE: 1 (Assumed)	DOC: Low

(6) Table H-46 provides toxicity estimates for mild effects on eyes from an ocular vapor exposure to HN-1.

Table H-46. HN-1 Profile Estimates (Mild Effects, Ocular)[1]

Ct Profile		
Exposure Duration (min)	ECt_{50} (mg-min/m^3)	Concentration (mg/m^3)
2	25	12.5
10	25	2.5
30	25	0.8
60	25	0.4
120	25	0.2
240	25	0.1
360	25	0.07
Selected Toxicology Information		
Probit Slope: Unknown	TLE: 1 (Assumed)	DOC: Low

c. HN-2 Profile.

(1) Table H-47 provides toxicity estimates for a lethal dose from an inhalation/ocular vapor exposure to HN-2.

Table H-47. HN-2 Profile Estimates (Lethal Dose, Inhalation/Ocular)[1]

Ct Profile (15L MV)				MV Profile (2-Minute Exposure)	
Exposure Duration (min)	LCt_{50} (mg-min/m^3)	Concentration (mg/m^3)		MV (L)	LCt_{50} (mg-min/m^3)
2	1000	500		10	1500
10	1000	100		15	1000
30	1000	35		30	500
60	1000	15		50	300
120	1000	10			
240	1000	5			
360	1000	2.5			
Selected Toxicology Information					
Probit Slope: Unknown	TLE: Unknown	DOC: Low			

(2) Table H-48 provides toxicity estimates for a lethal dose from a percutaneous vapor exposure to HN-2. Personnel are masked.

Table H-48. HN-2 Profile Estimates (Lethal Dose, Percutaneous)[1]

Ct Profiles						
Moderate Temperatures				Hot Temperatures		
Exposure Duration (min)	LCt_{50} (mg-min/m^3)	Concentration (mg/m^3)		Exposure Duration (min)	LCt_{50} (mg-min/m^3)	Concentration (mg/m^3)
2	10,000	5000		2	5000	2500
10	10,000	1000		10	5000	500
30	10,000	330		30	5000	165
60	10,000	165		60	5000	85
120	10,000	85		120	5000	40
240	10,000	40		240	5000	20
360	10,000	30		360	5000	15
Selected Toxicology Information						
Probit Slope: Unknown	TLE: 1 (Assumed)	DOC: Low		Probit Slope: Unknown	TLE: 1 (Assumed)	DOC: Low

(3) Table H-49 provides toxicity estimates for severe effects (vesication) for exposure to HN-2. Personnel are masked.

Table H-49. HN-2 Profile Estimates (Severe Effects, Percutaneous)[1]

Ct Profiles						
Moderate Temperatures				Hot Temperatures		
Exposure Duration (min)	ECt_{50} (mg-min/m^3)	Concentration (mg/m^3)		Exposure Duration (min)	ECt_{50} (mg-min/m^3)	Concentration (mg/m^3)
2	500	250		2	200	100
10	500	50		10	200	20
30	500	15		30	200	5
60	500	10		60	200	3
120	500	5		120	200	1.7
240	500	2		240	200	0.8
360	500	1.4		360	200	0.6
Selected Toxicology Information						
Probit Slope: Unknown	TLE: 1 (Assumed)	DOC: Low		Probit Slope: Unknown	TLE: 1 (Assumed)	DOC: Low

(4) Table H-50 provides toxicity estimates for severe effects on eyes from an ocular vapor exposure to HN-2.

Table H-50. HN-2 Profile Estimates (Severe Effects, Ocular)[1]

Ct Profile		
Exposure Duration (min)	**ECt_{50} (mg-min/m3)**	**Concentration (mg/m3)**
2	75	37.5
10	75	7.5
30	75	2.5
60	75	1.25
120	75	0.63
240	75	0.31
360	75	0.21
Selected Toxicology Information		
Probit Slope: Unknown	TLE: 1 (Assumed)	DOC: Low

(5) Table H-51 provides toxicity estimates for mild effects from a percutaneous vapor exposure to HN-2. Mild effects include erythema, itching and some pain. Personnel are masked.

Table H-51. HN-2 Profile Estimates (Mild Effects, Percutaneous)[1]

Ct Profiles						
Moderate Temperatures				**Hot Temperatures**		
Exposure Duration (min)	**ECt_{50} (mg-min/m^3)**	**Concentration (mg/m^3)**		**Exposure Duration (min)**	**ECt_{50} (mg-min/m^3)**	**Concentration (mg/m^3)**
2	50	25		2	25	12.5
10	50	5		10	25	2.5
30	50	1.7		30	25	0.8
60	50	0.8		60	25	0.4
120	50	0.4		120	25	0.2
240	50	0.2		240	25	0.1
360	50	0.1		360	25	0.07
Selected Toxicology Information						
Probit Slope: Unknown	TLE: 1 (Assumed)	DOC: Low		Probit Slope: Unknown	TLE: 1 (Assumed)	DOC: Low

(6) Table H-52 provides toxicity estimates for mild effects on eyes from an ocular vapor exposure to HN-2.

Table H-52. HN-2 Profile Estimates (Mild Effects, Ocular)[1]

Ct Profile		
Exposure Duration (min)	**ECt_{50} (mg-min/m^3)**	**Concentration (mg/m^3)**
2	25	12.5
10	25	2.5
30	25	0.8
60	25	0.4
120	25	0.2
240	25	0.1
360	25	0.07
Selected Toxicology Information		
Probit Slope: Unknown	TLE: 1 (Assumed)	DOC: Low

d. HN-3 Profile.

(1) Table H-53 provides toxicity estimates for a lethal dose from an inhalation/ocular vapor exposure to HN-3.

Table H-53. HN-3 Profile Estimates (Lethal Dose, Inhalation/Ocular)[1]

Ct Profile (15L MV)				MV Profile (2-Minute Exposure)	
Exposure Duration (min)	LCt_{50} (mg-min/m^3)	Concentration (mg/m^3)		MV (L)	LCt_{50} (mg-min/m^3)
2	1000	500		10	1500
10	1000	100		15	1000
30	1000	35		30	500
60	1000	15		50	300
120	1000	10			
240	1000	5			
360	1000	2.5			
Selected Toxicology Information					
Probit Slope: Unknown	TLE: 1 (Assumed)	DOC: Low			

(2) Table H-54 provides toxicity estimates given for a lethal dose from a percutaneous vapor exposure to HN-3. Personnel are masked.

Table H-54. HN-3 Profile Estimates (Lethal Dose, Percutaneous)[1]

Ct Profiles						
Moderate Temperatures				Hot Temperatures		
Exposure Duration (min)	LCt_{50} (mg-min/m^3)	Concentration (mg/m^3)		Exposure Duration (min)	LCt_{50} (mg-min/m^3)	Concentration (mg/m^3)
2	10,000	5000		2	5000	2500
10	10,000	1000		10	5000	500
30	10,000	330		30	5000	165
60	10,000	165		60	5000	85
120	10,000	85		120	5000	40
240	10,000	40		240	5000	20
360	10,000	30		360	5000	15
Selected Toxicology Information						
Probit Slope: Unknown	TLE: 1 (Assumed)	DOC: Low		Probit Slope: Unknown	TLE: 1 (Assumed)	DOC: Low

(3) Table H-55 provides toxicity estimates for severe effects (vesication) from a percutaneous vapor exposure to HN-3. Personnel are masked.

Table H-55. HN-3 Profile Estimates (Severe Effects, Percutaneous)[1]

Ct Profiles						
Moderate Temperatures				Hot Temperatures		
Exposure Duration (min)	ECt_{50} (mg-min/m^3)	Concentration (mg/m^3)		Exposure Duration (min)	ECt_{50} (mg-min/m^3)	Concentration (mg/m^3)
2	500	250		2	200	100
10	500	50		10	200	20
30	500	15		30	200	5
60	500	10		60	200	3
120	500	5		120	200	1.7
240	500	2		240	200	0.8
360	500	1.4		360	200	0.6
Selected Toxicology Information						
Probit Slope: Unknown	TLE: 1 (Assumed)	DOC: Low		Probit Slope: Unknown	TLE: 1 (Assumed)	DOC: Low

(4) Table H-56 provides toxicity estimates given for severe effects on eyes from an ocular vapor exposure to HN-3.

Table H-56. HN-3 Profile Estimates (Severe Effects, Ocular)[1]

Ct Profile		
Exposure Duration (min)	**ECt_{50} (mg-min/m^3)**	**Concentration (mg/m^3)**
2	75	37.5
10	75	7.5
30	75	2.5
60	75	1.25
120	75	0.63
240	75	0.31
360	75	0.21
Selected Toxicology Information		
Probit Slope: Unknown	TLE: 1 (Assumed)	DOC: Low

(5) Table H-57 provides toxicity estimates for mild effects from a percutaneous vapor exposure to HN-3. Personnel are masked.

Table H-57. HN-3 Profile Estimates (Mild Effects, Percutaneous)[1]

Ct Profiles					
Moderate Temperatures			**Hot Temperatures**		
Exposure Duration (min)	**ECt_{50} (mg-min/m^3)**	**Concentration (mg/m^3)**	**Exposure Duration (min)**	**ECt_{50} (mg-min/m^3)**	**Concentration (mg/m^3)**
2	50	25	2	25	12.5
10	50	5	10	25	2.5
30	50	1.7	30	25	0.8
60	50	0.8	60	25	0.4
120	50	0.4	120	25	0.2
240	50	0.2	240	25	0.1
360	50	0.1	360	25	0.07
Selected Toxicology Information					
Probit Slope: Unknown	TLE: 1 (Assumed)	DOC: Low	Probit Slope: Unknown	TLE: 1 (Assumed)	DOC: Low

(6) Table H-58 provides toxicity estimates for mild effects on eyes from an ocular vapor exposure to HN-3.

Table H-58. HN-3 Profile Estimates (Mild Effects, Ocular)[1]

Ct Profile		
Exposure Duration (min)	**ECt_{50} (mg-min/m^3)**	**Concentration (mg/m^3)**
2	25	12.5
10	25	2.5
30	25	0.8
60	25	0.4
120	25	0.2
240	25	0.1
360	25	0.07
Selected Toxicology Information		
Probit Slope: Unknown	TLE: 1 (Assumed)	DOC: Low

e. HT Profile.

(1) Table H-59 provides toxicity estimates for a lethal dose from an inhalation/ocular vapor exposure to HT.

Table H-59. HT Profile Estimates (Lethal Dose, Inhalation/Ocular)[1]

Ct Profile (15L MV)				MV Profile (2-Minute Exposure)	
Exposure Duration (min)	LCt_{50} (mg-min/m^3)	Concentration (mg/m^3)		MV (L)	LCt_{50} (mg-min/m^3)
2	1000	500		10	1500
10	1000	100		15	1000
30	1000	35		30	500
60	1000	15		50	300
120	1000	10			
240	1000	5			
360	1000	2.5			
Selected Toxicology Information					
Probit Slope: Unknown		TLE: 1 (Assumed)		DOC: Low	

(2) Table H-60 provides toxicity estimates for a lethal dose from a percutaneous vapor exposure to HT. Personnel are masked.

Table H-60. HT Profile Estimates (Lethal Dose, Percutaneous)[1]

Ct Profiles						
Moderate Temperatures				Hot Temperatures		
Exposure Duration (min)	LCt_{50} (mg-min/m^3)	Concentration (mg/m^3)		Exposure Duration (min)	LCt_{50} (mg-min/m^3)	Concentration (mg/m^3)
2	10,000	5000		2	5000	2500
10	10,000	1000		10	5000	500
30	10,000	330		30	5000	165
60	10,000	165		60	5000	85
120	10,000	85		120	5000	40
240	10,000	40		240	5000	20
360	10,000	30		360	5000	15
Selected Toxicology Information						
Probit Slope: Unknown	TLE: 1 (Assumed)	DOC: Low		Probit Slope: Unknown	TLE: 1 (Assumed)	DOC: Low

(3) Table H-61 provides toxicity estimates for severe effects (vesication) from a percutaneous vapor exposure to HT. Personnel are masked.

Table H-61. HT Profile Estimates (Severe Effects, Percutaneous)[1]

Ct Profiles						
Moderate Temperatures				Hot Temperatures		
Exposure Duration (min)	ECt_{50} (mg-min/m^3)	Concentration (mg/m^3)		Exposure Duration (min)	ECt_{50} (mg-min/m^3)	Concentration (mg/m^3)
2	500	250		2	200	100
10	500	50		10	200	20
30	500	15		30	200	5
60	500	10		60	200	3
120	500	5		120	200	1.7
240	500	2		240	200	0.8
360	500	1.4		360	200	0.6
Selected Toxicology Information						
Probit Slope: Unknown	TLE: 1 (Assumed)	DOC: Low		Probit Slope: Unknown	TLE: 1 (Assumed)	DOC: Low

(4) Table H-62 provides toxicity estimates for severe effects on eyes from an ocular vapor exposure to HT.

Table H-62. HT Profile Estimates (Severe Effects, Ocular)[1]

Ct Profile		
Exposure Duration (min)	ECt_{50} (mg-min/m^3)	Concentration (mg/m^3)
2	75	37.5
10	75	7.5
30	75	2.5
60	75	1.25
120	75	0.63
240	75	0.31
360	75	0.21
Selected Toxicology Information		
Probit Slope: Unknown	TLE: 1 (Assumed)	DOC: Low

(5) Table H-63 provides toxicity estimates for mild effects from a percutaneous vapor exposure to HT. Personnel are masked.

Table H-63. HT Profile Estimates (Mild Effects, Percutaneous)[1]

Ct Profiles						
Moderate Temperatures				Hot Temperatures		
Exposure Duration (min)	ECt_{50} (mg-min/m^3)	Concentration (mg/m^3)		Exposure Duration (min)	ECt_{50} (mg-min/m^3)	Concentration (mg/m^3)
2	50	25		2	25	12.5
10	50	5		10	25	2.5
30	50	1.7		30	25	0.8
60	50	0.8		60	25	0.4
120	50	0.4		120	25	0.2
240	50	0.2		240	25	0.1
360	50	0.1		360	25	0.07
Selected Toxicology Information						
Probit Slope: Unknown	TLE: 1 (Assumed)	DOC: Low		Probit Slope: Unknown	TLE: 1 (Assumed)	DOC: Low

(6) Table H-64 provides toxicity estimates for mild effects on eyes from an ocular vapor exposure to HT.

Table H-64. HT Profile Estimates (Mild Effects, Ocular)[1]

Ct Profile		
Exposure Duration (min)	ECt_{50} (mg-min/m^3)	Concentration (mg/m^3)
2	25	12.5
10	25	2.5
30	25	0.8
60	25	0.4
120	25	0.2
240	25	0.1
360	25	0.07
Selected Toxicology Information		
Probit Slope: Unknown	TLE: 1 (Assumed)	DOC: Low

f. L Profile.

(1) Table H-65 provides toxicity estimates for a lethal dose from an inhalation/ocular vapor exposure to L.

Table H-65. L Profile Estimates (Lethal Dose, Inhalation/Ocular)[1]

Ct Profile (15L MV)				MV Profile (2-Minute Exposure)	
Exposure Duration (min)	LCT_{50} (mg-min/m^3)	Concentration (mg/m^3)		MV (L)	LCt_{50} (mg-min/m^3)
2	1000	700		10	1500
10	1000	140		15	1000
30	1000	45		30	500
60	1000	25		50	300
120	1000	10			
240	1000	5			
360	1000	4			
Selected Toxicology Information					
Probit Slope: Unknown	TLE: 1 (Assumed)	DOC: Low			

(2) Table H-66 provides toxicity estimates for a lethal dose from a percutaneous vapor exposure to L. Personnel are masked.

Table H-66. L Profile Estimates (Lethal Dose, Percutaneous)[1]

Ct Profiles						
Moderate Temperatures				Hot Temperatures		
Exposure Duration (min)	LCt_{50} (mg-min/m^3)	Concentration (mg/m^3)		Exposure Duration (min)	LCt_{50} (mg-min/m^3)	Concentration (mg/m^3)
2	5000-10,000	2500-5000		2	2500-5000	1260-2500
10	5000-10,000	500-1000		10	2500-5000	250-500
30	5000-10,000	165-330		30	2500-5000	85-165
60	5000-10,000	85-165		60	2500-5000	40-85
120	5000-10,000	40-85		120	2500-5000	20-40
240	5000-10,000	20-40		240	2500-5000	10-20
360	5000-10,000	15-30		360	2500-5000	5-15
Selected Toxicology Information						
Probit Slope: Unknown	TLE: 1 (Assumed)	DOC: Low		Probit Slope: Unknown	TLE: 1 (Assumed)	DOC: Low
True human LCt_{50} may be lower.				True human LCt_{50} may be lower.		

(3) Table H-67 provides toxicity estimates for severe effects (vesication) from a percutaneous vapor exposure to L. Personnel are masked.

Table H-67. L Profile Estimates (Severe Effects, Percutaneous)[1]

Ct Profiles						
Moderate Temperatures				Hot Temperatures		
Exposure Duration (min)	ECt_{50} (mg-min/m^3)	Concentration (mg/m^3)		Exposure Duration (min)	ECt_{50} (mg-min/m^3)	Concentration (mg/m^3)
2	500	250		2	200	100
10	500	50		10	200	20
30	500	15		30	200	5
60	500	10		60	200	3
120	500	5		120	200	1.7
240	500	2		240	200	0.8
360	500	1.4		360	200	0.6
Selected Toxicology Information						
Probit Slope: Unknown	TLE: 1 (Assumed)	DOC: Low		Probit Slope: Unknown	TLE: 1 (Assumed)	DOC: Low

(4) Table H-68 provides toxicity estimates for severe effects on eyes from an ocular vapor exposure to L.

Table H-68. L Profile Estimates (Severe Effects, Ocular)[1]

Ct Profile		
Exposure Duration (min)	**ECt_{50} (mg-min/m^3)**	**Concentration (mg/m^3)**
2	75	37.5
10	75	7.5
30	75	2.5
60	75	1.25
120	75	0.63
240	75	0.31
360	75	0.21
Selected Toxicology Information		
Probit Slope: Unknown	TLE: 1 (Assumed)	DOC: Low

(5) Table H-69 provides toxicity estimates for mild effects from a percutaneous vapor exposure to L. Personnel are masked.

Table H-69. L Profile Estimates (Mild Effects, Percutaneous)[1]

Ct Profiles							
Moderate Temperatures				**Hot Temperatures**			
Exposure Duration (min)	**ECt_{50} (mg-min/m^3)**	**Concentration (mg/m^3)**		**Exposure Duration (min)**	**ECt_{50} (mg-min/m^3)**	**Concentration (mg/m^3)**
2	50	25		2	25	12.5
10	50	5		10	25	2.5
30	50	1.7		30	25	0.8
60	50	0.8		60	25	0.4
120	50	0.4		120	25	0.2
240	50	0.2		240	25	0.1
360	50	0.1		360	25	0.07
Selected Toxicology Information						
Probit Slope: Unknown	TLE: 1 (Assumed)	DOC: Low		Probit Slope: Unknown	TLE: 1 (Assumed)	DOC: Low

(6) Table H-70 provides toxicity estimates for mild effects on eyes from an ocular vapor exposure to L.

Table H-70. L Profile Estimates (Mild Effects, Ocular)[1]

Ct Profile		
Exposure Duration (min)	**ECt_{50} (mg-min/m^3)**	**Concentration (mg/m^3)**
2	25	12.5
10	25	2.5
30	25	0.8
60	25	0.4
120	25	0.2
240	25	0.1
360	25	0.07
Selected Toxicology Information		
Probit Slope: Unknown	TLE: 1 (Assumed)	DOC: Low

g. HL Profile.

(1) Table H-71 provides toxicity estimates for a lethal dose from an inhalation/ocular vapor exposure to HL.

Table H-71. HL Profile Estimates (Lethal Dose, Inhalation/Ocular)[1]

Ct Profile (15L MV)				MV Profile (2-Minute Exposure)	
Exposure Duration (min)	LCt_{50} (mg-min/m^3)	Concentration (mg/m^3)		MV (L)	LCt_{50} (mg-min/m^3)
2	1000	500		10	1500
10	1000	170		15	1000
30	1000	80		30	500
60	1000	50		50	300
120	1000	30			
240	1000	20			
360	1000	15			
Selected Toxicology Information					
Probit Slope: Unknown	TLE: 1 (Assumed)	DOC: Low			

(2) Table H-72 provides toxicity estimates for a lethal dose from a percutaneous vapor exposure to HL. Personnel are masked.

Table H-72. HL Profile Estimates (Lethal Dose, Percutaneous)[1]

Ct Profiles						
Moderate Temperatures				Hot Temperatures		
Exposure Duration (min)	LCt_{50} (mg-min/m^3)	Concentration (mg/m^3)		Exposure Duration (min)	LCt_{50} (mg-min/m^3)	Concentration (mg/m^3)
2	10,000	5000		2	5000	2500
10	10,000	1000		10	5000	500
30	10,000	330		30	5000	165
60	10,000	165		60	5000	85
120	10,000	85		120	5000	40
240	10,000	40		240	5000	20
360	10,000	30		360	5000	15
Selected Toxicology Information						
Probit Slope: Unknown	TLE: 1 (Assumed)	DOC: Low		Probit Slope: Unknown	TLE: 1 (Assumed)	DOC: Low

(3) Table H-73 provides toxicity estimates for severe effects (vesication) from a percutaneous vapor exposure to HL. Personnel are masked.

Table H-73. HL Profile Estimates (Severe Effects, Percutaneous)[1]

Ct Profiles						
Moderate Temperatures				Hot Temperatures		
Exposure Duration (min)	ECt_{50} (mg-min/m^3)	Concentration (mg/m^3)		Exposure Duration (min)	ECt_{50} (mg-min/m^3)	Concentration (mg/m^3)
2	500	250		2	200	100
10	500	50		10	200	20
30	500	15		30	200	5
60	500	10		60	200	3
120	500	5		120	200	1.7
240	500	2		240	200	0.8
360	500	1.4		360	200	0.6
Selected Toxicology Information						
Probit Slope: Unknown	TLE: 1 (Assumed)	DOC: Low		Probit Slope: Unknown	TLE: 1 (Assumed)	DOC: Low

(4) Table H-74 provides toxicity estimates for severe effects on eyes from an ocular vapor exposure to HL.

Table H-74. HL Profile Estimates (Severe Effects, Ocular)[1]

Ct Profile		
Exposure Duration (min)	**ECt_{50} (mg-min/m^3)**	**Concentration (mg/m^3)**
2	75	37.5
10	75	7.5
30	75	2.5
60	75	1.25
120	75	0.63
240	75	0.31
360	75	0.21
Selected Toxicology Information		
Probit Slope: Unknown	TLE: 1 (Assumed)	DOC: Low

(5) Table H-75 provides toxicity estimates for mild effects from a percutaneous vapor exposure to HL. Personnel are masked.

Table H-75. HL Profile Estimates (Mild Effects, Percutaneous)[1]

Ct Profiles						
Moderate Temperatures				**Hot Temperatures**		
Exposure Duration (min)	**ECt_{50} (mg-min/m^3)**	**Concentration (mg/m^3)**		**Exposure Duration (min)**	**ECt_{50} (mg-min/m^3)**	**Concentration (mg/m^3)**
2	50	25		2	25	12.5
10	50	5		10	25	2.5
30	50	1.7		30	25	0.8
60	50	0.8		60	25	0.4
120	50	0.4		120	25	0.2
240	50	0.2		240	25	0.1
360	50	0.1		360	25	0.07
Selected Toxicology Information						
Probit Slope: Unknown	TLE: 1 (Assumed)	DOC: Low		Probit Slope: Unknown	TLE: 1 (Assumed)	DOC: Low

(6) Table H-76 provides toxicity estimates for mild effects on eyes from an ocular vapor exposure to HL.

Table H-76. HL Profile Estimates (Mild Effects, Ocular)[1]

Ct Profile		
Exposure Duration (min)	**ECt_{50} (mg-min/m^3)**	**Concentration (mg/m^3)**
2	25	12.5
10	25	2.5
30	25	0.8
60	25	0.4
120	25	0.2
240	25	0.1
360	25	0.07
Selected Toxicology Information		
Probit Slope: Unknown	TLE: 1 (Assumed)	DOC: Low

h. PD Profile. Table H-77 provides toxicity estimates for a lethal dose from an inhalation/ocular vapor exposure to PD.

Table H-77. PD Profile Estimates (Lethal Dose, Inhalation/Ocular)[1]

Ct Profile (15L MV)				MV Profile (2-Minute Exposure)	
Exposure Duration (min)	LCt_{50} (mg-min/m^3)	Concentration (mg/m^3)		MV (L)	LCt_{50} (mg-min/m^3)
2	2600	1300		10	3900
10	2600	260		15	2600
30	2600	85		30	1300
60	2600	45		50	780
120	2600	20			
240	2600	10			
360	2600	5			
Selected Toxicology Information					
Probit Slope: Unknown	TLE: 1 (Assumed)	DOC: Low			

i. CX Profile. Table H-78 provides toxicity estimates for a lethal dose from an inhalation/ocular vapor exposure to CX.

Table H-78. CX Profile Estimates (Lethal Dose, Inhalation/Ocular)[1]

Ct Profile (15L MV)				MV Profile (2-Minute Exposure)	
Exposure Duration (min)	LCt_{50} (mg-min/m^3)	Concentration (mg/m^3)		MV (L)	LCt_{50} (mg-min/m^3)
2	3200	1600		10	4800
10	3200	320		15	3200
30	3200	105		30	1600
60	3200	55		50	960
120	3200	25			
240	3200	15			
360	3200	10			
Selected Toxicology Information					
Probit Slope: Unknown	TLE: 1 (Assumed)	DOC: Low			

6. Respiratory Irritants

Table H-79 provides toxicity estimates for a lethal dose from an inhalation/ocular vapor exposure to DM.

Table H-79. DM Profile Estimates (Lethal Dose, Inhalation/Ocular)[1]

Ct Profile (15L MV)				MV Profile (2-Minute Exposure)	
Exposure Duration (min)	LCt_{50} (mg-min/m^3)	Concentration (mg/m^3)		MV (L)	LCt_{50} (mg-min/m^3)
2	11,000	5500		10	16,500
10	11,000	1100		15	11,000
30	11,000	365		30	5500
60	11,000	185		50	3300
120	11,000	90			
240	11,000	45			
Selected Toxicology Information					
Probit Slope: Unknown	TLE: 1 (Assumed)	DOC: Low			

NOTES

[1]Sharon Reutter, et al., *Review and Recommendations for Human Toxicity Estimates for FM 3-11.9*, ECBC-TR-349, September 2003.

Appendix I

PROPERTIES OF SELECTED BIOLOGICAL AGENTS

This appendix provides information on the properties of selected biological agents.

Table I-1. Properties of Selected Biological Agents

BW Agents[1]	Likely Methods of Dissemination	Transmissibility Person-to-Person	Infectivity	Lethality[2]	Stability[2]
Anthrax (Inhalation)	Spores in aerosols	None	Moderate	High	Spores are highly stable
Brucellosis	1. Aerosol 2. Sabotage (food supply)	None	High	Low	Long persistence in wet soil and food
Cholera	1. Sabotage (food/water supply) 2. Aerosol	Negligible	Low	Moderate to high	Unstable in aerosol and pure water; more so in polluted water
Glanders	Aerosol	DNA	DNA	DNA	DNA
Melioidosis	Aerosol	Negligible	High	Variable	Stable
Plague (Pneumonic)	1. Aerosol 2. Infected Vectors	High	High	Very high	Less important because of high transmissibility
Psittacosis	Aerosol	Negligible	Moderate	Very low	Stable
Shigellosis	Sabotage (Food/Water Supply)	DNA	DNA	DNA	DNA
Tularemia	Aerosol	Negligible	High	Moderate if untreated	Not very stable
Typhoid Fever	1. Sabotage (food/water supply) 2. Aerosol	Negligible	Moderate	Moderate if untreated	Unknown
Q Fever	1. Aerosol 2. Sabotage (food supply)	None	High	Very low	Stable
Rocky Mountain Spotted Fever	1. Aerosol 2. Infected Vectors	None	High	High	Not very stable
Trench Fever	1. Aerosol 2. Vector	None	DNA	Low	DNA
Typhus Fever	1. Aerosol 2. Infected vectors	None	High	High	Not very stable
Chikungunya	Aerosol	None	High	Very low	Relatively stable
Crimean-Congo Hemorrhagic Fever	Aerosol	Moderate	High	High	Relatively stable
Dengue Fever	Aerosol	None	High	Low	Relatively unstable
Eastern Equine Encephalitis	Aerosol	None	High	High	Relatively unstable
Western Equine Encephalitis	Aerosol	None	High	Low	Relatively unstable
Ebola Fever	Aerosol	Moderate	High	High	Relatively unstable
Far Eastern Tick-borne Encephalitis	1. Aerosol 2. Milk	None	High	Moderate	Relatively unstable
Hantaan Virus (Korean HFV)	Aerosol	None	High	Moderate	Relatively stable
Juinn Hemorrhagic Fever	Aerosol	DNA	DNA	DNA	DNA

Table I-1. Properties of Selected Biological Agents (Continued)

BW Agents[1]	Likely Methods of Dissemination	Transmissibility Person-to-Person	Infectivity	Lethality[2]	Stability[2]
Machuo Hemorrhagic Fever	Aerosol	DNA	DNA	DNA	DNA
Lassa Fever	Aerosol	Low to moderate	High	Unknown	Relatively stable
Lymphocytic Choriomeningitis	Aerosol	None	DNA	Low	DNA
Monkeypox	Aerosol	DNA			
Rift Valley Fever	1. Aerosol 2. Infected vectors	Low	High	Low	Relatively stable
Smallpox	Aerosol	High	High	High	Stable
Venezuelan Equine Encephalitis	1. Aerosol 2. Infected vectors	Low	High	Low	Relatively unstable
Yellow Fever	Aerosol	None	High	High	Relatively unstable
Botulinum Toxin	1. Sabotage (food/water supply) 2. Aerosol	None	N/A	High	Stable
Clostridium Perfringens Toxins	1. Sabotage 2. Aerosol	None	N/A	Low	Stable
Conotoxins	1. Aerosol 2. Sabotage (food/water)	None	N/A	High	DNa
Shigatoxins	1. Aerosol 2. Sabotage (food/water)	DNA	N/A	DNA	DNA
Microcystin	1. Aerosol 2. Sabotage (food/water)	None	N/A	DNA	DNA
Ricin	Aerosol	None	N/A	High	Stable
Saxitoxin	1. Sabotage 2. Aerosol	None	N/A	High	Stable
Staphylococcal Enterotoxin B	1. Aerosol 2. Sabotage	None	N/A	Low	Stable
Tetrodotoxin	1. Sabotage 2. Aerosol	None	N/A	High	Stable
Trichothecene Mycotoxins	Aerosol	None	N/A	High	Stable

NOTES

[1]FM 8-9/NAVMED P-5059/AFJMAN 44-151, *NATO Handbook on the Medical Aspects of NBC Defense Operations AMEDP-6(B),* 1 February 1996.

[2]Lethality and stability information is based on studies of natural, endemic strains of these agents. There is potential that the lethality and/or stability of some agents may be changed through various laboratory alterations and/or the use of stabilizing chemicals.

Appendix J

SELECTED ANIMAL PATHOGENS

1. Background

This appendix provides summaries of animal diseases listed in *The Biological and Chemical Warfare Threat* (1999). The information generally addressed in this appendix provides descriptions, treatment, and control measures. The military role, if any, would likely be in a consequence management support role, and the brief descriptions provided furnish an awareness of the agents that could be used. For further information, contact the Army Medical Department (AMEDD).

2. Animal Diseases

a. African Swine Fever (ASF).[1]

(1) ASF is a tick-borne and contagious disease of swine caused by a virus and characterized by hemorrhages of internal organs, high fever, moderate anorexia, and leukopia. Mortality rates may reach 100 percent. It occurs in Africa and on the island of Sardinia.

(2) There is no treatment available.

(3) Control. Slaughter and disposal of all acutely infected pigs, widespread testing and elimination of all seropositive animals, and good herd isolation and sanitary practices can accomplish control and eradication of ASF in developed countries. There is no vaccine.

(4) Human beings are not susceptible to ASF.

b. Avian Influenza (AI) (Fowl Plague).[2]

(1) AI is a disease of viral etiology that ranges from a mild or even asymptomatic infection to an acute, fatal disease of chickens, turkeys, guinea fowls, and other avian species, especially migratory waterfowl. Highly pathogenic AI viruses have periodically occurred in recent years in Australia, England, South Africa, Scotland, Ireland, Mexico, Pakistan, and the US. It is generally accepted belief that waterfowl, sea birds, or shore birds are generally responsible for introducing the virus into poultry. Death may occur within 24 hours of first signs of disease, frequently within 48 hours, or can be delayed for as long as a week.

(2) The practice of accepted sanitation and biosecurity procedures in the rearing of the poultry is of utmost importance. In areas where waterfowl, shore birds, or sea birds are prevalent, the rearing of poultry on open range is incompatible with a sound AI prevention program. Cleaning and disinfection procedures are critical.

(3) The AI viruses are Type A influenza viruses, and the possibility exists that they could be involved in the development, through genetic reassortment, of new mammalian strains. The infection and deaths of 6 of 18 humans infected with an AI virus in Hong Kong in 1997 has resulted in a reconsideration of the role that the avian species have on the epidemiology of human influenza.

(4) The AI viruses are Type A influenza viruses, and the possibility exists that they could be involved in the development, through genetic reassortment, of new mammalian strains.

c. Bluetongue (Sore Muzzle).[3]

(1) Bluetongue is an acute, noncontagious, insect-borne disease of sheep, goats, cattle, and wild ruminants caused by a virus. Occurrence is probably worldwide. Cattle and goats with inapparent infections are important reservoirs of the virus.

(2) The only applicable treatment available is to minimize animal stress and administer broad-spectrum antibiotics to combat secondary infections.

(3) A live attenuated vaccine is available for use in sheep in the US.

(4) There is only one documented human infection, and that was a laboratory worker.

d. Foot-and-Mouth Disease (FMD).[4]

(1) FMD is a highly contagious disease of cloven-footed domestic and wild animals caused by a virus. The morbidity is essentially 100 percent, and the mortality is less than 1 percent. Great economic loss results from the effects of the disease, which include lameness, low milk production, weight loss, mastitis, debilitation, and abortion. FMD occurs in many major livestock-producing countries.

(2) A number of inactivated vaccines, including those prepared in cell cultures containing the appropriate types or subtypes, are used in countries where the disease is endemic. The duration of immunity may be as short as 4 months. Treatment to prevent and cope with secondary infection may be necessary.

(3) In countries free of FMD, a policy of quarantine and slaughter is usually practiced with the goal of complete eradication.

(4) In a review of the zoonotic aspects of FMD by K. Bauer in 1997, he reported that, since 1921, the FMD virus has been isolated and typed from slightly over 40 human cases. The cases occurred on three continents: Europe, Africa, and South America. Because infection is uncommon, FMD is not considered to be a public health problem.

e. Sheep Pox and Goat Pox.[5]

(1) Sheep pox and goat pox, in fully susceptible animals, are highly contagious and often fatal diseases caused by a capripoxvirus and characterized by fever and pox. The diseases are endemic in Africa, the Middle East, the Indian subcontinent, and much of Asia. Transmission is by direct and indirect contact, and the viruses can survive on contaminated premises for many months.

(2) There is no treatment available.

(3) When a new case is confirmed, the area should be quarantined, infected and exposed animals should be slaughtered, and the premises cleaned and disinfected. Vaccination of susceptible animals on premises surrounding the infected flock should be considered.

(4) There is no conclusive evidence that SGPV infects humans.

f. Aujeszky's Disease (AJD) (Pseudorabies).[6]

(1) AJD is caused by porcine herpesvirus-1. Pigs are the main host, but sporadic cases have occurred in cattle, sheep, goats, horses, dogs, cats, foxes, and rodents. AJD is highly contagious and is principally spread via the respiratory route. It is endemic in the swine of many countries. It occurs frequently in the US, but not in Canada and Australia.

(2) There is no treatment available.

(3) Live attenuated and inactivated virus vaccines are available to reduce losses in herds in which the disease is a continuing problem. Vaccination does not necessarily prevent infection or shedding of the virus.

(4) There is no transmission to humans.

g. Hog Cholera (Swine Fever).[7]

(1) Hog cholera, a highly contagious disease, is caused by a virus and is characterized in fully susceptible pigs by high mortality. The disease occurs in many countries of Europe, Africa, and Asia. It has been eradicated in Australia, Canada, and the US. The disease is spread by direct and indirect contact.

(2) There is no treatment available.

(3) A strict regimen of vaccination will reduce the number of outbreaks to a level at which complete eradication by sanitary measure alone will be feasible.

(4) Human beings are not susceptible to Hog Cholera infection.

h. Lyssa Virus (Rabies).[8]

(1) Rabies infections are usually established following introduction of virus-infected saliva into a bite or scratch, although animals can also be infected by the oral and nasal (olfactory) routes. Bat-infested caves may result in infectious aerosols; nonbite transmission of the disease from this source has been reported in a number of animal species, including humans. The disease generally takes one of two forms: "furious," with sporadic episodes of rage; or "dumb," in which there is an early progressive paralysis. Both forms almost invariably result in death. A normally lively and sociable dog, for example, may become anorexic, withdrawn, irritable, or restless. This behavior may suddenly change, with the animal becoming highly affectionate. At this stage, the dog may try repeatedly to lick the hands and face of its owner or handler. As the disease progresses, the animal may appear to have difficulty swallowing, as if a bone were caught in its throat. Any attempt to alleviate the problem manually exposes the handler to considerable risk, either through a bite or the deposition of virus-infected saliva on mucous membranes or minor scratches. The dog's bark becomes high pitched and hoarse, indicating the onset of paralysis. The animal drools saliva. Convulsive seizures and muscular incoordination become apparent, followed by progressive paralysis, usually terminating in death within 7 days of the onset of symptoms. In about 25 to 50 percent of cases, apparently as a result of limbic lobe dysfunction, dogs with rabies develop the furious form of the disease. Affected animals may eat abnormal objects, and during paroxysms of rage, will attack almost anything. Rabies is present in most of Europe, throughout Africa, the Middle East, and most of Asia and the Americas. The UK, Ireland, parts of Scandinavia, Japan, Singapore, Australia, New Zealand, Papua New Guinea, and the Pacific Islands are free of rabies.

(2) There is no treatment available.

(3) Once a virulent strain of rabies virus has established itself in the CNS of an infected animal, the outcome is almost always death. At a practical level, however, the only ways of preventing rabies are by preexposure immunization.

(4) All warm blooded animals, including humans, are susceptible.

i. Velogenic Newcastle Disease (VND) (Exotic Newcastle Disease, Asiatic Newcastle disease).[9]

(1) VND is likely the most serious disease of poultry throughout the world. In chickens it is characterized by morbidity rates near 100 percent and mortality rates as high as 90 percent in susceptible chickens.

(2) The establishment of a strict quarantine, destruction of all infected and exposed birds followed by thorough cleaning, and disinfection of the premises are the main actions necessary for eradication of VND virus. VND virus has been recovered from effluent water for as long as 21 days and from carcasses for 7 days when the daytime temperatures were over 90 degrees.

(3) Although people may become infected with VND virus, the resulting disease is typically limited to conjunctivitis. Recovery is usually rapid, and the virus is no longer present in eye fluids after 4 to 7 days. Infections have occurred mostly in laboratory workers and vaccinating crews, with rare cases in poultry handlers. No instance of transmission to humans through handling or consuming of poultry products is known. Individuals with conjunctivitis from VND virus should not enter poultry premises or come in contact with live avian species.

j. Peste Des Petits Ruminants (PPR) (Pest of Small Ruminants).[10]

(1) PPR is an acute or subacute disease of sheep and goats caused by a virus. The disease is not contagious, and transmission requires close contact. The mortality, depending on the form, varies form 50 to 80 percent. The disease has been reported from west and central Africa, the Middle East, and the Indian subcontinent.

(2) There is no specific treatment for PPR. Drugs that control bacterial and parasitic complications may decrease mortality.

(3) Eradication is recommended when PPR appears in new areas. Methods should include quarantine, slaughter, and proper disposal of carcasses and contact fomites; decontamination; and restrictions on importation of sheep and goats from affected areas.

(4) PPR is not infectious to humans.

k. Swine Vesicular Disease (SVD).[11]

(1) SVD is a contagious disease of swine. This disease produces lesions that are grossly indistinguishable from those of FMD. In the natural disease, the prognosis is favorable; however, in most countries, as soon as it is recognized, the pigs are slaughtered. This disease, which has been reported from several European countries and Asia, does not occur in North America.

(2) There is no treatment available.

(3) Preventive measures include control of animals imported from infected areas and sanitary disposal of garbage from international aircraft and ships. Eradication

measures consist of quarantining infected farms and areas, slaughtering and disposing of infected and exposed pigs, and cleaning and disinfecting infected premises.

(4) Human infection has been reported in laboratory personnel working with the virus. Caution should be taken when working with infected material.

l. Rinderpest.[12]

(1) Rinderpest is a contagious viral disease of cattle, domestic buffalo, and some species of wildlife. The disease is found only in the Indian subcontinent, Near East, and sub-Saharan Africa. Transmission is by direct and indirect contact. Aerosol transmission is not a significant means of transmission except in confined areas and over short distances. In fully susceptible cattle, the mortality may approach 100 percent with some virulent strains.

(2) There is no treatment available.

(3) If an outbreak occurs, the area should be quarantined, infected and exposed animals slaughtered and buried or burned, and ring vaccination considered.

(4) There are no reports of Rinderpest infection in humans.

m. Enterovirus Encephalomyelitis (Porcine Polioencephalomyelitis, Teschen Disease, Talfan Disease).[13]

(1) These diseases are caused by several closely related enteroviruses. They occur in pigs of all ages and are characterized by a fever and convulsions, stiffness, spasms, and paralysis. The highly virulent strain (Teschen Disease) has a mortality rate of 70 to 90 percent. Transmission is made by direct or indirect contact, swill feeding, and fomites. Teschen disease is found in central and eastern Europe, Madagascar, and Uganda.

(2) There is no treatment available for this pathogen.

(3) Live attenuated and inactivated vaccines are effective in controlling the disease. Quarantine and hygienic measures should be applied. In the US, strict import regulations must be followed for Teschen disease.

(4) Pigs are the only susceptible species.

n. Vesicular Stomatitis.[14]

(1) Vesicular stomatitis is a contagious disease of cattle, horses, and pigs. The disease occurs in North and Central America and the northern part of South America. The incubation period is usually 24 hours; and direct contact, biting insects, and fomites spread the disease rapidly. Humans may be infected by contact and by aerosol.

(2) There is no treatment except the use of mild antiseptics and astringents on the mucosa of the mouth.

(3) Infected and exposed animals should be quarantined. A vaccine is available in some countries.

(4) Vesicular stomitis (New Jersey and Indiana) infection frequently occurs in man and causes influenza-like symptoms, but rarely results in vesicles. Other vesicular stomitis viruses (Piry, Isfahan, and Chandipura) are much more infectious for man.

o. Contagious Bovine Pleuropneumonia (CBPP).[15]

(1) CBPP is caused by *Mycoplasma mycoides* and primarily affects cattle. The principal mode of infection is by inhalation. The disease occurs in Africa and some regions of Asia (especially India and China), with occasional outbreaks in Europe.

(2) Treatments are generally ineffective for this pathogen.

(3) CBPP is a notifiable disease in most countries. A live attenuated vaccine is employed in areas where eradication is not feasible.

(4) There is no evidence to indicate that humans are susceptible to this disease.

p. Contagious Equine Metritis (CEM).[16]

(1) CEM is a highly contagious disease of horses. It has been reported from Australia, Ireland, a number of European countries, Japan, Belgium, Denmark, Netherlands, Norway, Sweden, Switzerland, and Luxembourg. The disease has been eradicated from the US.

(2) Treatment of mares with antimicrobial drugs is not always effective, even though the disease-causing organism is generally susceptible to antibiotics.

(3) This is a reportable disease in North America. Quarantine or isolation with attempts to eliminate actively infected animals and positive carriers is generally used.

(4) There is no evidence that man is affected by the CEM.

q. Heartwater (Cowdriosis).[17]

(1) Heartwater is a disease of ruminants caused by rickettsia and transmitted by ticks. It occurs in Africa, Madagascar, Guadeloupe, and other islands of the Caribbean. Cattle, sheep, and goats are affected; and indigenous breeds are more resistant than imported ones.

(2) Treatment involves administering tetracyclines.

(3) Preventive measures include control of ticks. Simultaneous infection with infectious blood and treatment with tetracyclines provides protection against some strains.

(4) Humans are not known to be affected by Heartwater.

r. Screwworm Myiasis.[18]

(1) Myiasis is the infestation of live vertebrate animals with larvae, which for at least a certain period, feed on the host's dead or living tissue, liquid body substances, or ingested food. Screwworm larvae penetrate deeply into a wound of a warm-blooded animal and feed on living tissue and body fluid.

(2) Screwworm myiasis is treated with topical application of an approved larvacide directly into the infested wound.

(3) Where screwworm is endemic, animals must be inspected at least every 3 to 4 days to discover and treat cases of screwworm myiasis.

(4) Humans are susceptible to screwworm myiasis.

s. Enterovirus Infections of Swine (Virus Infection of Swine).[19]

(1) This virus causes stillbirths, mummification, embryonic death, and infertility. These infections can be widespread where swine are raised intensively.

(2) There is no treatment for this virus.

(3) Enterovirus can be controlled by comingling new replacement boars or gilts by fence-line contact or exchange of manure for 30 days before breeding begins.

NOTES

[1] C.A. Mebus, "African Swine Fever (Peste porcine Africaine, fiebre porcina Africana, maladie de Montgomer)," *Foreign Animal Diseases "The Gray Book"*, 1998 ed., http://www.vet.uga.edu/vpp/gray_book/FAD/asf.htm (31 March 2003).

[2] C.W. Beard, "Avian Influenza (Fowl Plague)," *Foreign Animal Diseases "The Gray Book"*, 1998 ed., http://www.vet.uga.edu/vpp/gray_book/FAD/avi.htm (31 March 2003).

[3] O.L Stott, "Bluetongue and Epizootic Hemorrhagic Disease (Sore muzzle, pseudo foot-and-mouth disease, muzzle disease)," *Foreign Animal Diseases "The Gray Book"*, 1998 ed., http://www.vet.uga.edu/vpp/gray_book/FAD/blt.htm (31 March 2003).

[4] James House and C.A. Mebus, "Foot-and-Mouth Disease (Afta epizootica, Bek-en-klouseer, Fiebra aftosa, Fievre aphteuse, Maul-und-Klauenseuche)," *Foreign Animal Diseases "The Gray Book"*, 1998 ed., http://www.vet.uga.edu/vpp/gray_book/FAD/fmd.htm (31 March 2003).

[5] James A. House, "Sheep and Goat Pox", *Foreign Animal Diseases "The Gray Book"*, 1998 ed., http://www.vet.uga.edu/vpp/gray_book/FAD/sgp.htm (31 March 2003).

[6] Alexandre Fediaevsky, "B052-Aujeszky's Disease," *Manual for the Recognition of Exotic Diseases of Livestock*, 17 December 2002, http://www.spc.int/rahs/Manual/Porcine/AUJESZKYSE.HTM (1 April 2003). (Developed by G. Garner and Peter Saville).

[7] Gilles C. Dulac, "Hog Cholera (Classical swine fever, peste du porc, colera porcina, Virusschweinepest)," *Foreign Animal Diseases "The Gray Book"*, 1998 ed., http://www.vet.uga.edu/vpp/gray_book/FAD/hoc.htm (31 March 2003).

[8] Alexandre Fediaevsky, "B058-Rabies," *Manual for the Recognition of Exotic Diseases of Livestock*, 17 December 2002, http://www.spc.int/rahs/Manual/Multiple_Species/RABIESE.HTM (31 March 2003). (Developed by G. Garner and Peter Saville).

[9] Charles W. Beard, "Velogenic Newcastle Disease (Exotic Newcastle disease, Asiatic Newcastle disease)," *Foreign Animal Diseases "The Gray Book"*, 1998 ed., http://www.vet.uga.edu/vpp/gray_book/FAD/vnd.htm (31 March 2003).

[10] J.T. Saliki, "Peste Des Petits Ruminants (Pest of Small Ruminants, stomatitis-pneumoenteritis complex or syndrome pseudorinderpest of small ruminants and kata [Pidgin English for catarrh])," *Foreign Animal Diseases "The Gray Book"*, 1998 ed., http://www.vet.uga.edu/vpp/gray_book/FAD/pdp.htm (31 March 2003).

[11] C.A. Mebus, "Swine Vesicular Disease," *Foreign Animal Diseases "The Gray Book"*, 1998 ed., http://www.vet.uga.edu/vpp/gray_book/FAD/svd.htm (31 March 2003).

[12] C.A. Mebus, "Rinderpest," *Foreign Animal Diseases "The Gray Book"*, 1998 ed., http://www.vet.uga.edu/vpp/gray_book/FAD/rin.htm (31 March 2003).

[13] Alexandre Fediaevsky, "B256-Enterovirus Encephalomyelitis (Previsously Teschen Disease)," *Manual for the Recognition of Exotic Diseases of Livestock,* 17 December 2002, http://www.spc.int/rahs/Manual/Porcine/PEVE.HTML (1 April 2003).(Developed by G. Garner and Peter Saville).

[14] C.A. Mebus, "Vesicular Stomatitis," *Foreign Animal Diseases "The Gray Book"*, 1998 ed., http://www.vet.uga.edu/vpp/gray_book/FAD/vst.htm (31 March 2003).

[15] Corrie Brown, "Contagious Bovine Pleuropneumonia," *Foreign Animal Diseases "The Gray Book"*, 1998 ed., http://www.vet.uga.edu/vpp/gray_book/FAD/CBP.htm (31 March 2003).

[16] T.W. Swerczek, "Contagious Equine Metritis," *Foreign Animal Diseases "The Gray Book"*, 1998 ed., http://www.vet.uga.edu/vpp/gray_book/FAD/cem.htm (31 March 2003).

[17] C. John Mare, "Heartwater (Cowdriosis)," *Foreign Animal Diseases "The Gray Book"*, 1998 ed., http://www.vet.uga.edu/vpp/gray_book/FAD/hea.htm (31 March 2003).

[18] James E. Novy, "Screwworm Myiasis (Gusanos, Mosca Verde, Gusano barrendor, Gusaneras)," *Foreign Animal Diseases "The Gray Book"*, 1998 ed., http://www.vet.uga.edu/vpp/gray_book/FAD/SCM.htm (8 April 2003).

[19] Alex Hogg and Donald G Levis, "Swine Reproductive Problems: Infectious Causes)," *NebGuide,* February 1997, http://www.ianr.unl.edu/pubs/swine/g926.htm (8 April 2003).

Appendix K

SELECTED PLANT PATHOGENS

1. Background

The purpose of this appendix is to provide a brief summary of select plant diseases listed in *The Biological and Chemical Warfare Threat* (1999). The military could potentially be involved in a consequence management support role, and brief descriptions are provided to furnish an awareness of these agents. For further information, contact AMEDD.

2. Bacterial Diseases

a. Xanthomonas albilineans.[1]

(1) Description. Leaf Scald (Sugar Cane).

(2) Specific Bacteria. Xanthomonas albilineans.

(3) Symptom. The initial characteristic symptom is one or more narrow, white "pencil lines" running longitudinally down the leaf blade into the sheath. Under severe disease conditions, entire plants may die.

(4) Transmission. The bacterium lives from year to year in infected plants. It is spread by the harvester and possibly by other cultivation practices that cause plant wounding. The disease can be spread aerially in windblown rain.

(5) Control Measures. The main control measures are use of disease-free or treated seed and crop rotation. Additionally, the disease is kept out of production areas through quarantine of varieties introduced from other areas.

b. Xanthomonas campestris pv. citri.[2]

(1) Description. Citrus Canker (citrus plants).

(2) Specific Bacteria. Bacterium-Xanthomonas campestris pv. citri.

(3) Symptom. On leaves, first appearance is as oily-Looking, 2- to 10-mm circular spots. The lesions are often similarly sized. Later, they become white or yellow spongy pustules. The pustules darken and thicken into tan to brown corky cankers. Sunken craters are noticeable on fruits. Defoliation occurs on severely infected trees.

(4) Transmission. The bacterium is spread primarily by wind-driven rain, overhead irrigation, and contaminated equipment. Citrus canker seems to be much more severe in areas in which the periods of high rainfall coincide with the period of high mean temperature (such as Florida and other Gulf Coast states).

(5) Control Measures. In canker-free, citrus-producing areas, strict quarantine measures are practiced. When the canker bacterium is found in such an area, control is attempted by burning all infected and adjacent trees to prevent the spread of the pathogen.

c. Xanthomonas campestris pv. oryzae.[3]

(1) Description. Bacterial Leaf Blight (rice).

(2) Specific Bacteria. Bacterium-Xanthomonas campestris pv. oryzae.

(3) Symptom. The first symptom of the disease is a water-soaked lesion on the edges of the leaf blades near the leaf tip.

(4) Transmission. High rainfall and strong winds are thought to provide conditions for the bacteria to multiply and enter the leaf through injured tissue.

(5) Control Measures. Fall plowing or rolling of stubble to hasten decay of the rice debris should help to manage the disease by destroying the tissue in which the bacterium is maintained.

d. Xylella fastidiosa (grapevines).[4]

(1) Description. Pierce's disease (PD) (grapevines).

(2) Specific Bacteria. Bacterium-Xylella fastidiosa.

(3) Symptom. In grapes, symptoms appear as a sudden drying and scorching of much of the margin area of the leaf while the rest of the leaf is still green. Grape clusters on vines with leaf symptoms stop growing, wilt, and dry up. The bacterium also causes leaf scorching in the American elm, maple, mulberry, and plum. In the peach and alfalfa plants, the bacterium slows and stunts growth.

(4) Transmission. The pathogen is transmitted by certain kinds of leafhoppers known as sharpshooters. The leafhopper vectors transmit the bacteria from diseased to healthy plants.

(5) Control Measures. There is no practical control of Pierce's disease of grapes in the field. All commercial grape varieties are susceptible to the disease.

3. Fungal Diseases

a. Colletotrichum Coffeanum.[5]

(1) Description. Coffee Berry Disease (CBD) (coffee).

(2) Specific Fungi. Colletotrichum Coffeanum.

(3) Symptom. CBD attacks the green tissues at the beginning stage of berry development, often penetrating into the interior of the berry and destroying the bean.

(4) Transmission. Infected berries are the major source of transmission. Rain is also a major factor responsible for spreading spores. Windblown rain also results in local dispersal from tree to tree or over relatively short distances. If an infected seed is planted, systemic seed-borne inoculum could easily infect the bark of young seedlings and become established in mature orchards.

(5) Control Measures. Since CBD is limited to Africa, precautions should be taken with coffee seeds from this region.

b. Cochiliobolus miyabeans (Helminthosporium oryzae).[6]

(1) Description. Brown Spot (rice).

(2) Specific Fungi. Cochliobolus miyabeanus (Helminthosporium oryzae).

(3) Symptom. Leaf spots initially appear as small circular to oval spots on the first seedling leaves. Leaf spots are observed throughout the growing season and can vary in size, shape, and color. Small spots are dark brown to reddish brown while large spots have a light, reddish-brown or gray center surrounded by a dark to reddish-brown margin. Older spots may have a bright yellow halo surrounding the lesion. Severely infected leaves will produce lightweight or chalky kernels.

(4) Transmission. The disease spreads from plant to plant in the field by airborne spores. Disease development is favored by high relative humidity (86 to 100 percent) and temperatures between 68 and 78 degrees F. Leaves must be continuously wet for 8 to 24 hours for infection to occur.

(5) Control Measures. The best management strategy is balanced nutrition because plants that grow in soils with nutritional deficiencies or in soils where nutrient uptake is hindered are more susceptible to infection. Fungicidal seed treatment has proven very effective in reducing seedling brown spot disease.

c. Microcyclus ulei.[7]

(1) Disease. South American Leaf Blight (Hevea species only, yields latex).

(2) Specific Fungi. Microcyclus ulei *(M. ulei)*.

(3) Symptom. The symptoms vary with the age of the leaves; however, on young leaves up to 10 days old, slightly colored hypertrophic deformations are visible 3 to 4 days after inoculation.

(4) Transmission. Plants older than 4 to 5 years normally change leaves once a year at the onset of the dry season. This change behavior is very important for an epidemic of M. ulei because leaves are only susceptible when they are less than 10 to 15 days old. The spores of M. ulei are disseminated mainly by rain splash or wind.

(5) Control Measures. The disease threatens the rubber cultivation in the tropical regions of the globe.

d. Puccinia graminis tritici.[8]

(1) Disease. Wheat Stem Rust (wheat; some varieties of barley, oats, and rye; wild barley; and goat grass).

(2) Specific Fungi. Puccinia graminis.

(3) Symptom. Wheat stem rust is encountered during growing season and may occur on any aboveground parts. Rust spots are very small, circular or elongated, and vivid orange-red in color. Later in the year, the rust pustules darken because of the production of dark brown spores that are the over-wintering stage of the rust.

(4) Transmission. Virtually all of the stem rust infections come by way of spores blown in from infected fields of other regions.

(5) Control Measures. Plant-resistant wheat varieties are available. If necessary, apply a foliar fungicide at the very early stage of disease development.

e. Puccinia striiformis.[9]

(1) Disease. Stripe Rust of Wheat (mainly wheat, can occasionally infect barley, triticale, and cereal rye).

(2) Specific Fungi. Puccinia striiformis.

(3) Symptom. Initially, small areas 1 to 10 meters in diameter will appear as yellow patches in paddocks. Infected leaves develop yellow-orange spore masses (pustules) in long stripes on the leaves, but rarely on the stems and heads. As the crop matures, the spore masses turn from yellow to black stripes.

(4) Transmission. The disease, like leaf rust, develops from spores blown in from other wheat-growing areas. Stripe rust spores can be spread on contaminated clothing.

(5) Control Measures. Resistant varieties can be selected if this disease becomes serious. Foliar fungicides also are labeled for stripe rust.

f. Pyricularia grisea.[10]

(1) Disease. Rice Blast (rice).

(2) Specific Fungi. Pyricularia grisea.

(3) Symptom. Lesions that occur on the leaf are usually diamond-shaped with a gray or white center and brown or reddish brown border. Panicle lesions are usually brown, but may also be black.

(4) Transmission. Shortly after the fungus infects and produces a lesion on rice, fungal strands called "conidiophores" grow and produce spores called "conidia." These conidia are dispersed in the air. The disease is favored by long periods of free moisture, high humidity, little or no wind at night, and night temperatures between 63 and 73 degrees F. Leaf wetness from dew or other sources is required for infection. Sporulation is greatest when relative humidity is above 93 percent.

(5) Control Measures. Resistant varieties are available. Continuous flooding is recommended to limit blast development. Avoid field drainage, especially for extended periods because it allows the formation of nitrate and may cause drought stress.

g. Deuterophoma tracheiphila.[11]

(1) Disease. Citrus.

(2) Specific Fungi. Deuterophoma tracheiphila.

(3) Symptom. The first symptoms appear in the spring, followed by a dieback of twigs and branches. Gradually, the pathogen affects the entire tree, and it eventually dies.

(4) Transmission. Prunings containing affected twigs or branches can be a source of inoculum for several weeks. The fungus can survive within infected twigs in the soil for more than 4 months. The disease can also be transmitted by rain, hail, and wind. Dissemination by birds and contaminated insects is also suspected.

(5) Control Measures. Resistant varieties of lemon trees are available.

h. Moniliophthora roreri.[12]

(1) Disease. Monilia Pod Rot (cocoa).

(2) Specific Fungi. Moniliophthora roreri.

(3) Symptom. Moniliophthora roreri completes its entire life cycle on the pods in the tree. There will be conspicuous bumpy swellings on the pod surfaces. Sporulation begins over the pod causing a tan discoloration within 12 days after pod swellings.

(4) Transmission. To monitor for Moniliophthora Pod Rot, "sanitation sweeps" are made through the planting after pod set and should be continued on 7- to 10-day schedules. This is well within infected pod surface sporulation, which disperses spores throughout the trees during rainy periods.

(5) Control Measures. Detection of cocoa-infected pods and their removal is the real and only key to pest management. Good surface drainage and removal of weeds should be practiced on a regular basis.

4. Viral Diseases

a. Barley Yellow Dwarf (BYD) Virus.[13]

(1) Disease. BYD (wheat, barley, and oat).

(2) Specific Virus. BYD.

(3) Symptom. Symptoms include uneven, blotchy leaf discoloration in various shades of yellow, red, or purple, progressing from leaf tip to base and margin to midrib. The most striking symptoms occur on older leaves Wheat and barley leaves usually turn yellow, while oat leaves are more red.

(4) Transmission. Transmitted by more than 20 aphid species (e.g., corn leaf aphid).

(5) Control Measures. The main hope of control of BYD is the use of resistant varieties. Most of the commercial varieties of oats, barley, and wheat are susceptible to BYD, but some are less susceptible than others. A number of varieties have been found or developed that show some tolerance or resistance to BYD. Also, avoid very early or very late planting dates during active aphid populations.

b. Banana Bunchy Top Virus (BBTV).[14]

(1) Disease. BBTV (bananas).

(2) Specific Virus. BBTV.

(3) Symptom. New leaves of infected plants develop dark green streaks. On mature plants, new leaves emerge with difficulty, are narrower than normal, are wavy rather than flat, and have yellow leaf margins. They appear to be "bunched" at the top of the plant.

(4) Transmission. BBTV is transmitted from plant to plant by aphids and transmitted from place to place by people transporting planting materials obtained from infected plants.

(5) Control Measures. The most important factors in controlling this disease are killing the aphid vector and removing and destroying infected banana plants.

NOTES

[1]Kenneth Witam, et al., "Field Crops: Sugarcane" *2001 Plant Disease Control Guide*, 2001, http://www.lsuagcenter.com/Subjects/guides/plantdisease/01disease.html (31 March 2003).

[2]Dean W. Gabriel, *Citrus Canker Disease*, 5 August 2002, http://www.biotech.ufl.edu/PlantContainment/canker.htm (31 March 2003).

[3]Texas Extension Plant Pathologists, "Bacterial Leaf Blight Symptoms on Rice," February 1996, http://plantpathology.tamu.edu/Texlab/Grains/Rice/riceblb.html (8 April 2003).

[4]A.H. Purcell, "An Introduction to Pierce's Disease," *Xylella Fastidiosa Web Site*, November 1999, http://www.cnr.berkeley.edu/xylella/page2.html (31 March 2003).

[5]Stephen A. Ferreira and Rebecca A. Boley, "Colletrotrichum Coffeanum," *Crop Knowledge Master*, November 1991, http://www.extento.hawaii.edu/kbase/crop/Type/c_coffee.htm (April 1, 2003).

[6]Lawrence E. Datnoff and Richard S. Lentini, "Brown Spot in Florida Rice," May 1994, http://edis.ifas.ufl.edu/BODY_RH007 (2 April 2003).

[7]The International Rubber Research and Development Board, "South American Leaf Blight," http://www.irrdb.com/irrdb/NaturalRubber/Diseases/salb.htm (1 April 2003).

[8]J.E. Partridge, "Stem Rust of Wheat," *University of Nebraska-Lincoln, Department of Plant Pathology*, 1997, http://plantpath.unl.edu/peartree/homer/disease.skp/agron/wheat/WhStRst.html (8 April 2003).

[9]Department of Agriculture-Western Australia, "Stripe Rust of Wheat: Puccinia striiformis f.sp. tritici," August 2002, http://www.agric.wa.gov.au/agency/Pubns/factsheets/2002/fs002_2002.rtf (8 April 2003).

[10]R.K. Webster, "Rice: Rice Blast," *UC Pest Management Guidelines*, July 2000, http://www.ipm.ucdavis.edu/PMG/r682100611.html (2 April 2003).

[11]Data Sheets on Organisms Cuarentenarios for the Paises Members of the Cosave Card Cuarentenaria, "Xanthomonas campestris (Pammel) Dowson pv. Oryzae (Ishiyama) Dye," July 1999, http://www.cosave.org.py/lpcdeuterophomatracheiphila.htm (1 April 2003).

[12] "Monilia (Moniliophthora roreri)," http://www.oardc.ohio-state.edu/cocoa/monilia.htm (1 April 2003).

[13]R.M. Davis, "Small Grains: Barley Yellow Dwarf," *University of California Statewide Integrated Pest Management Program*, December 2002, http://www.ipm.ucdavis.edu/PMG/r730101911.html (1 April 2003).

[14]College of Tropical Agriculture & Human Resources, University of Hawaii, "Banana Bunchy Top Virus," *Plant Disease*, December 1997, http://www2.ctahr.hawaii.edu/oc/freepubs/pdf/PD-12.pdf (8 April 2003).

Appendix L

DISSEMINATION OF BIOLOGICAL AGENTS

1. Background

The term "dissemination" refers to the intentional release of a biological agent by an adversary so that it will reach the portals of entry of target personnel in a viable and virulent state. Based on the portals of entry, the characteristics of agents used, and the results desired, certain methods of dissemination are feasible for biological attacks. The effectiveness of these methods is determined by physical and environmental factors that limit the ability of the agent to establish infection. Dissemination methods are related to the routes of entry through which pathogens may be introduced into the body to establish infection.[1] These routes of entry are inhalation, percutaneous, and oral.

2. Inhalation or Aerosol Route of Entry

The primary route of exposure to biological agents is through the respiratory tract. This would be accomplished by disseminating the agent as an aerosol. An aerosol is comprised of finely divided particles, either liquid or solid, suspended in a gaseous medium. Examples of common aerosols are dust, fog, and smoke. A biological agent aerosol is defined as an airborne suspension of particles containing biological agents.

a. Characteristics of Aerosol Dissemination.

(1) Difficulty of Detection. A biological agent aerosol in field concentrations cannot be detected by the physical senses.

(2) Capability of Penetration. Aerosol particles tend to diffuse in much the same manner as a gas. The aerosol cloud travels with the wind and is capable of diffusing into nonairtight structures that are not equipped with adequate filtering devices.

(3) Difficulty of Diagnosis. The classical symptoms of a disease associated with a particular agent can mimic the symptoms of diseases, such as the flu.

(4) Increased Severity and Mortality Rate. Certain diseases have altered incubation periods and incapacitation times and increased mortality rates when the agent enters the body through the respiratory tract. This is the result of the organism diffusing directly into the bloodstream and being carried by the blood directly to the body tissues.

(5) Massive Overdoses. Personnel might be exposed to massive overdoses of an agent through the use of an aerosol. Thus, the acquired immunities of target personnel might be overcome by the use of selected agents.

(6) Increased Susceptibility. Man has a constant requirement for oxygen; therefore, he is breathing continually. This increases the probability of contacting an airborne organism.

b. Particle Size. For the biological agent aerosol to be effective, it must reach target personnel. Particle size is a critical factor in lung infections. Particles in the size range from 1 to 5 microns are much more capable of passing through the defensive barriers of the upper respiratory tract and of being retained in the lungs than those below or above this size range.

c. Formation of Aerosol Particles. Particles in the proper size range can be formed by physically breaking up a substance. This can be accomplished by release through a nozzle or a spray or by an explosive force. There are three general methods of forming biological agent aerosols—

(1) Generator. Particles can be formed by forcing the wet (slurry form) agent through a nozzle at a regulated pressure. The amount of pressure, the size of the orifices, the viscosity of the agent, and the relative humidity determine particle size. Size control of solid particles (dry form of agent) can be achieved by presizing before dissemination.

(2) Spray. Releasing the agent in slurry form into a high-velocity air stream can produce aerosol particles in the proper size range.

(3) Explosive Force. Biological agents can be disseminated by explosive means. The use of an explosive means for aerosol production is feasible because large numbers of organisms can be packaged in a munition. The total volume and agent concentration will support low efficiencies and still produce aerosols that result in adequate area coverage with high infective dose concentrations.

d. Agent Aerosol Stability. From the instant an aerosol is created, certain physical and environmental factors affect its stability. Aerosols eventually diffuse and become too dilute to be effective; environmental conditions cause the agent in the aerosol to gradually lose its ability to establish infection. This decline in aerosol effectiveness is called the "aerosol decay rate," which is usually expressed as percent decay (death) of microorganisms per minute. The decay rate differs from agent to agent and for meteorological conditions. Some of the factors that determine aerosol stability are—

(1) Settling. The rate of fall of the particle is directly related to its size. The terminal velocity of a 1- to 5-micron particle is relatively small (5 inches per hour for a 1-micron-diameter particle in still air). This slow settling rate and the presence of convection currents within the target area cause fallout of 1- to 5-micron particles to be negligible.

(2) Impaction. As the aerosol cloud moves downwind from its release point, particles within the cloud strike and stick to objects in their path. While a number of particles will impact, the overall effect is negligible.

(3) UV Radiation. The UV radiation in sunlight kills microorganisms. In spite of the low penetrating power of UV radiation, its killing effect for most pathogens is complete and takes place in a relatively short period of time upon direct contact.

(4) Wind Direction and Speed. These determine the direction in which the aerosol cloud will travel and the size of the area that it will cover. Aerosols of biological agents with a high decay rate can be employed effectively at high wind speeds (8 to 18 knots). At these speeds, the aerosol may be carried over extensive areas during the agent's survival period. Low wind speeds decrease downwind travel, which reduces area coverage. However, low wind speeds also tend to lengthen the time the aerosol is on the target and thereby increase the inhaled dose in target personnel.

(5) Relative Humidity and Evaporation. Liquid particles in a biological aerosol may be reduced in size by evaporation. A decrease in the amount of liquid in the particle creates a corresponding increase in the percentage of salts remaining in the liquid

surrounding the agent. This results in increased osmotic pressure, which tends to draw fluids out through the cell membrane and results in dehydration of the living microorganisms. The rate of evaporation is dependent upon the relative humidity and the temperature in the environment surrounding the particle. Disseminating the agents affected during conditions of high relative humidity reduces the rate of evaporation. Low relative humidity is conducive to the stability of some biological agents. These agents would be disseminated during conditions of low relative humidity.

(6) Temperature. Temperature has little direct effect on the living portion of a biological aerosol. Indirectly, however, an increase in temperature is normally followed by an increase in evaporation rate. High temperatures (170 to 180 degrees F) tend to kill most vegetative bacteria as well as the viral and rickettsiae agents; however, these temperatures are not normally encountered under field conditions. Subfreezing temperatures tend to freeze the aerosol (if it is in liquid form) after its dissemination. This freezing tends to preserve the agent and decrease its rate of decay.

(7) Air Stability. The temperature gradient conditions of lapse, inversion, and neutral affect the biological agent aerosol in much the same manner as they affect a chemical agent cloud. Inversion and neutral conditions are most effective for aerosol travel because the cloud is kept at a height conducive to inhalation by target personnel. Turbulence, which occurs during lapse conditions, will cause vertical diffusion of the cloud with a resulting loss of agent to higher altitudes and a reduction of area coverage.

(8) Precipitation. Heavy and prolonged precipitation will substantially reduce the number of agent particles in the air. The high relative humidity associated with very light rain makes it less important in aerosol effectiveness than the rainout effect of heavy rainfall.

e. Effects of Terrain on Cloud Travel. Terrain affects cloud travel of biological agent aerosols in the same general manner as it affects chemical agent clouds. The ground contour of rough terrain creates wind turbulence, which in turn influences the vertical diffusion of the aerosol cloud. Soil will have an effect only as related to heat absorption and reflection, which aid in determining temperature gradients.

3. Percutaneous Route of Entry

A second portal of entry that can be utilized for biological agent employment is the skin. Penetration of the skin can be accomplished by the bite of an arthropod vector (carrier), injection, or absorption.

a. Arthropod Vector. These insects are capable of transferring pathogens to man through breaks in the skin. For the purpose of this manual, the definition of "vector" is limited to the arthropods. The spread of pathogens by arthropod vectors to man is well-established in history. Some examples of vectors and the pathogens that they have shown to be capable of transmitting are as follows:

(1) Mosquitoes. The virus of yellow fever is transmitted from man to man by the bite of a mosquito. Other important mosquito-borne viral diseases are dengue fever and several types of encephalitis.

(2) Flies. Most varieties of true flies have sucking mouthparts, but those few that have mouthparts capable of piercing the skin of man or animals carry pathogens that cause some of the most feared human diseases. The sucking flies introduce pathogens

through previously injured body surfaces or mechanically transport them on their body surfaces to exposed food and water. Typhoid fever, bacillary and amoebic dysentery, and Asiatic cholera are examples of diseases that may be spread mechanically by nonbiting flies. Pathogens transported by the biting flies include those that cause the dreaded African sleeping sickness—an infection of man, domestic animals, and wild game. The vector responsible for this disease is the tsetse fly (*Glossina morsitans*). Tularemia, a bacterial disease of man and wild animals, is sometimes transmitted by the deer fly.

(3) Lice. Lice are sucking, dorsoventrally flattened, wingless insect parasites of the skin of mammals and birds. The human body louse, *Pediculis humanus*, is the vector for the rickettsiae that cause epidemic typhus and trench fever.

(4) Fleas. Fleas are small, wingless insect parasites of the skin of mammals and birds. Their bodies are flattened laterally, and they have mouthparts for piercing the skin. While different species show preferences for certain hosts, when hungry they will attack any warm-blooded animal. This habit increases their potential to transmit disease to man. The common rat flea, *Xenopsylla cheopis*, is the vector of endemic typhus.

(5) Ticks and Mites. These arthropod vectors are known as acarids and are not true insects. As adults, they possess eight legs while insects have six. Most of these are merely parasitic skin pests of land vertebrate animals, but a few are important disease vectors. Certain mites transport the causative organism of scrub typhus. The wood tick, *Dermacentor andersoni*, indigenous to the western US, is known to transmit to man the rickettsia of Rocky Mountain spotted fever, the bacterium of tularemia, and the virus of Colorado tick fever.

b. Injection. Biological agents can be injected through the skin. In 1978, Bulgarian exile Georgi Markov, was attacked in London, England, with a device disguised as an umbrella. The weapon discharged a tiny pellet in the subcutaneous tissue of his leg while he was waiting for a bus. He died days later. The pellet, which contained Ricin, was found during autopsy.[2] A Flechette is another penetrating device.

c. Absorption. Biological agents may be absorbed through the skin or placed on the skin to do damage to the integument.

4. Oral Route of Entry

Another possibility is through the oral route by ingestion of contaminated food or water supplies. Contamination with toxins of chlorinated water, rivers, lakes, or reservoirs would be difficult because of dilution effects.[5]

5. Covert Dissemination[1]

a. Characteristics. Biological agents lend themselves well to covert or hidden operations because of detection difficulties, the variety of potential agents, the ways they might be employed, and the small amounts of materials required to cause infection. Sabotage is the direct application, by a person, of material to the target. It is generally covert in nature.

b. Targets. Covert use of biological agents might be aimed primarily at the respiratory tract and secondly at the digestive tract. Since many pathogens are spread naturally in food and water, these provide proven vehicles in which the saboteur could employ an agent. The respiratory tract is an excellent target for the small-scale employment of a biological antipersonnel agent aerosol.

NOTES

[1]TM 3-216/AFM 355-6, *Technical Aspects of Biological Defense,* January 1971.

[2]BG Russ Zajtchuk et al. (eds), *Textbook of Military Medicine: Medical Aspects of Chemical and Biological Warfare,* Office of the Surgeon General, 1997, Chapter 28, "Viral Encephalitides."

[3]BG Russ Zajtchuk et al. (eds), *Textbook of Military Medicine: Medical Aspects of Chemical and Biological Warfare,* Office of the Surgeon General, 1997, Chapter 32, "Ricin Toxin."

[4]BG Russ Zajtchuk et al. (eds), *Textbook of Military Medicine: Medical Aspects of Chemical and Biological Warfare,* Office of the Surgeon General, 1997, Chapter 20, "Use of Biological Weapons."

[21]BG Russ Zajtchuk et al. (eds), *Textbook of Military Medicine: Medical Aspects of Chemical and Biological Warfare,* Office of the Surgeon General, 1997, Chapter 30, "Defense Against Toxic Weapons."

REFERENCES

Office of the US President

Office of the US President, *The Biological and Chemical Warfare Threat*, 1999.

Department of Defense

DOD Chemical and Biological Defense Program Annual Report to Congress, Vol. I, April 2002.

DOD Chemical and Biological Defense Program Annual Report to Congress, Vol. I, April 2003.

Office of the Secretary of Defense, *Proliferation: Threat and Response*, ISBN: 0-16-042727-4, US Government Printing Office, November 1997.

Office of the Secretary of Defense, *Proliferation: Threat and Response,* US Government Printing Office, January 2001.

Joint

Joint Publication 1-02, *Department of Defense Dictionary of Military and Associated Terms*, as amended through 05 September 2003.

Joint Publication 3-11, *Joint Doctrine for Operations in Nuclear, Biological, and Chemical (NBC) Environments*, 11 July 2000.

Multiservice

FM 3-5/MCWP 3-37.3, *NBC Decontamination*, 28 July 2000.

FM 3-6/FMFM 7-11-H/AFM 105-7, *Field Behavior of NBC Agents (Including Smoke and Incendiaries)*, 3 November 1986.

FM 3-11.4/MCWP 3-37.2/NTTP 3-11.27/AFTTP (I) 3-2.46, *Multiservice Tactics, Techniques, and Procedures for Nuclear, Biological, and Chemical (NBC) Protection*, 2 June 2003.

FM 3-11.11/MCRP 3-3.7.2, *Flame, Riot Control Agent and Herbicide Operations*, 19 August 1996 (Renumbered from FM 3-11).

FM 3-11.21/MCRP 3-37.2C/NTTP 3-11.24/AFTTP (I) 3-2.37, *Multiservice Tactics, Techniques, and Procedures for Nuclear, Biological, and Chemical Aspects of Consequence Management*, 12 December 2001.

FM 8-9/NAVMED P-5059/AFJMAN 44-151, *NATO Handbook on the Medical Aspects of NBC Defense Operations AMEDP-6(B),* 1 February 1996.

FM 8-284/NAVMED P-5042/AFMAN (I) 44-156/MCRP 4-11.1C, *Treatment of Biological Warfare Agent Casualties*, 17 July 2000.

FM 8-285/NAVMED P-5041/AFJMAN 44-149/FMFM 11-11, *Treatment of Chemical Agent Casualties and Conventional Military Chemical Injuries*, 22 December 1995.

TM 3-215/AFM 355-7, *Military Chemistry and Chemical Agents*, Washington, DC, December 1963, UNCLASSIFIED Technical Manual (ADA292141).

TM 3-216/AFM 355-6, *Technical Aspects of Biological Defense*, 12 January 1971.

Army

BG Russ Zajtchuk, et al. (eds), *Textbook of Military Medicine: Medical Aspects of Chemical and Biological Warfare*, Office of the Surgeon General, 1997.

DA Form 12-99-R, *Initial Distribution (ID) Requirements for Publications (LRA)*, 1 April 1996.

DA Pam 40-8, *Occupational Health Guidelines for the Evaluation and Control of Occupational Exposure to Nerve Agents GA, GB, GD, AND VX*, 4 December 1990.

DA Pam 40-173. *Occupational Health Guidelines for the Evaluation and Control of Occupational Exposure to Mustard Agents H, HD, AND HT*, 3 June 2003.

FM 3-11.22, *Weapons of Mass Destruction Civil Support Team Tactics, Techniques, and Procedures*, 6 June 2003.

FM 3-50, *Smoke Operations*, 4 December 1990.

USACHPPM TG 204, *Glossary of Terms for Nuclear, Biological, and Chemical Agents and Defense Equipment*, December 2001.

USA Corps of Engineers, Construction Engineering Research Laboratory, *Methods for Field Studies of the Effects of Military Smokes, Obscurants, and Riot-control Agents on Threatened and Endangered Species, Vol. 4: Chemical Analytical Methods*, USACERL Technical Report 99/56, July 1999.

Air Force

AFMAN 10-2602, *Nuclear, Biological, Chemical, and Conventional (NBCC) Defense Operations and Standards (Operations)*, 29 May 2003.

Other Sources

2000 Emergency Response Guidebook, Guide 153, Substances-Toxic and/or Corrosive (Combustible).

Abercrombie, P., ECBC Notebook # NB 98-0079 (U).

Abercrombie, P. L., and Butrow, A.B., *Selected Physical Properties of Ton Container HD (Mustard) and VX*, ERDEC-TR-450, U.S. Army Edgewood Research, Development, and Engineering Center, Aberdeen Proving Ground, MD, July 1998, UNCLASSIFIED Report (ADA350462).

A.H. Purcell, "An Introduction to Pierce's Disease," *Xylella Fastidiosa Web Site*, November 1999, http://www.cnr.berkeley.edu/xylella/page2.html (31 March 2003).

A.K. Steumpfle et al., *Final Report of International Task Force-25: Hazard From Toxic Industrial Chemicals,* March 18, 1996.

Alex Hogg and Donald G Levis, "Swine Reproductive Problems: Infectious Causes)," *NebGuide,* February 1997, http://www.ianr.unl.edu/pubs/swine/g926.htm (8 April 2003).

Allan, C.R., *The Relationship Between Oxygen Index and the Flashing Propensity of Explosively Disseminated Liquids,* ARCSL-TR-77061, USA Armament Research and Development Command, Chemical Systems Laboratory, Aberdeen Proving Ground. MD, October 1977, UNCLASSIFIED Report (ADA045976).

Asbury Graphite Mills, Inc., MSDS, "Synthetic Graphite," CAS No. 7782-42-5, January 2003.

ASTM Method D 167, Standard Test Method for Apparent and True Specific Gravity and Porosity of Lump Coke.

Atkinson, R.H., et al., "The Preparation and Physical Properties of Carbonyl Chloride," *J. Chem. Soc.*, Vol. 117, 1920.

Balson, E.W., *Determination of the Vapor Pressure of T.2104,* A.3804/3, Military Intelligence Division, Chemical Defence Experimental Establishment, Porton, England, April 1945, UNCLASSIFIED Report.

Bartlett, P.D., and Swain, C.G., "Kinetics of Hydrolyisis and Displacement Reactions of '-(Dichlorodiethyl Sulfide (Mustard Gas) and of -Chloro- '-hyroxidediethyl Sulfide (Mustard Chlorohydrin)," *J. Chem. Soc.*, Vol. 71, 1949.

Baskerville, C., and Cohen, P.W., "Solvents for Phosgene," *J. Ind. Eng. Chem.*, Vol. 13, 1921.

Beebe, C.H, *Important Constants of Fourteen Common Chemical Warfare Agents,* EACD 328, Chemical Warfare Service, Edgewood Arsenal, Edgewood, MD, December 1924, UNCLASSIFIED Report (ADB958296).

Belkin, F., and Brown, H.A., *Vapor Pressure Measurements of Some Chemical Agents Using Differential Thermal Analysis. Part III,* ECTR-75032, Edgewood Arsenal, Aberdeen Proving Ground, MD, June 1975, UNCLASSIFIED Report (ADA010666).

Bodenstein, M., and Durant, G., "Die Dissociation des Kohlenoxychlorids," *Z. Physik. Chem.*, Vol. 61, 1908.

Bossle, P.C., et al., *Determination of Lewisite Contamination in Environmental Waters by High Performance Liquid Chromatography,* CRDEC-TR-042, USA Chemical Research Development and Engineering Center, Aberdeen Proving Ground, MD, January 1989, UNCLASSIFIED Report (ADA206000).

Brookfield, K.J., et al., *The Kinetics of the Hydrolysis of Vesicants Part II.–2:2'-Dichlorodiethylsulphide (H),* SO/R/576, Military Intelligence Division, Great Britain, March 1942, UNCLASSIFIED Report.

Brooks, M. E. and Parker, G.A., *Incineration/Pyrolysis of Several Agents and Related Chemical Materials Contained in Identification Sets,* ARCSL-TR-79040, October 1979, UNCLASSIFIED Report (ADB042888).

Brooks, M.E., et al., *Corrosion, Compatibility and Other Physicochemical Studies (U),* DA18-108-CML-6602 (A), Final Report – Task I, W.R. Grace and Company, Washington Research Center, Clarksville, MD, May 1964, UNCLASSIFIED Report (AD350755).

Brooks, M.E. et al, *Final Report – Task VII, Contract DA18-108-CML-6602 (A), Corrosion, Compatibility and Other Physicochemical Studies (U)*, Final Report – Task VII RES-64-86, W. R. Grace & Co., Washington Research Center, Clarksville, Maryland, June 1964, UNCLASSSIFIED Report (AD352753).

Brown, H.A., Jr., et al., *Modified NM: An Improved Liquid Binary VX Reactant (U)*, EC-TR-76075, USA Armament Command, Edgewood Arsenal, Aberdeen Proving Ground, MD, November 1976, UNCLASSIFIED Report (ADC008561).

Buchanan, J.H., et al., *Vapor Pressure of VX*, ECBC-TR-068, USA Soldier and Biological Chemical Command, Aberdeen Proving Ground, MD, November 1999, UNCLASSIFIED Report (ADA371297).

Buchi, K.M., *Environmental Overview of Common Industrial Chemicals with Potential Application in the Binary Munitions Program*, CRDEC-TR-87041, USA Chemical Research Development and Engineering Center, Aberdeen Proving Ground, MD, July 1987, UNCLASSIFIED Report (ADA186083).

Buchi, K.M., *Environmental Overview of Intermediates, By-Products, and Products in the Production of QL, DC, and DF*, CRDEC-TR-076, USA Chemical Research Development and Engineering Center, Aberdeen Proving Ground, MD, May 1991, UNCLASSIFIED Report (ADB155651).

Buckles, L.C., *The Hydrolysis Rate of GD*, TCIR 373, Chemical Corps Technical Command, Army Chemical Center, MD, March 1947, UNCLASSIFIED Report (ADB966291).

Buckles, L.C., *The Hydrolysis Rate of G Agents*, TCIR 393, USA Chemical Research and Development Laboratories, Army Chemical Center, MD, December 1947, UNCLASSIFIED Report (ADB966236).

Buckles, M.F., *CW Vesicants: Selected Values for the Physical Properties of H, T, and Q (U)*, Special Report CRLR 542, Chemical Corps Chemical and Radiological Laboratories, Army Chemical Center, MD, May 1956, UNCLASSIFIED Report (AD108272).

Buswell, A.M., et al., *The Chemistry of Certain Arsenical Chemical Warfare Agents as Water Contaminants*, OSRD 4193, Division 9 National Defense Research Committee of the Office of Scientific Research and Development, June 1944, UNCLASSIFIED Report.

Butrow, A.B., ECBC Notebook # NB 03-0025 (U).

Butrow, A., Chemical Research and Development Center Notebook #NB 83-0155 (U).

Butrow, B., ECBC Notebook # NB 97-0109 (C).

Carter, R.H., and Knight, H.C., *Fundamental Study of Toxicity: Solubility of Certain Toxics in Water and in Olive Oil*, EACD 445, Chemical Warfare Service, Edgewood Arsenal, MD, May 1928, UNCLASSIFIED Report (ADB955216).

CDC, Division of Bacterial and Mycotic Diseases, "Disease Information: Glanders (General Information), 20 June 2001, <http://www.cdc.gov/ncidod/dbmd/diseaseinfo/glanders_g.htm>, 29 August 2003.

CDC, Division of Bacterial and Mycotic Diseases, "Disease Information: Glanders (Technical Information), 7 March 2003, <http://www.cdc.gov/ncidod/dbmd/diseaseinfo/glanders_t.htm>, 11 August 2003.

CDC, HHS, “Regulatory Impact Analysis: 42 CFR Part 73: Select Biological Agents and Toxins, Interim Final Rule (Draft),” 9 December 2002.

CDC, Monkeypox, “Fact Sheet: Basic Information About Monkeypox,” 12 June 2003, http://www.cdc.gov/ncidod/monkeypox/factsheet.htm, 2 September 2003.

CDC, Monkeypox, “Fact Sheet: Smallpox Vaccine and Monkeypox,” 9 July 2003, http://www.cdc.gov/ncidod/monkeypox/smallpoxvaccine_mpox.htm, 2 September 2003.

CDC, Monkeypox, “Updated Interim Infection Control and Exposure Management Guidance in the Health-Care and Community Setting for Patients with possible Monkeypox Virus Infection,” 18 July 2003, http://www.cdc.gov/ncidod/monkeypox/infectioncontrol.htm, 2 September 2003.

CDC, NIOSH, *Documentation for Immediately Dangerous to Life or Health Concentrations*, NTIS Publication No. PB-94-195047, May 1994.

CDC, OHS, “BMBL Section VII: Agent Summary Statements, Section VII-D: Prions,” 17 June 1999,. http://www.cdc.gov/OD/OHS/BIOSFTY/bmb14/bmbl4s7d.htm, 19 August 2003.

CDC, Special Pathogens Branch, “Ebola Hemorrhagic Fever,” 6 August 2003, http://www.cdc.gov/ncidod/dvrd/spb/mnpages/dispages/ebola.htm, 2 September 2003.

Cheicante, R.L., et al., “Investigation for the Determination of Nitrogen Mustard and Related Compounds in Air by Gas Chromatography Using Solid Sorbent Collection and Thermal Desorption,” *In Proceedings of the 1998 ERDEC Scientific Conference on Chemical and Biological Defense Research 17-20 November 1998,* UNCLASSFIED Paper, ERDEC-SP-004, USA Edgewood Chemical Biological Center, Aberdeen Proving Ground, MD, July 1999, UNCLASSFIED Report (ADA375171).

Chemical Agent Data Sheets Volume I, Edgewood Arsenal Special Report EO-SR-74001, Edgewood Arsenal, Aberdeen Proving Ground, MD, December 1974, UNCLASSIFIED Report (ADB028222).

Chemical Agent Data Sheets Vol. II, Edgewood Arsenal Special Report EO-SR-74002, USA Armament Command, Edgewood Arsenal, Aberdeen Proving Ground, MD, December 1974, CONFIDENTIAL Report (AD000020).

Chinn, Kenneth, S. K., *Joint CB Technical Data Source Book, Volume III, G Nerve Agents, Part Three: Agents GD and GF (U),* DPG-TR-82-004, USA Dugway Proving Ground, Utah, August 1983, SECRET Report (ADC032927).

Chris Heilman, *The pictorial Periodic Table,* 23 December 2002, http://chemlab.pc.maricopa.edu/periodic/default.html, 20 May 2004.

Clark, D.N, *Review of Reactions of Chemical Agents in Water, Final Report to USA Biomedical Research and Development Laboratory*, Battelle, Columbus, OH, January 1989, UNCLASSIFIED Report (ADA213287).

College of Tropical Agriculture & Human Resources, University of Hawaii, “Banana Bunchy Top Virus,” *Plant Disease*, December 1997, http://www2.ctahr.hawaii.edu/oc/freepubs/pdf/PD-12.pdf (8 April 2003).

CNN.com./World, “Russia names Moscow siege gas,” 30 October 2002, http://www.cnn.com/2002/WORLD/europe/10/30/moscow.gas/, 27 August 2003.

Coates, J.E., and Davies, R.H., "Studies on Hydrogen Cyanide. Part XVIII. Some Physical Properties of Anhydrous Hydrogen Cyanide," *J. Chem. Soc.*, 1950.

Code of Federal Regulations

Comparison of GA and GB as Chemical Warfare Agents (U), CWL-PS-1, USA Chemical Warfare Laboratories, Army Chemical Center, MD, November 1957, UNCLASSIFIED Report.

Cone, N. M. and Roviller, C.A., *HQ & HT Review of British & U.S. Literature,* TDMR 575, USA Chemical Research and Development Laboratories, Army Chemical Center, MD, February 1943, UNCLASSIFIED Report.

Conoco International, Inc., MSDS, "No. 1 Diesel Fuel, No. 1 Fuel Oil."

Cook, R.P., and Robinson, P.L., "Certain Physical Properties of Cyanogen and its Halides," *J. Chem. Soc.*, 1935.

Costal Corporation, MSDS, "Diesel fuel 2."

Coulter, P.B., et al., *Physical Constants of Thirteen V Agents,* CWLR 2346, USA Chemical Warfare Laboratories, Army Chemical Center, MD, December 1959, UNCLASSIFIED Report (AD314520).

Dahl, A.R., et al., *Acute Toxicity of Methylphosphonic Difluoride (DF) Methylphosphonic Dichloride (DC) and their Hydrolysis Products by Inhalation and other Routes in Mice, Rats and Guinea Pigs,* CRDEC-CR-86049, USA Chemical Research Development and Engineering Center, Aberdeen Proving Ground, MD, June 1986, UNCLASSIFIED Report (ADB105158).

Daroff, P.M., et al., *Oleoresin Capsicum: An Effective Less-Than Lethal Riot Control Agent,* DPG/JCP-097-002, Chemical Biological Defense, USA Dugway Proving Ground, UT, January 1997, UNCLASSIFIED Report (ADB225032).

Data Sheets on Organisms Cuarentenarios for the Paises Members of the Cosave Card Cuarentenaria, "Xanthomonas campestris (Pammel) Dowson pv. Oryzae (Ishiyama) Dye," July 1999, http://www.cosave.org.py/lpcdeuterophomatracheiphila.htm (1 April 2003).

Davies, C. N., "The Density and Thermal Expansion of Liquid Phosgene," *J. Chem. Phys.*, Vol. 14, 1946.

Dawson, T., and Witten, B., *A Memorandum Report New Compounds 2,2' Dichlorotriethylamine,* Technical Division Memorandum Report 552, USA Chemical Research and Development Laboratories, Army Chemical Center, MD, February 1943, UNCLASSFIED Report (ADB960467).

Dawson, T.P., *A Memorandum Report: New Compounds Bis(B-Chloroethylthioethyl) Ether (T) and its Mixtures with Mustard (HT),* TDMR 534, USA Chemical Research and Development Laboratories, Army Chemical Center, MD, January 1943, UNCLASSIFIED Report (ADB960651).

Dawson, T.P., and Witten, B., *Bis (2-chloroethyl) methylamine, Preparation, Decontamination, and Stability,* Technical Division Memorandum Report 442, Chemical Warfare Center, Edgewood Arsenal, MD, September 1942, UNCLASSIFIED Report (ADB960331).

Dawson, T.P., *Ethyldichloroarsine (ED): Preliminary Investigation (1939)*, EATR 325, Chemical Warfare Service, Edgewood Arsenal, MD, November 1941, UNCLASSIFIED Report (ADB957078).

Dean W. Gabriel, *Citrus Canker Disease*, 5 August 2002, http://www.biotech.ufl.edu/PlantContainment/canker.htm (31 March 2003).

Defense General Supply Center, MSDS MIL-F-12070C, "SGF-2 Type; Fog Oil."

Department of Agriculture-Western Australia, "Stripe Rust of Wheat: Puccinia striiformis f.sp. tritici," August 2002, http://www.agric.wa.gov.au/agency/Pubns/factsheets/2002/fs002_2002.rtf (8 April 2003).

Douglas, D.E., and Winkler, C.A., "The Preparation, Purification, Physical Properties and Hydrolysis of Cyanogen Chloride," *Ca. J. Research,* Vol. 25B, (1947).

EA 2277 (U): A Summary Report as of 15 March 1961, CRDL-SP-4-28, USA Chemical Research and Development Laboratories, Army Chemical Center, MD, March 1961, CONFIDENTAL Report.

Eakle, B.F., *Chemical Agent GF (U)*, Technical Study 69-C4, USA Deseret Test Center, Fort Douglas, Utah, January 1969, UNCLASSIFIED Report (AD509689).

Eckhaus, S.R., et al., *Resistance of Various Materials of Construction in Contact with Transester Process*, CWL Technical Memorandum 31-73, USA Chemical Research and Development Laboratories, Army Chemical Center, MD, February 1959, UNCLASSIFIED Report (ADB963125).

Edwards, J.O., and Sauer, M., *Chemical Reactivity of Cyanogen Chloride in Aqueous Solution, Quarterly Status Report (March through May 1972)*, Report No. III, DAAA15-71-C-0478-QSR 3, USA Chemical Laboratories, Edgewood Arsenal, MD, February 1973, UNCLASSIFIED Report (ADA090556).

Ellzy, M., et al., *Difluor (DF) – Flooring Compatibility Studies*, CRDEC-TR-229, USA Chemical Research Development and Engineering Center, Aberdeen Proving Ground, MD, May 1991, UNCLASSIFIED Report (ADB154752).

Epstein, J., *Studies on Hydrolysis of GB I. Effect of pH and Temperature on Hydrolysis Rates. II. Observations on Hydrolysis of GB in Sodium Bicarbonate Buffered Waters*, MDR 132, Chemical Warfare Laboratories, Army Chemical Center, MD, February 1948, UNCLASSIFIED Report.

Exxon Company, MSDS, "Jet Fuel Grade JP-8, 27020-00079."

Federal Register, Department of Commerce, Bureau of Export Administration, "15 CFR Part 710 et al., Chemical Weapons Convention Regulations; Final Rule," December 30, 1999.

Felsing, W.A., et al., "The Melting Point of Mustard Gas," *J. Amer. Chem. Soc.*, Vol. 70, 1948.

Fielder, D., Chemical Systems Laboratories Notebook #NB-CSL-82-0213 (U).

Fielder, D, USA Chemical Warfare Laboratories Notebook # NB 6695 (C).

Food & Agriculture Organization of the United Nations, Secretariat of the Pacific Community, *Manual for the Recognition of Exotic Diseases of Livestock*, Last Updated 12 December 2002. (http://www.spc.int/rahs/Manual).

Franke, S., *Manual of Military Chemistry Volume I- Chemistry of Chemical Warfare Agents,* ACSI-J-3890, Chemie der Kampfstoffe, East Berlin, April 1968, UNCLASSIFIED Technical Manual (AD849866).

Furukawa, G.T., et al., "Thermodynamic Properties of Some Methylphosphonyl Dihalides From 15 to 335°K," J. *Rsch. NBS Phy. & Chem.*, Vol. 68A, No. 4, 1964.

Germann, A.F.O. and Taylor, Q.W., "The Critical Constants and Vapor Tension of Phosgene," *J. Amer. Chem. Soc.,* Vol. *48(*5), 1928.

Giauque, W. F., and Jones, W.M., "Carbonyl Chloride. Entropy. Heat Capacity. Vapor Pressure. Heats of Fusion and Vaporization. Comments on Solid Sulfur Dioxide Structures," *J. Amer. Chem. Soc.,* Vol. *70,* 1948.

Giauque, W.F. and Ruehrwein, R.A., "The Entropy of Hydrogen Cyanide. Heat Capacity, Heat of Vaporization and Vapor Pressure. Hydrogen Bond Polymerization of the Gas in Chains of Indefinite Length." J. *Am. Chem. Soc.*, Vol. 61 1939.

Grula, R.J., et al., *Compatibility Studies with Candidate Binary VX2 Components,* EC-TM-76009, USA Armament Command, Edgewood Arsenal, Aberdeen Proving Ground, MD, February 1976, UNCLASSIFIED Report.

Grula, R.S., et al., *Storage Stability of GD, GF and EA 1356 (U),* CRDLR 3342, USA Chemical Research and Development Laboratories, Edgewood Arsenal, MD, December 1965, CONFIDENTIAL Report (AD369299).

Hall, R.W., *An Investigation on the Solubility and Rate of Hydrolysis of Phosgene in Water,* Porton Report 2663, Chemical Defence Experimental Establishment, Porton, England, 19 December 1944, UNCLASSIFIED Report.

Harris, B.L. and Macy, R., *Storage Stability of German GA in Uncoated and Lacquered 75mm Shell at 50°C and 65°C. Corrosion Rate of Steel by GA at 65°C,* Technical Division Memorandum Report 1299, USA Chemical Research and Development Laboratories, Army Chemical Center, MD, December 1946, UNCLASSIFIED Report (ADB964902).

Harris, B.L. and Macy, R., *Corrosion by Vesicants: Rate of Corrosion of Steel and Other Metals by H, HQ, HN-3, HN-1, and L, Mostly at 65°C,* Technical Division Memorandum Report 1031, USA Chemical Research and Development Laboratories, Army Chemical Center, MD, April 1945, UNCLASSIFIED Report (ADB963161).

Harris, B.L. and Macy R., *Memorandum Report Storage Stability of HL, Mixtures of Mustard and Lewisite,* Technical Division Memorandum Report 1302, USA Chemical Research and Development Laboratories, Army Chemical Center, MD, February 1947, UNCLASSIFIED Report (ADB96498).

Harris, B.L., et al., *Thickened Vesicants: Storage Stability of Unthickened and Thickened Nitrogen Mustards and Their Mixtures with Levinstein Mustard,* Technical Division Memorandum Report 706, USA Chemical Research and Development Laboratories, Army Chemical Center, MD, July 1943, UNCLASSIFIED Report (ADB962153).

Harris, B.L., *Physical Constants of MCE,* Technical Division Memorandum Report 1094, USA Chemical Research and Development Laboratories, Army Chemical Center, MD, July 1945, UNCLASSIFIED Report (ADB964103).

Henley, F.M., *Surveillance Tests on 75 MM. Steel Gas Shell Extending Over a Period of One Year*, EACD 11, Chemical Warfare Service, Edgewood Arsenal, MD, June 1920, UNCLASSIFIED Report (ADB959731).

Henry F. Holtzclaw, Jr., et al., *General Chemistry with Qualitative Analysis,* 9th ed., D.C. Heath and Company, Lexington, MA, 1991.

Herbst, V.H., "Uber die Fluchtigkeit und Vernebelung einer Reihe organischer Stoffe," *Kolloid Beihefte,* Vol. 23, 1927.

HHS, *NIOSH Recommendations for Occupational Safety and Health: Compendium of Policy Documents and Statements*, January 1992.

Hood, H.P., and Murdock, H.R., "Superpalite," *J. Phys. Chem.,* Vol. 23, 1919.

Holzclaw, Jr. H.F. and Robinson, W.R., *College Chemistry with Qualitative Analysis,* 8th ed., D.C. Heath and Company, MA, 1988.

Hormats, S., et al., *Storage Stability in Steel at 65°C of Pure GD. Corrosion Rate of Steel at 65°C,* TDMR 1346, Chemical Corps Technical Compound, Army Chemical Center, MD, March 1948, UNCLASSIFIED Report (ADB964759).

Hutchcraft, A.S. Jr., et al., *Special Report: Corrosion Resistance of Metals Toward Isopropyl-Methylphosphonofluoridate (GB*), CRLR 510, USA Chemical and Radiological Laboratories, Army Chemical Center, MD, May 1955, UNCLASSIFIED Report (AD474404).

Hyttinen, L.J., et al., *Mixed Binary Agents New Approach Toward Meeting Expanded Chemical Munitions Effectiveness Requirements,* ARCSL-TR-83080, USA Armament Research and Development Command, APG, MD, June 1983, CONFIDENTIAL Report (ADC033576).

ICSC 0126, "Chlorine (Cl_2)."

ISCS 0128, "2-Chloroacetophenone."

ICSC 0701, "Magnesium (Pellets)."

ICSC 1064, "Zinc Chloride."

Ingrid Chorus and Jamie Bartram ed., *Toxic Cyanobacteria in Water: A guide to their public health consequences, monitoring and management,* Geneva, WHO, 1999.

IUPAC Commission on Atomic Weights and Isotopic Abundances: Atomic Weights of the Elements 2001, http://www.chemgmw.ac.uk/uipac/atwt/>, 25 May 2004.

Jackson, A.M., and Semiatin, W.J., *Long-Term Storability of the M20 DF Canister Used in the M687 Binary Projectile,* CRDC-TR-84104, USA Chemical Research Development and Engineering Center, Aberdeen Proving Ground, MD, January 1985, UNCLASSIFIED Report (ADB092563).

James Chin (ed.), *Control of Communicable Diseases Manual,* 17th ed., United Book Press, Inc., Baltimore, MD, 2000.

Jeffrey H. Grotte and Lynn I Yang, *Report of the Workshop on Chemical Agent Toxicity for Acute Effects: Institute for Defense Analyses, May 11-12, 1998,* IDA Document D-2176, June 2001.

J.E. Partridge, "Stem Rust of Wheat," *University of Nebraska-Lincoln, Department of Plant Pathology*, 1997, http://plantpath.unl.edu/peartree/homer/disease.skp/agron/wheat/WhStRst.html (8 April 2003).

Johnson, W. and Pechukas, A., "Hydrogen Compounds of Arsenic. I. Preparation of Arsine in Liquid Ammonia Some Physical Properties of Arsine," *J. Am. Chem. Soc.*, Vol. 59, 1937.

Kaiser, W.A., *Summary of Information on Agent GF*, CRLR 164, USA Chemical Research and Development Laboratories, Army Chemical Center, MD, March 1954, UNCLASSIFIED Report (ADB969120).

Kay Lau, Tony Man, *Glass or Polymer Etching Due to the Reaction of Methylphosphonic Difluoride (DF) with Water (U),* CRDEC-TR-86074, USA Chemical Research Development and Engineering Center, Aberdeen Proving Ground, MD, August 1986, CONFIDENTIAL Report (ADC039896).

Kenneth Witam, et al., "Field Crops: Sugarcane" *2001 Plant Disease Control Guide,* 2001, http://www.lsuagcenter.com/Subjects/guides/plantdisease/01disease.html (31 March 2003).

Kibler, A.L., *Data on Chemical Warfare,* Technical Division Memorandum Report 456, Chemical Warfare Center, Edgewood Arsenal, MD, November 1942, UNCLASSIFIED Report (ADB969725).

Kibler, A.L., *Fundamental Study of Toxicity Miscellaneous Data,* EACD459, Chemical Warfare Service, Edgewood Arsenal, MD, 1928 UNCLASSIFIED Report (ADB955210).

Kinkead, E.R., *Evaluation of the Acute Toxicity of Four Compounds Associated with the Manufacture of O-Ethyl-O'- (2-Diisopropyaminoethyl) Methylphosphonite*, CRDEC-CR-87077, USA Chemical Research, Development & Engineering Center, Aberdeen Proving Ground, MD, June 1987, UNCLASSIFIED Report (ADB113969).

Klosky, S. and Stricker, P.F., *Physico Chemical Constants of Diphenylaminochlorarsine*, EATR 58, Chemical Warfare Service Edgewood Arsenal, MD, July 1921, UNCLASSIFIED Report (ADB955024).

Klosky, S., and Stricker, P.F., *The Physico Chemical Properties of Methyldichloroarsine and Arsenic Trichloride*, EACD 63, Chemical Warfare Service, Edgewood Arsenal, Edgewood, MD, August 1921, UNCLASSIFIED Report (ADB955049).

Lau, T.M.K., *Brief Evaluation of the Possibilities of Using Arsenicals as Incapacitating Agents*, CRDEC-TR-87061, USA Chemical Research Development and Engineering Center, Aberdeen Proving Ground, MD, July 1987, UNCLASSIFIED Report (ADB114319).

Lawrence E. Datnoff and Richard S. Lentini, "Brown Spot in Florida Rice," May 1994, http://edis.ifas.ufl.edu/BODY_RH007 (2 April 2003).

L. Fishbein and S. Czerczak, "Concise International Chemical Assessment Document 47: Arsine: Human Health Aspects," WHO, 2002.

Lewis, R.J., *Sax's Dangerous Properties of Industrial Materials*, 10th ed., Volume 2, John Wiley & Sons, Inc., New York, NY, 2001.

Lewis, R.J., *Sax's Dangerous Properties of Industrial Materials,* 10th ed., Volume 3, John Wiley & Sons, Inc., New York, NY, 2001.

Lide, D.R., *CRC Handbook of Chemistry and Physics,* 82nd ed., CRC Press, Washington, DC, 2001.

Lochboehler, C.M., *The Physical Properties of the Glycolates (U),* EASP-100-61; USA Munitions Command, Chemical Research Laboratory; Edgewood Arsenal, MD, 1970, CONFIDENTIAL Report (AD508308).

Macintire, B.G., et al., *Methyldichloroarsine and Methyldifluoroarsine Field Tests,* EACD 410, Chemical Warfare Service, Edgewood Arsenal, MD, March 1931, UNCLASSIFIED Report (ADB955243).

Macy, R., *Constants and Physiological Action of Chemical Warfare Agents,* EATR 78, Chemical Warfare Service Edgewood Arsenal, MD, July 1932, UNCLASSIFIED Report (ADB956574).

Macy, R, *Freezing Point and Volatilities of Mustard and Lewisite Mixtures,* TCIR 512, USA Chemical Research and Development Laboratories, Army Chemical Center, MD, March 1935, UNCLASSIFIED Report (ADB9670444).

Mallinckroct Baker, Inc., MSDS Number T0080, "Tantalum, 1,000 µg/mL or 10,000 µg/mL," effective date 29 October 2001.

Mallinckroct Baker, Inc., MSDS Number T3627, "Titanium Dioxide," effective date: 15 February 1998.

Mark Davis, *Baseline Study on the Problem of Obsolete Pesticide Stocks,* FAO Pesticide Disposal Series N.9, 2001.

Marsh, D.J. et al., *Kinetics of the Hydrolysis of Ethyl Dimethylamino Cyanophosphonate (and certain other related compounds) in Water,* Proton Technical Paper No. 85 (PTP-85), Chemical Defense Experimental Establishment, Porton, England, December 1948, UNCLASSIFIED Report.

Matheson Gas Data Book, 4th ed., The Matheson Company, Inc. East Rutherford, NJ, 1966.

Matt T. Roberts and Don Etherington, "Vapor Density," Bookbinding and the Conservation of Books: A Dictionary of Descriptive Terminology, 7 January 2002, http://palimpsest.stanford.edu/don/dt/dt3670.html (8 April 2003).

Mead, W.P. *Freezing Points of Mixtures of Methyldichloroarsine and Mustard Gas and of Lewisite and Mustard Gas,* EACD 170, Chemical Warfare Service, Edgewood Arsenal, Edgewood, MD, May 1922, UNCLASSIFIED Report (ADB95011).

Miller, C.E., *Thermal Studies on MCE,* Technical Division Memorandum Report 1132, USA Chemical Research and Development Laboratories, Army Chemical Center, MD, September 1945, UNCLASSIFIED Report (ADB964104).

Moelwyn-Hughes, E.A., and Owens, R., *The Surface Tension, The Molecular Surface Energy and the Parachor of Toxic Compounds and of Certain Chlorides Used in Their Manufacture, Part XV of the Thermal Decomposition of the Secondary Alkylflurophosphonites,* Sutton Oak Report 544, Sutton Oak, England, September 1941, UNCLASSIFIED Report.

"Monilia (Moniliophthora roreri)," http://www.oardc.ohio-state.edu/cocoa/monilia.htm (1 April 2003).

Mumford, S.A. and Perry, G.A., *Report on Physical Properties of Mixtures of H and Lewisite I*, PR-1342, USA Chemical Research and Development Laboratories, Army Chemical Center, MD, March 1935, UNCLASSIFIED Report.

NCI Nomination Submitted to the NTP, "Blue-Green Algae," September 2000.

Nelson, E.K., and Dawson, L.E., "The Constitution of Capsaicin, The Pungent Principle of Capsicum. III," *J. Am. Chem. Soc.*, Vol. 45, 1923.

Newman, J.H., Edgewood Arsenal Notebook # NB 9298 (U).

Newman, J.H., et al., *A Thickener for GD (U)*, EC-TR-77016, Edgewood Arsenal, Aberdeen Proving Ground, MD, April 1977, CONFIDENTIAL Report (ADC009719).

NIOSH-DOD-OSHA Sponsored Chemical and Biological Respiratory Protection Workshop Report, February 2000.

NIOSH Pocket Guide to Chemical Hazards, "a-Chloroacetophenone," CAS 532-27-4.

NIOSH Pocket Guide to Chemical Hazards, "o-Chlorobenzylidene malononitrile," CAS 2698-41-1.

NIOSH Pocket Guide to Chemical Hazards, "Phosphorus (yellow)," CAS 7723-14-0.

NIOSH Pocket Guide to Chemical Hazards, "Zinc Chloride fume," CAS 7646-85-7.

Nowlin, T.E., et al., *A New Binary VX Reaction-Two-Liquid System (U)*, EATR 4700, USA Munitions Command, Edgewood Arsenal, MD, November 1972, UNCLASSIFIED Report (AD524088).

NRC, *Toxicity of Military Smokes and Obscurants,* Vol.1, National Academy Press, 1997.

NRC, *Toxicity of Military Smokes and Obscurants,* Vol. 2, National Academy Press, 1999.

National Toxicity Program, "NTP Chemical Repository: Chloroacetophenone."

Owens, R., *Diphenylcyanoarsine Part III – The Pope-Turner Process*, SO/R 488, Sutton Oak, England, December 1940, UNCLASSIFIED Report.

Owens, R., *Diphenylcyanoarsine: Part V – The Physical Properties of M.A., D.A. T.A., and D.C.*, SO/R 492, Sutton Oak, England, December 1940, UNCLASSIFIED Report.

Parker, D.H., *Vapor Pressure of D.M. (Diphenylaminechloroarsine)*, EATR 46, Chemical Warfare Service Edgewood Arsenal, MD, June 1921, UNCLASSIFIED Report (ADB955053).

Patrice L. Abercrombie, *Physical Property Data Review of Selected Chemical Agents and Related Compounds*, September 2003.

Patten, H.E., and Bouder, N.M., *Chemical Properties of Phosgene*, EACD 124, Chemical Warfare Service, Edgewood Arsenal, MD, March 1923, UNCLASSIFIED Report (ADB955133).

Penski, E.C., *Properties of Di-(2-Chloroethyl) Sulfide I. Vapor Pressure Data Review and Analysis*, ERDEC-TR-043, USA Edgewood Research, Development and Engineering

Center, Aberdeen Proving Ground, MD, April 1993, UNCLASSIFIED Report (ADA267059).

Penski, E. C., *Vapor Pressure Data Analysis of Dichloroformoxime*, ERDEC-TR-042, USA Chemical and Biological Defense Agency, Aberdeen Proving Ground, MD, March 1993, UNCLASSIFIED Report (ADA265873).

Penski, E.C., *Vapor Pressure Data Analysis Methodology, Statistics, and Applications,* CR DEC-TR-386, Chemical Research, Development, and Engineering Center, Aberdeen Proving Ground, MD, 1992, UNCLASSIFIED Report (AD-A255090).

Penski, Elwin C., *The Properties of 2-Propyl Methylfluorophosphonate (GB) I. Vapor Pressure Data Review and Analysis,* ERDEC-TR-166, USA Chemical and Biological Defense Command, Aberdeen Proving Ground, MD, June 1994, UNCLASSIFED Report (ADB187225).

Perry, B.J., et al., *The Chemistry of the Alkylfluorophosphonites and Related Compounds,* Porton Technical Paper No. 258, Chemical Defense Experimental Establishment, Porton, England, 31 August 1951, UNCLASSIFIED Report.

Petersen, T.G., *Agent CX (Phosgene Oxime) Summary Report (U),* CRDL Special Publication 7, USA Chemical Research and Development Laboratories, Edgewood Arsenal, MD, October 1965, CONFIDENTIAL Report (AD367890).

Potts, A.M., *The Physical and Chemical Properties of Phosgene and Diphosgene,* OEMCMR-114, 1945, UNCLASSIFIED Report.

Prandtl, W. and Sennewald, K., "Trichloronitrosomethane, Dichloroformoxime (Phosgene Oxime) and Their Derivatives," *Chemische Berichte,* Vol. 62, 1929.

Price, C.C., et al., "Hydrolysis and Chlorinolysis of Cyanogen Chloride", *J. Amer. Chem. Soc.*, Vol. 69, 1947.

Properties of War Gases Volume II: Blood and Nettle Gases (U), ETF 100-41/Vol-2, Chemical Corps Board, Army Chemical Center, MD, December 1956, CONFIDENTIAL Report (AD108457).

Properties of War Gases Volume III: Vomiting & Choking Gases & Lacrimators (U), ETF 100-41/Vol-3, Chemical Corps Board, Army Chemical Center, MD, December 1944, CONFIDENTIAL Report (AD108458).

Properties of War Gases Volume IV: Vesicants (U), ETF 100-41/Vol-4, Chemical Corps Board, Army Chemical Center, MD, December 1956, CONFIDENTIAL Report (AD108459).

Richard J. Lewis, Sr., *Hawley's Condensed Chemical Dictionary,* 13th ed., John Wiley & Sons, Inc., New York, NY, 1997.

Riordan, M.B., *Pilot-Scale Operations of Process for Manufacture of VX Binary Intermediate NM (U),* EM-TR-76055, USA Armament Command, Edgewood Arsenal, Aberdeen Proving Ground, MD, November 1976, CONFIDENTIAL Report (ADC008383).

R.K. Webster, "Rice: Rice Blast," *UC Pest Management Guidelines,* July 2000, http://www.ipm.ucdavis.edu/PMG/r682100611.html (2 April 2003).

R.M. Davis, "Small Grains: Barley Yellow Dwarf," *University of California Statewide Integrated Pest Management Program*, December 2002, http://www.ipm.ucdavis.edu/PMG/r730101911.html (1 April 2003).

Rohrbaugh, D.K., *Detection and Identification of QL Impurities by Electron and Chemical Ionization Gas Chromatography/ Mass Spectrometry*, USA Chemical Research, Development, & Engineering Center, Aberdeen Proving Ground, MD, July 1989, UNCLASSIFIED Report (ADB136428).

Rohrbaugh, D.K., et al., *Studies in Support of SUPLECAM (Surveillance Program for Lethal Chemical Agents and Munitions) II, 1. Thermal Decomposition of VX*, CRDEC-TR-88056, USA Chemical Research, Development and Engineering Center, Aberdeen Proving Ground, MD, May 1988, UNCLASSIFIED Report (ADB124301).

Rosenberg, H.R., and Sharp, S.S., *Evaluation and Synthesis of Chemical Compounds Volume II, DA-18-108-CML-6673 (A), Final Comprehensive Report November 1961 – February 1965*, USA Chemical Research Laboratories, Edgewood Arsenal, MD, February 1965, UNCLASSIFIED Report (ADB253543).

Rosenblatt, D.H., et al., *Problem Definition Studies on Potential Environmental Pollutants VIII: Chemistry and Toxicology of BZ (3-Quinuclidinyl Benzilate),* USAMBRDL-TR 7710, USA Medical Bioengineering Research and Development Laboratory, Fort Detrick, MD, 1977, UNCLASSIFIED Report (ADB030349).

Ruth Levy Guyer, "Research in the News: Prions: Puzzling Infectious Proteins," http://science-education.nih.gov/nihHTML/ose/snapshots/multimedia/ritn/prions/prions1.html, 8 August 2003.

Salamon, M.K., *Agent VX*, CWL Special Publication 4-10, USA Chemical Warfare Laboratories, Army Chemical Center, MD, June 1959, UNCLASSIFIED Report.

Samuel, J.B., et al., *Physical Properties of Standard Agents, Candidate Agents, and Related Compounds at Several Temperatures (U)*, ARCSL-SP-83015, USA Armament Research and Development Command, Aberdeen Proving Ground, MD, June 1983, UNCLASSIFIED Report (ADC033491).

Sass, S., et al., *Basic Esters of Glycolic Acids (U): Part III Analysis and Chemical Properties of Microgram and Larger Quantities of EA 2277 and Related Compounds*, CRDLR 3088, USA Chemical Research and Development Laboratories, Army Chemical Center, MD, August 1961, UNCLASSIFIED Report (AD325351).

Satu M. Somani and James A. Romano, Jr. (eds), *Chemical Warfare Agents: Toxicity at Low Levels*," CRC Press, 2001.

Savage, J. J., and Fielder, D., *The Vapor Pressure of Chemical Agents GD, VX, EA2223, EA 3547, EA 3580, EA 5365, and EA 5533*, EC-TR-76058, Aberdeen Proving Ground, MD, August 1976, UNCLASSIFIED Report (ADB013164).

Sax, N.I., *Dangerous Properties of Industrial Materials*, 3rd ed., Reinhold Book Corporation, Albany, NY, 1968.

Schweitzer, P.A., *Corrosion Resistance Tables: Metals, Plastics, Nonmetallics, and Rubbers*, 2nd ed., Ed., Marcel Dekker, INC., Chester, NJ, 1986.

Scientific Section (Laboratory), Policy Development and Analysis Branch, Division for Operations and Analysis, United Nations Office on Drugs and Crime, *Terminology and Information on Drugs: Part I,* "Opioids: Fentanyls," October 1998, http://www.unodc.org/unodc/report_1998-10-01_1_page016.html, 8 August 2003.

Scott, D.W., et al., "2,3-Dithiabutane: Low Temperature Heat Capacity, Heat of Fusion, Heat of Vaporization, Vapor Pressure, Entropy and Thermodynamic Functions," *J. Amer. Chem. Soc.*, Vol. 72, 1950.

Sharon Reutter, et al., *Review and Recommendations for Human Toxicity Estimates for FM 3-11.9*, ECBC-TR-349, September 2003.

Sherrill, M.L., *Investigation of the Synthesis of Capsaicin and Related Compounds*, EACD 307, USA Chemical Research Laboratories, Edgewood Arsenal, MD, April 1924, UNCLASSIFIED Report (ADB955292).

Siegel, M., *The Corrosive Effect of War Gases on Metals and Materials*, EACD 113, Chemical Warfare Service, Edgewood Arsenal, MD, December 1921, UNCLASSIFIED Report (ADB955153).

Spectrum Chemical Fact Sheet, "Phosphorus," CAS 7723-14-0.

Steadman, A., *Isolation of Capsaicin from Capsicum*, EACD 188, USA Chemical Research Laboratories, Edgewood Arsenal, MD, June 1922, UNCLASSIFIED Report (ADB955131).

Steadman's Medical Dictionary, 25th Ed., Williams & Wilkins, Baltimore, MD, 1990.

Stephen A. Ferreira and Rebecca A. Boley, "Colletrotrichum Coffeanum," *Crop Knowledge Master*, November 1991, http://www.extento.hawaii.edu/kbase/crop/Type/c_coffee.htm (April 1, 2003).

Stern, R.A., USA Chemical Research and Development Laboratories Notebook # NB 7265 (C).

Sumner, J.F., et al., *The Vapour Pressure of Arsenious Chloride and of Lewisite I,* Sutton Oak Report 561(SO/R/561), Military Intelligence Division, Great Britain, December 1941, UNCLASSIFIED Report. (ADB956574).

Szafraniec, L.J., et al., *On the Stoichiometry of Phosphonothiolate Ester Hydrolysis*, CRDEC-TR-212, USA Chemical Research Development & Engineering Center, Aberdeen Proving Ground, MD, July 1990, UNCLASSIFIED Report (ADA225952).

Sze, J.M., and Simak, R.S., *Binary GB: A Compilation of Relevant Data*, ARCSL-TR-82019, USA Armament Research and Development Command, APG, MD, March 1983, CONFIDENTIAL Report (ADC030931).

Tannenbaum, H. and Zeffert, B.M., *Crystallization of GB*, TCIR-513, USA Chemical and Radiological Laboratories, Army Chemical Center, MD, November 1949, UNCLASSIFIED Report (AD474404).

Tarantino, P.A., *Electrochemical Corrosion Study of Miscellaneous Metals/Alloys with Methylphosphonic Difluoride*, CRDEC-TR-88032, USA Chemical Research Development and Engineering Center, Aberdeen Proving Ground, MD, November 1987, UNCLASSIFIED Report (ADB117574).

Tevault, D.E., et al., *Vapor Pressure of GF, TR-304S*, USA ECBC, Aberdeen Proving Ground, MD, submitted for publication 2 May 2003, UNCLASSIFIED Report.

Texas Extension Plant Pathologists, "Bacterial Leaf Blight Symptoms on Rice," February 1996, http://plantpathology.tamu.edu/Texlab/Grains/Rice/riceblb.html (8 April 2003).

The International Rubber Research and Development Board, "South American Leaf Blight," http://www.irrdb.com/irrdb/NaturalRubber/Diseases/salb.htm (1 April 2003).

The Merck Index, An Encyclopedia of Chemicals, Drugs, and Biologicals, 12th ed., Merck Research Laboratories, Whitehouse Station, NJ, 1996.

The Merck Index: An Encyclopedia of Chemicals, Drugs, and Biologicals, 13th ed., Merck & Company, Inc., Whitehouse Station, MJ, 2001.

Thomas, M.T., *Research and Development for Candidate Materials for Use as a DF Containment Vessel,* CRDC-CR-85058, USA Chemical Research Development and Engineering Center, Aberdeen Proving Ground, MD, September 1985, UNCLASSIFIED Report (ADB096058).

US Animal Health Association, *Foreign Animal Diseases: "The Gray Book (Part IV),"* Pat Campbell & Associates and Carter Printing Company, Richmond, VA, Revised 1998.

USDA, APHIS, "Protocol for Military Clearance," 18 June 2001.

US EPA, "Pesticides: Health and Safety: Human Health Issues," 19 May 2003, http://www.epa.gov/pesticides/health/human.htm, 19 August 2003.

US HHS, CDC, "Compendium of Measures to Control *Chlamydia psittaci* Infection Among Humans (Psittacosis) and Pet Birds (Avian Chlamydiosis), 1998," *Morbidity and Mortality Weekly Report,* Vol. 47, N0. RR-10, 10 July 1998.

US HHS, Public Health Service, National Toxicology program, *10th Report on Carcinogens,* December 2002.

Walpole, J.L., *Determination of the Flash Points of GA and GB,* Porton Technical Paper No. 45 (PTP 45), Chemical Defence Experimental Establishment, Porton, England, March 1948, UNCLASSIFIED Report.

Wardrop, A.W.H., and Bryant, P.J.R., *Physico-Chemical Properties of Phosphorus Esters Part II: Some Constants of Isopropyl methylfluorophosphinate (GB),* Porton, Technical Paper No. 278 (PTP-278), Chemical Warfare Laboratories, Army Chemical Center, MD, March 1952, UNCLASSIFIED Report.

Watson, P.D., *Determination of the Vapor Pressure of D.C.,* EACD 79, Chemical Warfare Service, Edgewood Arsenal, MD, December 1921, UNCLASSIFIED Report (ADB959611).

Watson, P.D., *Determination of the Vapor Pressure of Methyldichloroarsine,* EACD 176, Chemical Warfare Service, Edgewood Arsenal, Edgewood, MD, May 1922, UNCLASSIFIED Report (ADB959625).

Weast, R.C., *CRC Handbook of Chemistry and Physics* 50th ed., CRC, Cleveland, OH, 1969.

Welchman, R.M.A., *Preliminary Report on the Potential Value of Nerve Gases as C.W. Agents,* Porton Report No. 2747 (PR 2747), Chemical Defence Experimental Establishment, Porton, England, January 1947, UNCLASSIFIED Report.

Williams, A.H., "The Thermal Decomposition of 2:2'-Dichlorodiethyl Sulphide," *J. Chem. Soc.,* 1947.

Witten, Benjamin, *The Search for Toxic Chemical Agents (U)*, EATR 4210, Edgewood Arsenal Research Laboratories, MD, November 1969, UNCLASSIFIED Report (AD507852).

Witten, B., *The Hydrolysis of MCE,* Technical Division Memorandum Report 1121, USA Chemical Research and Development Laboratories, Army Chemical Center, MD, August 1945, UNCLASSIFIED Report (ADB964102).

W.R. Kirner, *Summary Technical Report of Division 9, NDRC Volume 1, Chemical Warfare Agents, and Related Chemical Problems Part I-II,* Office of Scientific Research and Development, Washington, DC, 1946, UNCLASSIFIED Report (AD234270).

W.R. Kirner, *Summary Technical Report of Division 9,* NDRC Volume 1, *Chemical Warfare Agents, and Related Chemical Problems Parts III-VI,* NDRC-DIV-9-VOL-1-PT1-2, Office of Scientific Research and Development Washington, DC, 1946, UNCLASSIFIED Report (AD234249).

Yang, Y., et al., "Characterization of HD Heels and the Degradation of HD in Ton Containers," *In Proceedings of the 1996 ERDEC Scientific Conference on Chemical and Biological Defense Research 19-22 November 1996,* UNCLASSFIED Paper, ERDEC-SP-048, USA Edgewood Research, Development and Engineering Center, Aberdeen Proving Ground, MD, October 1997, UNCLASSFIED Report (ADA334105).

Yang, Y., et al., "Decontamination of Chemical Warfare Agents," *Chem. Rev.*, Vol. 92, 1992.

Yang, Y, et al., "Hydrolysis of VX: Activation Energies and Autocatalysis," *In Proceedings of the 1994 ERDEC Scientific Conference on Chemical Biological Defense Research 15-18 November1994,* UNCLASSIFIED Paper, ERDEC-SP-036, USA Edgewood Research, Development and Engineering Center, Aberdeen Proving Ground, MD, May 1996, UNCLASSFIED Report (ADA313080).

Yang, Y., et al., "Hydrolysis of VX with Equimolar Water at Elevated Temperatures: Activation Parameters of VX, CV and EA 2192", *In Proceedings of the 1996 ERDEC Scientific Conference on Chemical Biological Defense Research 19-22 November1996,* UNCLASSIFIED Paper, ERDEC-SP-048, USA Edgewood Research, Development and Engineering Center, Aberdeen Proving Ground, MD, October 1997, UNCLASSFIED Report (ADA334105).

Yang, Y., et al., "Perhydrolysis of Nerve Agent VX," *J. Org. Chem.,* Vol. 58, 1993.

Yaws, C.L., *Matheson Gas Data Book,* 7th ed., McGraw-Hill Companies, New York, NY, 2001.

Yvan J.F. Hutin (CDC), et al., "Research: Outbreak of Human Monkeypox, Democratic Republic of Congo, 1996-1997," *Emerging Infectious Diseases*, Vol. 7, No. 3, May-June 2001.

Zeffert, B.M., and Coulter, P.B., *Physical Constants of G-Series Compounds: Compounds EA 1210, EA 1211, EA 1212, EA 1213, EA 1214,* Technical Division Memorandum Report 1292, USA Chemical Research and Development Laboratories, Army Chemical Center, MD, July 1947, UNCLASSIFIED Report (ADB964904).

Zeffert, B.M. et al., "Properties, Interaction and Esterification of Methylphosphonic Dihalides," *J. Am. Chem. Soc.*, Vol. 82, 1960.

Zeffert, B.M. et al., *Slow Fractional Crystallization of GB*, CRLR-2, USA Chemical and Radiological Laboratories, Army Chemical Center, MD, April 1951, UNCLASSIFIED Report (AD498968).

GLOSSARY

PART I – ABBREVIATIONS AND ACRONYMS

A	
AC	hydrogen cyanide
ACAA	automatic chemical agent alarm
ACADA	automatic chemical agent detector/alarm
ACh	acetylcholine
AChE	acetycholinesterase
AFJMAN	Air Force Joint Manual
AFM	Air Force Manual
AI	area of interest, Avian Influenza (fowl plague)
AJD	Aujeszky's disease
AMAD	automatic mustard agent alarm
AMEDD	Army Medical Department
AO	area of operations
APHIS	Animal and Plant Health Inspection Service
ASF	African swine fever
ASTM	American Society for Testing and Materials
atm	atmosphere
ATSDR	Agency for Toxic Substance and Disease Registry
avdp	avoirdupois weight
B	
BBTV	banana bunchy top virus
BG	Brigadier General
BMBL	Biosafety in Microbiological and Biomedical Labs
BW	biological warfare
BWC	Biological Weapons Convention
BYD	Barley Yellow Dwarf (plant disease)
BZ	3-Quinuclidinyl benzilate (an incapacitating agent)
C	
C	Celsius
CA	bromobenzylcyanide
CADK	chemical agent detector kit
CAM	chemical agent monitor
CANA	convulsant antidote for nerve agents
CAPDS	Chemical Agent Point Detection System
CARC	chemical agent-resistant coating
CAS	Chemical Abstract Service
CB	chemical-biological
CBD	coffee berry disease
CBPP	contagious bovine pleuropneumonia
CBR	chemical, biological, and radiological
CBW	chemical and biological warfare
Cd	Cadmium
CDC	Centers for Disease Control and Prevention
CEM	contagious equine metritis

CFR	Code of Federal Regulations
CG	commanding general, phosgene
ChE	cholinesterase
CHPPM	US Army Center for Health Promotion and Preventive Medicine
CHRIS	Chemical Hazard Response Information System
CIC	chloroacetophenone in chloroform
CK	cyanogen chloride
Cl_2	chlorine
CMPF	cyclohexyl methylphosphonafluoridate
CN	chloroacetophenone (tear gas or mace)
CNB	chloroacetophenone-in benzene
CNC	chloroacetophenenone-in chloroform
CNOH	cyanic acid
CNS	central nervous system
cP	centipoises
CR	dibenz (b,f)-1:4-oxazepine
CS	o-chlorobenzylidene malononitrile (a tear agent)
Ct	concentration time
CW	chemical warfare
CWA	chemical warfare agent
CWC	Chemical Weapons Convention
CWDD	Chemical Warfare Directional Detector
CX	phosgene oxime
CZ	combat zone
D	
DA	diphenylchloroarsine (a vomiting agent), Department of the Army
DC	diphenylcyanoarsine (a vomiting agent), District of Columbia
DF	methylphosphonic difluoride
DM	diphenylaminochloroarsine (Adamsite) (a vomiting agent)
DMDS	dimethyl disulfide
DNA	deoxyribonucleic acid
DOC	degree of confidence
DOD	Department of Defense
DOT	Department of Transportation
DP	diphosgene
dynes/cm	dynes per centimeter
E	
ECBC	Edgewood Chemical and Biological Center
ECt	effective dosage of an aerosol
ED	ethyldichloroarsine, effective dose
EEE	eastern equine encephalitis
EHEC	enterohemorrhagic
EHS	extremely hazardous substances
EMPA	ethyl methylphosphonate
EMT	emergency medical technician
EPA	Environmental Protection Agency
ERG	Emergency Response Guide

F

F	Fahrenheit
F_2	fluorine
FAO	Food and Agriculture Organization
FM	field manual
FMD	foot and mouth disease
FMFM	Fleet Marine Field Manual
FP	freezing point

G

g	gram(s)
GA	tabun (a nerve agent)
GB	sarin (a nerve agent)
g/cm^3	grams per cubic centimeter
GD	soman (a nerve agent)
GF	cyclosarin (a nerve agent)
GI	gastrointestinal
g/ml	grams per milliliter

H

H	Levinstein mustard (a blister agent)
H_2S	hydrogen sulfide
H_2SO_4	sulphuric acid
HAZMAT	hazardous material
HBr	hydrogen bromide
HC	hexachloroethane
HCl	hydrogen chloride
HCN	hydrogen cyanide
HD	distilled mustard (a blister agent)
HF	hydrogen fluoride
HFV	hemorrhagic fever virus
H/HD	sulfur mustards
HHS	health and human services
HL	mustard-lewisite mixture
HN	nitrogen mustard (HN-1, HN-2, HN-3)
HNO_3	nitric acid
hr	hour
HSBD	Hazardous Substance Data Bank
HT	mustard-T mixture
HTH	high test hypochlorite (calcium hypochlorite)

I

ICAM	improved chemical agent monitor
IDA	Institute for Defense Analysis
IDLH	immediately dangerous to life and health
IEDK	individual equipment decontamination kit
ILO	International Labor Office
in	inch(es)
IND	investigational new drug
IPB	intelligence preparation of the battlespace
IPCS	International Program on Chemical Safety
IPDS	improved point detection system

IR	infrared
ISBN	International Standard Book Number
ISCS	International Chemical Safety Cards
ITF	International Task Force
J	
JE	Japanese encephalitis
JP	joint publication
JP-8	jet fuel grade
K	
kg	kilogram(s)
KHV	Korean hemorrhagic fever
L	
L	lewisite, liters
LCt_{50}	medial lethal dosage of a chemical agent vapor or aerosol
LD	lethal dose
LD_{50}	median lethal dosage of a liquid chemical agent
L/min	liters per minute
LSD	d-lysergic acid diethylamide
M	
M	molarity (moles per liter)
MA	Massachusetts
MCCDC	Marine Corps Combat Development Command
MCPDS	Marine Corps Publication Distribution System
MCRP	Marine Corps Reference Publication
MD	Maryland, methyldichloroarsine
MDMA	3, 4-methylene dioxymethamphetimine (ecstasy)
MDMP	military decision making process
MFPA	methylfluorophosphonic acid
mg	milligrams
mg/kg	milligrams per kilogram
mg/m^3	
$mg\text{-}min/m^3$	milligram-minute(s) per cubic meter
mg/mL	milligram per milliliter
MILSTRIP	military standard requisitioning and issue procedures
min	minute
MOPP	mission oriented protective posture
MP	melting point
MPA	methylposphonic acid
MSHA	Mine Safety and Health Administration
MV	minute volume
MV(L)	minute volume (in liters)
MW	molecular weight
N	
N/A	not applicable
NAERG	North American Emergency Response Guide
NATO	North Atlantic Treaty Organization
NAVMED	Navy medical
NBC	nuclear, biological and chemical
NBCC	nuclear, biological, chemical, and conventional

NCI	National Cancer Institute
NCO	noncommissioned officer
NE	sulphur with small amounts of silica gel
ng	nanograms
NH_3	ammonia
NIOSH	National Institute for Occupational Safety and Health
NJ	New Jersey
NM	Dimethylpolysulfide
NPG	NIOSH Pocket Guide
NPL	National Priorities List
NR	none recommended
NRC	National Response Center
NTIS	National Technical Information Service
NTP	National Toxicity Program
NWDC	Navy Warfare Development Command
NTRP	Navy Tactical Reference Publication
NY	New York
O	
OC	capsaicin
OCONUS	outside the continental United States
OH	Ohio
OPA	isopropylamine and isopropyl alcohol
OPR	office of primary responsibility
OSD	Office of the Secretary of Defense
OSHA	Occupational Safety and Health Administration
P	
PB	pyridostigmine bromide
PCB	polychlorinated biphenyls
PCl_3	phosphorus trichloride
PCP	phencyclidine
PD	phenyldichloroarsine, Pierce's disease
PEL	permissible exposure limit
pH	potential of hydrogen
PHS	Public Health Service
PIC	prior informed consent
PMPA	pinacolyl methylphosphonic acid
ppm	parts per million
PPR	peste des petits ruminants (pest of small ruminants)
PS	chloropicrin (a choking agent)
PSP	paralytic shellfish poisoning
PTP	Porton Technical Paper
PVNTMED	preventive medicine
Q	
QL	2-diisopropylanminoethyl
R	
RCA	riot control agent
REL	recommended exposure limit
RF	Russian Federation
RMP	risk management program

ROA	rate of action
ROD	rate of detoxification
ROE	route of exposure
RP	red phosphorous
RTECS	Registry of Toxic Effects of Chemical Substances
RVF	Rift Valley Fever
S	
SA	arsine
SCP	Standards Completion Program
SCT	Secretariat of Communications and Transportation (Mexico)
SDK	skin decontamination kit
SEB	staphylococcal enterotoxin B
SGF	smoke generation fuels
SGF-2	fog oil
SME	subject matter expert
SMEDI	stillbirths, mummification, embryonic death and infertility
SO_2	sulphur dioxide
SOF	special operations forces
SRC	State Research Center
STB	supertropical bleach
STEC	Shiga toxin-producing *E. coli*
STEL	short-term exposure limit
SVD	swine vesicular disease
SWO	staff weather officer
T	
$t_{½}$	half-life of a reaction
TIB	toxic industrial biological
TIC	toxic industrial chemical
TIM	toxic industrial material
TIR	toxic industrial radiological
TLE	toxic load exponent
TLV	threshold limit value
TM	technical manual
TO	theater of operation
TOF	trioctylophosphite
TRADOC	United States Army Training and Doctrine Command
TWA	time-weighted average
U	
UK	United Kingdom
UN	United Nations
UNEP	United Nations Environment Program
UNEPO	United Nations Environmental Protection Organization
US	United States
USA	United States of America
USACMLS	United States Army Chemical School
USAF	United States Air Force
USAMRIID	United States Army Medical Research Institute of Infectious Diseases
USCG	United States Coast Guard

USDA	United States Department of Agriculture
USEPA	United States Environmental Protection Agency
USMC	United States Marine Corps
USN	United States Navy
USSR	Union of Soviet Socialist Republics
UV	ultraviolet
V	
VEE	Venezuelan equine encephalitis
VHF	viral hemorrhagic fever
VND	velogenic Newcastle disease
vol	volume
FP	vapor pressure
VTEC	vertoxin-producing *E. coli*
VX	a persistent nerve agent
Vx	a persistent nerve agent
W	
WEE	western equine encephalitis
WF_6	tungsten hexafluoride
WHA	World Health Assembly
WHO	World Health Organization
WMD	weapons of mass destruction
WP	white phosphorous
WWI	World War I
WWII	World War II
Z	
$ZnCl_2$	zinc chloride

Symbols

μg	microgram
μg/kg	micrograms per kilogram

PART II – TERMS AND DEFINTIONS

acetylcholine – The neurotransmitter of the cholinergic portion of the nervous system. (The Textbook of Military Medicine)

acetylcholinesterase – An enzyme that hydrolyzes acetylcholine very rapidly (thereby stopping its activity). Acetylcholinesterase is found at the receptor sites of tissue supplied with nerves by the cholinergic nervous system. Most cholinesterase-inhibiting compounds are either organophosphates or carbamates. (The Textbook of Military Medicine)

acid – 1. A compound yielding a hydrogen ion in a polar solvent (e.g., in water);a.'s form salts by replacing all or part of the ionizable hydrogen with an electropositive element or radical. An a. containing one ionizable atom of hydrogen in the molecule is called monobasic; one containing two such atoms, dibasic; and one containing more than two, polybasic. 2. In popular language, any chemical compound that has a sour taste (given by the hydrogen ion). 3. Sour; sharp to the taste. 4. Relating to a.; giving an a. reaction. For individual acids, see specific names. Bilea.'s, taurocholic and glyocholic a.'s, used when bilary secretion is inadequate and for biliary colic. (Stedman's Medical Dictionary)

acidosis – A state characterized by actual or relative decrease of alkali in body fluids in relation to the acid content. (Stedman's Medical Dictionary)

acute – Of short and sharp course, not chronic; said of a disease. (Stedman's Medical Dictionary)

acute toxicity – Toxic effects occurring within moments to a few days of toxic exposure. (Review and Recommendations for Human Toxicity Estimates for FM 3-11.9, ECBC-TR-349)

aerosol – A liquid or solid composed of finely divided particles suspended in a gaseous medium. Examples of common aerosols are mist, fog, and smoke. (JP 1-02)

alga, algae (p) – A division of eukaryotic, photosynthetic, nonflowering organisms that includes many seaweeds. (Stedman's Medical Dictionary)

alkali – A strongly basic substance yielding hydroxide ions (OH^-) in solution; e.g., sodium hydroxide, potassium hydroxid. (Stedman's Medical Dictionary)

ambient temperature - Temperature of the surrounding air or other medium. (EPA, Terms of Environment)

amino acids – An organic acid in which one of the CH hydrogen atoms has been replaced by NH_2. An α-amino acid is an amino acid of the general formula R-$CHNH_2COOH$; the L forms of these are the hydrolysis products of proteins. (Stedman's Medical Dictionary)

analogue – A compound that resembles another in structure but is not necessarily as isomer. (Stedman's Medical Dictionary)

anthrax – An acute bacterial disease caused by *Bacillus anthracis.* (Control of Communicable Diseases Manual)

antibiotics – a soluble substance derived from a mold or bacterium that inhibits the growth of other microorganisms. (Stedman's Medical Dictionary)

antibody – A protein made by vertebrates as the immune response to a foreign macromolecule or antigen. (The Biological & Chemical Warfare Threat)

antigen – a molecule capable of eliciting a specific antibody or T-cell response. (USACHPPM TG 204)

antiplant (biological) – Living organisms that cause disease or damage to plants. (Textbook of Military Medicine)

antitoxin – An antibody formed in response to and capable of neutralizing a biological poison; an animal serum containing antitoxins. (Medical Management of Biological Casualties Handbook)

aqueous – Watery; of, like, or containing water. (Stedman's Medical Dictionary)

arrhythmia – Loss of rhythm; denoting especially an irregularity of the heartbeat. (Stedman's Medical Dictionary)

arthralgia – Severe pain in a joint, especially one not inflammatory in character. (Medical Management of Biological Casualties Handbook)

arthropods – A member of the phylum Arthropoda that includes the classes Crustacea (crabs, shrimps, crayfish, lobsters), Insecta, Arachnida (spiders, scorpions, mites, ticks), Chilopoda (centipedes), Diplopoda (millipedes), Merostomata (horseshoe crabs), and various other extinct or lesser known groups. (Stedman's Medical Dictionary)

arsenical – Drug or agent, the effect of which depends o its arsenic content. (Stedman's Medical Dictionary)

ataxia – An inability to coordinate muscle activity during voluntary movement, so that smooth movements occur. (Medical Management of Biological Casualties Handbook)

atropine – Is used as an antidote for nerve agent poisoning. It inhibits the action of acetylcholine at the muscle junction by binding to acetylcholine receptors. (Textbook of Military Medicine)

bacteria – Single-celled, microscopic, plant-like organisms. (TM 3-216/AFM 355-6)

base – Any molecule or ion that combines with a hydrogen ion; e.g., OH^-, CN^-, NH_3. (Stedman's Medical Dictionary)

binary chemical munition – A munition in which chemical substances, held in separate containers, react when mixed or combined as a result of being fired, launched, or otherwise initiated to produce a chemical agent. (JP 1-02)

binary precursors - The component chemicals that combine to produce binary chemical agents. Examples of two common binary chemical agent components are as follows: a. The components for binary GB (GB2) are methylphosphonic difluoride (DF) and isopropyl alcohol with an amine added (OPA). b. The components for binary VX (VX2) are ethyl 2-didsopropyl aminoethyl methylphosphonite (QL) and dimethylpolysulfide (NM). (AR 50-6)

boiling point – The temperature at which the vapor pressure of a liquid equals the pressure of the gas above it. Normal boiling point is the temperature at which vapor pressure of a liquid equals one atmosphere (atm). (General Chemistry with Qualitative Analysis, 9th ed.)

biological agent – A microorganism that caused disease in personnel, plants, or animals or causes the deterioration of materiel. (JP 1-02)

biological ammunition – a type of ammunition, the filler of which is primarily a biological agent. (JP 1-02)

biological defense –The methods, plans, and procedures involved in establishing and executing defensive measures against attacks using biological agents. (JP 1-02)

biological environment – Conditions found in an area resulting from direct or persisting effects of biological weapons. (JP 1-02)

biological operation – employment of biological agents to produce casualties in personnel or animals or damage to plants. (JP 1-02)

biological threat – A threat that consists of biological material planned to be deployed to produce casualties in personnel or animals or damage plants. (JP 1-02)

biological weapon – An item of materiel which projects, disperses, or disseminates a biological agent including arthropod vectors. (JP 1-02)

bioregulator – Organic chemicals that regulate cell processes. (The Biological & Chemical Warfare Threat)

biovar – A group of bacterial strains distinguishable from other strains of the same species o the basis of physiological characters. Formerly called biotype. (Stedman's Medical Dictionary)

blepharospasm – Spasmodic winking, or contraction of the orbicularis oculi muscle. (Stedman's Medical Dictionary)

blister agent – A chemical agent which injures the eyes and lungs, and burns or blisters the skin. Also called vesicant agent. (JP 1-02)

blood agent – A chemical compound, including the cyanide group, the affects bodily functions by preventing the normal utilization of oxygen by body tissues. (JP 1-02)

botulism – Poisoning by toxin derived from the microorganism *Clostridium botulinum*. (Control of Communicable Diseases Manual)

brucellosis – A systemic bacterial disease characterized by irregular fever . (Control of Communicable Diseases Manual)

bubo – Inflammatory swelling of one or more lymph nodes in the groin. (Stedman's Medical Dictionary)

carcinogen – Any cancer-producing substance. (Stedman's Medical Dictionary)

case-fatality rate – Usually expressed as the percentage of persons diagnosed as having a specified disease who die as a result of that illness within a given period. This term is most frequently applied to a specific outbreak of acute disease in which all patients have been followed for an adequate period of time to include all attributable deaths. (Control of Communicable Diseases Manual)

casualty – Any person who is lost to the organization by having been declared dead, duty status – whereabouts unknown, missing, ill, or injured. (JP 1-02)

catalyst – A substance that accelerates a chemical reaction but is not consumed or changed permanently thereby. (Stedman's Medical Dictionary)

central nervous system – The nervous system consisting of the brain and spinal cord. (Hazardous Chemicals in Human and Environmental Health, WHO)

chemical agent – Any toxic chemical intended for use in military operations. (JP 1-02)

chemical agent cumulative action – the building up, within the human body, of small ineffective doses of certain chemical agents to a point where eventual effect is similar to one large dose. (JP 1-02)

chemical ammunition – a type of ammunition, the filler of which is primarily a chemical agent. (JP 1-02)

chemical ammunition cargo – cargo such as white phosphorous munitions (shell and grenades). (JP 1-02)

chemical, biological, and radiological operation – a collective term used only when referring to a combined chemical, biological, and radiological operation (JP 1-02)

chemical defense – the methods, plans and procedures involved in establishing and executing defensive measures against attack utilizing chemical agents. (JP 1-02)

chemical dose – the amount of chemical agent, expressed in milligrams, that is taken or absorbed by the body. (JP 1-02)

chemical environment – conditions found in an area resulting from direct or persisting effects of chemical weapons. (JP 1-02)

chemical monitoring – the continued or periodic process of determining whether or not a chemical agent is present. (JP 1-02)

chemical operations – employment of chemical agents to kill, injure, or incapacitate for a significant period of time, man or animals, and deny or hinder the use of areas, facilities, or material; or defense against such employment. (JP 1-02)

chemical survey – the directed effort to determine the nature and degree of chemical hazard in an area and to delineate the perimeter of the hazard area. (JP 1-02)

chemical warfare – All aspects of military operations involving the employment of lethal and incapacitating munitions/agents and the warning and protective measures associated with such offensive operations. Since riot control agents and herbicides are not considered to be chemical warfare agents, those two items will by referred to separately or under the broader term "chemical," which will be used to include all types of chemical munitions/agents collectively. Also called CW. (JP 1-02)

Chikungunya virus disease – A febrile viral disease characterized by arthalgia or arthritis. (Control of Communicable Diseases Manual)

cholera – An acute bacterial disease of serious consequence. (Control of Communicable Diseases Manual)

cholinergic – Relating to nerve cells or fibers that employ acetylcholine as their neurotransmitter. (Medical Management of Biological Casualties Handbook)

coagulate – To convert a fluid or a substance in solution into a solid or gel. (Stedman's Medical Dictionary)

compound – A substance formed by the covalent or electrostatic union of two or more elements, generally differing entirely in physical characteristics from any of its components. (Stedman's Medical Dictionary)

concentration – The quantity of a substance per unit volume or weight. (Stedman's Medical Dictionary)

conjunctiva, pl. conjunctivae – The mucous membrane inventing the anterior surface of the eyeball and the posterior surface of the lids. (Medical Management of Biological Casualties Handbook)

contamination – 1. The deposit, absorption, or absorption of radioactive material, or of biological or chemical agents on or by structures, areas, personnel, or objects. 2. Food and/or water made unfit for consumption by humans or animals because of the presence of environmental chemicals, radioactive elements, bacteria or organisms, the byproduct of the growth of bacteria or organisms, the decomposing material (to include the food substance itself), or waste in the food or water. (JP 1-02)

contamination control – Procedures to avoid, reduce, remove, or render harmless, temporarily or permanently , nuclear, biological, and chemical contamination for the purpose of maintaining or enhancing the efficient conduct of military operations. (JP 1-02)

convulsion – A violent spasm or series of jerkings of the face, trunk, or extremities. (Stedman's Medical Dictionary)

coronary – Specifically, denoting the blood vessels of the heart. (Stedman's Medical Dictionary)

corrosive – Causing gradual deterioration or consummation of a substance by another, especially by biochemical or chemical reaction. (Stedman's Medical Dictionary)

cutaneous – Relating to the skin. (Medical Management of Biological Casualties)

cyanosis – A dark bluish or purplish coloration of the skin and mucous membrane due to deficient oxygenation of the blood, evident when reduced hemoglobin I the blood exceeds 5 g/mL. (Medical Management of Biological Casualties)

cytotoxin – A specific substance, usually with reference to antibody, that inhibits or prevents the functions of cells, causes destruction of cells, or both. (Stedman's Medical Dictionary)

decomposition temperature – The temperature at which a chemical breaks down into two or more substances. (College Chemistry with Qualitative Analysis)

decontamination – The process of making any person, object, or area safe by absorbing, destroying, neutralizing, making harmless, or removing chemical or biological agents, or by removing radioactive material clinging to or around it. (JP 1-02)

decontamination station – A building or location suitably equipped and organized where personnel and material are cleansed of chemical, biological or radiological contaminants. (JP 1-02)

defoliant operation - The employment of defoliating agents on vegetated areas in support of military operations. (JP 1-02)

defoliating agent – A chemical which causes trees, shrubs, and other plants to shed their leaves prematurely. (JP 1-02)

degree of confidence – An indication of the level of confidence in each toxicity estimate. It is a subjective evaluation based on the quality and quantity of the underlying data and the method(s) by which the estimate was derived. (Review and Recommendations for Human Toxicity Estimates for FM 3-11.9, ECBC-TR-349).

dehydrate – To extract water from; to lose water. (Stedman's Medical Dictionary)

density (liquid/solid) – Is the mass per unit volume of the substance. (General Chemistry with Qualitative Analysis, 9th ed.)

desiccate – Exsiccation; to dry thoroughly; to render free from moisture. (Stedman's Medical Dictionary)

detection – The act of locating NBC hazards by use of NBC detectors or monitoring and/or survey teams. (JP 1-02)

diathesis – The constitutional or inborn state disposing to a disease, group of diseases, or metabolic or structural anomaly. (Stedman's Medical Dictionary)

dilate – To perform or undergo a physiologic or artificial enlargement of a hollow structure or opening. (Stedman's Medical Dictionary)

dilute solution – Solutions of chemical agents in concentration and quantities reduced by admixture (dilution) to levels that present significantly reduced hazards. Also called RDTE dilute solution. (AR 50-6)

dimer – A compound or unit produced by the combination of two like molecules; in the strictest sense, without loss of atoms (thus nitrogen tetroxide, N_2O_4, is the dimmer of nitrogen dioxide, NO_2). (Stedman's Medical Dictionary)

disease – Morbus; illness; sickness; an interruption, cessation, or disorder of body functions, systems, or organs. (Stedman's Medical Dictionary)

deoxyribonucleic acid - the genetic material of all organisms and viruses (except for a small class of RNA-containing viruses) that code for structures and materials used in normal metabolism. (The Biological & Chemical Warfare Threat)

dosage – The amount of substance administered (or received) per body weight. (USACHPPM TG 204)

dose – The amount of substance or energy that is taken into or absorbed by the body; the amount of substance, radiation, or energy absorbed in a unit volume, an organ, or an individual. (USACHPPM TG 204)

dysentery – A disease marked by frequent watery stools, often with blood and mucus, and characterized clinically by pain, tenesmus, fever, and dehydration. (Stedman's Medical Dictionary)

dyspnea – Shortness of breath, a subjective difficulty or distress in breathing, usually associated with disease of the heart or lungs. (Medical Management of Biological Casualties Handbook)

Eastern and Western Equine Encephalitis, Japanese Encephalitis (EEE, WEE, JE) – A group of acute inflammatory disease involving the brain, spinal cord, and meninges. (Control of Communicable Diseases Manual)

edema – An accumulation of an excessive amount of watery fluid in cells, tissues, or serous cavities. (Medical Management of Biological Casualties Handbook)

endemic –Present in a community or among a group of people; said of a disease prevailing continually in a region. (Stedman's Medical Dictionary)

end point – A biological effect used as an index of the effect of a chemical on an organism. (Hazardous Chemicals in Human and Environmental Health, WHO, 2000)

endogenous – Originating or produced within the organism or one of its parts. (Stedman's Medical Dictionary)

endotoxin – A bacterial toxin not freely liberated into the surrounding medium, in contrast to exotoxin. (Stedman's Medical Dictionary)

enterotoxin – Intestinotoxin; a cytotoxin specific for the cells of the intestinal mucosal. (Stedman's Medical Dictionary)

environment – The milieu; the aggregate of all the external conditions and influences affecting the life and development of an organism. (Stedman's Medical Dictionary)

enzyme – Organic catalyst; a protein, secreted by cells, that acts as a catalyst to induce chemical changes in other substances, itself remaining apparently unchanged by the process. (Stedman's Medical Dictionary)

epidemic – A disease whose frequency of occurrence is in excess of the expected frequency in a population during a given time interval; distinguished from endemic, since the disease is not continuously present but has been introduced from outside. (Stedman's Medical Dictionary)

eruption – A breaking out, specially the appearance of lesions on the skin. (Stedman's Medical Dictionary)

erythema – Inflammatory redness of the skin. (Stedman's Medical Dictionary)

exotoxin – Extracellular toxin; a specific, soluble, antigenic, usually heat labile, injurious substance elaborated by certain gram-positive bacteria (rarely by gram-negative species); it is formed within the cell, but is released into the environment where it is rapidly active in extremely small amounts. (Stedman's Medical Dictionary)

Far Eastern Tickborne Encephalitis – A tick-born viral diseases often associated with focal epilepsy and flaccid paralysis. (Control of Communicable Diseases Manual)

febrile – Denoting or relating to a fever. (Medical Management of Biological Casualties)

flaccid – Relaxed, flabby, or without tone. (Stedman's Medical Dictionary)

flash point – The temperature at which a liquid or volatile solid gives off sufficient vapor to form an ignitable mixture near the surface of the liquid. (Hawley's Condensed Chemical Dictionary, 13th ed.)

foliar – Pertaining to or resembling a leaf or leaflet. (Stedman's Medical Dictionary)

freezing point/melting point – The temperature at which the solid and liquid phases of a given substance are in equilibrium and is generally equivalent to the melting point. (General Chemistry with Qualitative Analysis, 9th ed.)

gene – A functional unit of heredity which occupies a specific place or locus on a chromosome, is cable of reproducing itself exactly at each cell division, and is capable of directing the formation of an enzyme or other protein. (Stedman's Medical Dictionary)

genetic engineering – The directed alteration or manipulation of genetic material. (The Biological & Chemical Warfare Threat)

glanders – Communicable disease of horses, mules, and donkeys. (Control of Communicable Diseases Manual)

G-series nerve agents – Chemical agents of moderate to high toxicity developed in the 1930s that act by inhibiting a key nervous system enzyme. Examples are Tabun (GA), sarin (GB), soman (GD), and GF. (The Biological & Chemical Warfare Threat)

half-life – Of a reaction-time required for half of the original concentration of the limiting reactant to be consumed Also called $t_{1/2}$. (College Chemistry with Qualitative Analysis, 8th ed.)

hazard – A condition with the potential to cause injury, illness, or death of personnel; damage to or loss of equipment or property; or mission degradation. (JP 1-02)

hemolysis – Alteration, dissolution, or destruction of red blood cells in such a manner that hemoglobin is liberated into the medium in which the cells that are suspended, e.g., by specific complement-fixing antibodies, toxins, various chemical agents, tonicity, alteration of temperature. (Stedman's Medical Dictionary)

hemorrhage – Bleeding; an escape of blood through ruptured or unruptured vessel walls. (Stedman's Medical Dictionary)

hepatitis – Inflammation of the liver; usually from a viral infection, but sometimes from toxic agents. (Stedman's Medical Dictionary)

herbicide – A chemical compound that will kill or damage plants. (JP 1-02)

hydrolysis – The reaction of a compound with water whereby decomposition of the substance occurs. (General Chemistry with Qualitative Analysis, 9th ed.)

hygiene – Cleanliness that promotes health and well being, especially of a personal nature. (Stedman's Medical Dictionary)

ileus – Mechanical, dynamic, or adynamic obstruction of the bowel; may be accompanied by severe colicky pain, abdominal distention, vomiting, absence of passage of stool, and often fever and dehydration. (Stedman's Medical Dictionary)

incapacitating agent – An agent that produces temporary physiological or mental effects, or both, which will render individuals incapable of concerted effort in the performance of their assigned duties. (JP 1-02)

incapacitating illness or injury – The casualty status of a person (a) whose illness or injury requires hospitalization but medical authority does not classify as very seriously ill or injured; or (b) seriously ill or injured and the illness or injury makes the person physically or mentally unable to communicate with the next of kin. Also called III. (JP 1-02)

incubation period – The interval (in hours, days, or weeks) between the initial, effective exposure to an infectious organism and the first appearance of symptoms of the infections. (Control of Communicable Diseases Manual)

indigenous – Native; natural to the country where found. (Stedman's Medical Dictionary)

industrial chemicals – Chemicals developed or manufactured for use in industrial operations or research by industry, government, or academia. These chemicals are not primarily manufactured for the specific purpose of producing human casualties or rendering equipment, facilities, or areas dangerous for human use. Hydrogen cyanide, cyanogen chloride, phosgene, and chloropicrin are industrial chemicals that also can be military chemical agents. (JP 1-02)

infectivity – The infectivity of an agent reflects the relative ease with which microorganisms establish themselves in a host species. Pathogens with high infectivity cause disease with relatively few organisms. (FM 8-9)

inflammation – aA fundamental pathologic process consisting of a dynamic complex of cytologic and histologic reactions that occur in the affected blood vessels and adjacent tissues in response to an injury or abnormal stimulation caused by a physical, chemical, or biologic agent. (Stedman's Medical Dictionary)

insecticide – An agent that kills insects. (Stedman's Medical Dictionary)

intoxication – Poisoning. (Stedman's Medical Dictionary)

intravenous – Within a vein or veins. (Stedman's Medical Dictionary)

inversion – An increase of air temperature with increase in altitude (the ground being colder than the surrounding air). When an inversion exists, there are no convection currents and wind speeds are below 5 knots. The atmosphere is stable and normally is considered the most favorable state for ground release of chemical agents. (FM 3-6)

ion – An atom or group of atoms carrying an electric charge by virtue of having gained or lost one or more valence electrons. (Stedman's Medical Dictionary)

jaundice – A yellowish staining of the integument, sclerae, and deeper tissues and the excretions with bile pigments, which are increased in the plasma. (Stedman's Medical Dictionary)

lachrymator (lacrimator) – An agent (such as tear gas) that irritates the eyes and produces tears. (Stedman's Medical Dictionary)

lapse – A marked decrease in air temperature with increasing altitude because the ground is warmer than the surrounding air. This condition usually occurs when skies are clear and between 1100 and 1600 hours, local time. Strong convection currents exist during lapse conditions. For chemical operations, the state is defined as unstable. This condition is normally considered the most unfavorable for the release of chemical agents. (FM 3-6)

latent heat of vaporization – The quantity of energy absorbed or given off as a substance undergoes a change in state with no change in temperature. (Hawley's Condensed Chemical Dictionary, 13th ed.) Also called Enthalpy of vaporization (ΔH_v).

lesion – One of the individual points or patches of a multifocal disease. (Stedman's Medical Dictionary)

Lymphocytic Choriomeningitis – A viral infection of animals, transmissible to humans. (Control of Communicable Diseases Manual)

macular – Relating to or marked by a small spot, perceptibly different in color from the surrounding tissue. (Stedman's Medical Dictionary)

malaise – A feeling of general discomfort or uneasiness, an out-of-sorts feeling, often the first indication of an infection or other disease. (Stedman's Medical Dictionary)

mass casualty – Any large number of casualties produced in a relatively short period of time, usually as the result of a single incident such as a military aircraft accident, hurricane, flood, earthquake, or armed attack that exceeds local logistical support capabilities. (JP 1-02)

mean lethal dose – The dose of chemical agent that would kill 50 percent of exposed, unprotected and untreated personnel. (JP 1-02)

median incapacitating dose – The amount or quantity of chemical agent which when introduced into the body will incapacitate 50 percent of exposed, unprotected personnel. (JP 1-02)

Melioidosis – An uncommon bacterial infection with manifestations ranging from benign to fatal septicemia. (Control of Communicable Diseases Manual)

membrane – A thin sheet or layer of pliable tissue, serving as a covering or envelope of a part, the lining of a cavity, as a partition or septum or to connect two structures. (Stedman's Medical Dictionary)

meningitis – Inflammation of the membranes of the brain or spinal cord. (Stedman's Medical Dictionary)

microbes – Any very minute organism. (Stedman's Medical Dictionary)

minute volume – The amount of air expelled from the lungs in a minute that is assumed to be 15 L, unless otherwise stated. This amount represents mild activity. (USACHPPM TG 204)

miosis – The excessive smallness or contraction of the pupil of the eye. The pupil is unable to dilate and remains contracted; thus, performance of tasks, navigating on foot, identifying or engaging targets, or driving vehicles is practically impossible. Miosis is often accompanied by pain, headaches, and pinpointing of the pupils. (USACHPPM TG 204)

miscible – Capable of being mixed and remaining so after the mixing process ceases. (Stedman's Medical Dictionary)

molecular weight – The value represented by the sum of the atomic weights of all the atoms in a molecule. (General Chemistry with Qualitative Analysis, 9th ed.)

molecule – The smallest possible quantity of a di-, tri-, or polyatomic substance that retains the chemical properties of the substance. (Stedman's Medical Dictionary)

Monkeypox – A sporadic zoonosic infection resembling smallpox. (Control of Communicable Diseases Manual)

morbidity – The ratio of sick to well individuals in a community; sick rate. (USACHPPM TG 204)

myalgia – Muscular pain. (Medical Management of Biological Casualties Handbook)

mycotoxin – A fungal toxin. They can cause illness or death upon ingestion skin contact, or inhalation. They exhibit great stability and heat resistance. Mycotoxins are difficult to detect, to identify, and to decontaminate. (USACHPPM TG 204)

mydirasis – Dilation of the pupil. (Steadman's Medical Dictionary)

myiasis – Any infection due to invasion of tissues or cavities of the body by larvae of dipterous insects. (Stedman's Medical Dictionary)

myocardia – The middle layer of the heart, consisting of cardiac muscle. (Steadman's Medical Dictionary)

neat chemical agent – A nondiluted, full-strength (as manufactured) chemical agent. A chemical agent manufactured by the binary synthesis route will also be considered a neat agent regardless of purity. (AR 50-6)

necrosis – Pathologic death of one or more cells, or of a portion of tissue or organ, resulting from irreversible damage. (USACHPPM TG 204)

nerve agent – A potentially lethal chemical agent which interferes with the transmission of nerve impulses. (JP 1-02)

neural – Relating to any structure composed of nerve cells or their processes, or that o further development will evolve into nerve cells. (Stedman's Medical Dictionary)

neuro- - Combining form denoting a nerve or relating to the nervous system. (Stedman's Medical Dictionary)

neuron – The functional unit of the nervous system; conducts or transmits nerve impulses. (Hazardous Chemicals in Human and Environmental Health, WHO)

neurotoxin – Any substance that is capable of destroying or adversely affecting nerve tissue. (Hazardous Chemicals in Human and Environmental Health, WHO)

neurotransmitter – Chemical responsible for the transfer of information along the nervous system. (Hazardous Chemicals in Human and Environmental Health, WHO)

neutralization – The act of altering chemical, physical, and toxicological properties to render the chemical agent ineffective for use as intended. (AR 50-6).

nonpersistent agent – A chemical agent that when released dissipates and/or loses its ability to cause casualties after 10 to 15 minutes. (JP 1-02)

nuclear, biological, and chemical capable nation – A nation that has the capability to produce and employ one or more types of nuclear, biological, and chemical weapons across the full range of military operations and at any level of war in order to achieve political and military objectives. (JP 1-02)

odor – Emanation from any substance that stimulates the olfactory cells in the organ of smell. (Stedman's Medical Dictionary)

operational decontamination – Decontamination carried out by an individual and/or a unit, restricted to specific parts of operationally essential equipment, material and/or working areas, in order to minimize contact and transfer hazards and to sustain operations. This may include decontamination of the individual beyond the scope of immediate decontamination, as well as decontamination of mission-essential spares and limited terrain decontamination. (JP 1-02)

organic solvent – An organic chemical compound that dissolves another to form a solution. (General Chemistry with Qualitative Analysis)

organophosphorus compound – A compound, containing phosphorus and carbon, whose physiological effects include inhibition of cholinesterase; many pesticides and virtually all nerve agents are organophosphorus combpounds. (Stedman's Medical Dictionary)

oxidize – To combine or cause an element or radical to combine with oxygen or lose electrons. (Stedman's Medical Dictionary)

oxime – A compound resulting from the action of hydroxylamine, NH_2OH, on a ketone or an aldehyde to yield the group =N-OH attached to the former carbonyl carbon atom. (Stedman's Medical Dictionary)

parenteral – By some other means that through the GI tract; referring particularly to the introduction of substances into an organism by intravenous, subcutaneous, intramuscular, or intramedullary injection. (Stedman's Medical Dictionary)

pathogen – A disease-producing microorganism. (JP 1-02)

pathogenic – Causing disease or abnormality. (Steadman's Medical Dictionary)

peptide – Molecules formed from two or more amino acids linked by peptide bonds. (General Chemistry with Qualitative Analysis, 9th ed.)

percutaneous – Through the skin; when applied to chemical agents, refers to route of entry into the body. (The Biological & Chemical Warfare Threat)

persistency – In biological or chemical warfare, the characteristics of an agent which pertains to the duration of its effectiveness under determined conditions after its dispersal. (JP 1-02)

persistent agent – A chemical agent that, when released, remains able to cause casualties for more than 24 hours to several days or weeks. (JP 1-02)

pesticide – General term for an agent that destroys fungi, insects, rodents, or any other pest. (Stedman's Medical Dictionary)

pH – A measure of the acidity or alkalinity of a solution. A value of seven is neutral; low numbers are acid, large numbers are alkaline. Strictly speaking, pH is the negative logarithm of the hydrogen-ion concentration. (FM 10-52)

physical state - Chemical agents may exist as solids, liquids, or gases. (General Chemistry with Qualitative Analysis, 9th ed.)

plague – A zoonosis involving rodents and their fleas, which transfer the bacterial infection to various animals and to people. (Control of Communicable Diseases Manual)

plasma – The fluid (noncellular) portion of the circulating blood, as distinguished from the serum obtain after coagulation. (Stedman's Medical Dictionary)

polymerize – To bring about a reaction in which a high molecular-weight product is produced by successive addition to or condensations of a simpler compound. (Stedman's Medical Dictionary)

polypeptide – A petide formed by the union of an indefinite (usually large) number of amino acids by peptide likes (-NH-CO-). (Stedman's Medical Dictionary)

postmortem – Pertaining to or occurring during the period after death. (Stedman's Medical Dictionary)

potable – Water that is free from disease-producing organisms, poisonous substances, and chemical or biological agents and radioactive contaminants that make it unfit for human consumption and many other uses. (FM 10-52)

prophylaxis – Prevention of disease or of a process which can lead to disease. (Stedman's Medical Dictionary)

prostration – A marked loss of strength, as in exhaustion. (Medical Management of Biological Casualties Handbook)

proteins – Macromolecules consisting of long sequences of α-amino acids in peptide (amide) linkage. (Stedman's Medical Dictionary)

protective clothing – Clothing especially designed, fabricated, or treated to protect personnel against hazards cause by extreme changes in physical environment, dangerous working conditions, or enemy action. (JP 1-02)

Psittacosis – An acute, generalized chlamydial disease caused by Chlamydia psittaci. (Control of Communicable Disease Manual)

pulmonary – Relating to the lungs. (Stedman's Medical Dictionary)

pyridostigmine bromide – An antidote enhancer that blocks acetylcholinesterase, protecting it from nerve agents. When taken in advance of nerve agent exposure, PB increases survival provided that atropine and oxime (Mark I NAAK) and other measures are taken. Also called PB. (Textbook of Military Medicine)

Q fever – An acute febrile rickettsial disease caused by *Coxiella burnetii.* (Control of Communicable Disease Manual)

quarantine – The isolation of a person with a known or possible contagious disease. (Stedman's Medical Dictionary)

rate of action – The rate at which the body reacts to or is affected by agent. Also called ROA. (DA PAM 385-61)

rate of detoxification – The rate at which the body is able to counteract the effects of poisonous substance (DA PAM 385-61)

ratification – The declaration by which a nation formally accepts with or without reservation the content of a standardization agreement. (JP 1-02)

reaction – In chemistry, the intermolecular action of two or more substances upon each other, whereby these substances are caused to disappear, new ones being formed in their place. (Stedman's Medical Dictionary)

reagent – Any substance added to a solution of another substance to participate in a chemical reaction. (Stedman's Medical Dictionary)

receptor – A structural protein molecule on the cell surface or within the cytoplasm that binds to a specific factor, such as hormone, antigen, or neurotransmitter. (Stedman's Medical Dictionary)

recombinant DNA – DNA resulting from the insertion into the chain, by chemical or biological means, of a sequence (a whole or partial chain of DNA) not originally present in that chain. (Stedman's Medical Dictionary)

rhabdovirus – Any virus of the family Rhabdoviridae, a family of rod- or bullet-shaped viruses of vertebrates, insects, and plants, including radies virus and vesicular stomatitis virus (of cattle). (Stedman's Medical Dictionary)

rickettsiae – Intracellular parasitic organisms in size between bacteria and viruses. (TM 3-216/AFM 355-6)

riot control agent – Any chemical that is not listed in the Chemical Weapons Convention, which can produce rapidly in humans sensory irritate or disabling physical effects which disappear within a short time following termination of exposure. Also called RCA. (JP 1-02)

riot control operations – The employment of riot control agents and/or special tactics, formations and equipment in the control of violent disorders. (JP 1-02)

Rocky Mountain Spotted Fever – A disease of the spotted fever group rickettsiae caused by *Rickettsia rickettsii.* (Control of Communicable Disease Manual)

rodenticide – An agent legal to rodents. (Stedman's Medical Dictionary)

ruminant – An animal that chews the cud, material regurgitated from the rumen for rechewing; e.g., the sheep, cow, deer, or antelope. (Stedman's Medical Dictionary)

Salmongellosis – A bacterial disease manifested by acute enterocolitis. (Control of Communicable Disease Manual)

septicemia – Systemic disease caused by the spread of microorganisms and their toxins via the circulating blood. (Stedman's Medical Dictionary)

Shigellosis – An acute bacterial disease involving the large and distal small intestine. (Control of Communicable Disease Manual)

slurry – A thin semifluid suspension of a solid in a liquid. (Stedman's Medical Dictionary)

Smallpox – A systemic viral disease caused by the Variola virus. (Control of Communicable Disease Manual)

smoke screen – A cloud of smoke used to conceal ground maneuver, obstacle breaching, recovery operations, and amphibious assault operations as well as key assembly areas, supply routes, logistic facilities. (JP 1-02)

solubility – The quantity of solute that will dissolve in a given amount of solvent to produce a saturated solution. (General Chemistry with Quyalitative Analysis, 9th ed.)

specific heat – The quantity of heat required to raise the temperature of 1 gram of a substance one degree C. (General Chemistry with Qualitative Analysis, 9th ed.)

spore – Resistant, dormant cell of some bacteria; primitive reproductive bodies of fungi. (USACHPPM TG 204)

sternutator – A substance, such as a gas, that induces sneezing. (Stedman's Medical Dictionary)

stupor – A state of impaired consciousness in which the individual shows a marked diminution in his reactivity to environmental stimuli. (Stedman's Medical Dictionary)

symptoms – Information related by an individual about himself/herself that may indicate illness or injury. Signs or observations are made about an individual or an animal that may indicate illness or injury. (USACHPPM TG 204)

synapse – The functional membrane-to-membrane contact of the nerve cell with another nerve cell, an effector (muscle, gland) cell, or a sensory receptor cell. (Stedman's Medical Dictionary)

synergistic – Acting together to enhance the effect of another force or agent. (USACHPPM TG 204)

synthesize – The formation of compounds by the union of simpler compounds or elements. (Stedman's Medical Dictionary)

systemic – Spread throughout the body, affecting all body systems and organs, not localized in one spot or area. (USACHPPM TG 204)

tenesmus – A painful spasm of the anal sphincter with an urgent desire to evacuate the bowel or bladder, involuntary straining, and the passage of but little fecal matter or urine. (Stedman's Medical Dictionary)

terrorism – The calculated use of unlawful violence or threat of unlawful violence to inculcate fear; intended to coerce or to intimidate governments or societies in the pursuit of goals that are generally political, religious, or ideological. (JP 1-02)

terrorist – An individual who uses violence, terror, and intimidation to achieve a result. (JP 1-02)

terrorist groups – Any element regardless of size or espoused cause, that commits acts of violence or threatens violence in pursuit of its political, religious, or ideological objectives. (JP 1-02)

tetrodotoxin – A highly poisonous fugu toxin. (Textbook of Military Medicine)

thickened agent – An agent to which a polymer or plastic has been added to minimize vaporization prior to its deposition on the target. (Joint CB Technical Data Source Book, Volume V, Part One: Agent H)

toxemia – A condition caused by the circulation of toxins in the blood. (USACHPPM TG 204)

toxic – Harmful to living organisms; oisonous. (USACHPPM TG 204)

toxic chemical, biological, or radiological attack – An attack directed at personnel, animals, or crops, using injurious agents of chemical, biological, or radiological origin. (JP 1-02)

toxicity – The degree to which a substance or mixture of substances can harm humans or animals. Acute toxicity involves harmful effects in an organism through a single or short-term exposure. Chronic toxicity is the ability of a substance or mixture of substances to cause harmful effects over an extended period, usually upon repeated or continuous exposure sometimes lasting for the entire life of the exposed organism. Subchronic toxicity is the ability of the substance to cause effects for more than one year but less than the lifetime of the exposed organism. (EPA, Terms of Environment)

toxins – Poisonous substances produced by living organisms. (The Biological & Chemical Warfare Threat)

toxin agent – A poison formed as a specific secretion product in the metabolism of a vegetable or animal organism, as distinguished from inorganic poisons. Such poisons can also be manufactured by a synthetic process. (JP 1-02)

toxoid – A toxin biologically inactivated by chemical or physical means, usually for vaccine production purposes. Because a prerequisite for toxoid generation is toxin production, the technology involved has applicability to BW. (The Biological & Chemical Warfare Threat)

Trench fever – A nonfatal bacterial disease. (Control of Communicable Disease Manual)

trichothecene mycotoxin – A very large family of chemically related toxins produced by various species of mold to include *Fusarium*. (Textbook of Military Medicine)

Tularemia – A zoonotic bacterial disease. (Control of Communicable Disease Manual)

Typhus fever – A rickettsial disease transmitted by body lice. (Control of Communicable Disease Manual)

ultraviolet – Denoting electromagnetic rays beyond the violet end of the visible spectrum. (Stedman's Medical Dictionary)

urticant – The third group listed under blister agent, although not a true vesicant. Causes erythema, wheals, and urticaria. (Textbook of Military Medicine)

V-series nerve agents – A class of chemical agents developed in the 1950s that act by inhibiting a key nervous system enzyme. The are generally persistent and have a moderate to high toxicity. Examples are VE, VG, VM, VS, and VX. (The Biological & Chemical Warfare Book)

vaccine – A substance administered to induce immunity in the recipient. (The Biological & Chemical Warfare Threat)

vapor density – The ratio of the weight of a given volume of a gaseous substance and that of the same volume of another gas measured under the same conditions of pressure and temperature. (Hawley's Condensed Chemical Dictionary, 13th ed.)

vapor pressure – The pressure exerted by a vapor when a state of equilibrium exists between the vapor and its liquid (or solid) state. (General Chemistry with Qualitative Analysis, 9th ed.)

vector – An animal, insect, or other organism that carries and transmits a virus or other microorganism. (USACHPPM TG 204)

Venezuelan Equine Encephalomyelitis Virus Disease – A viral infection involving a cycle with horses, mosquitoes and man. (Control of Communicable Disease Manual)

venom – A poisonous fluid secreted by snakes, spiders, scorpions, etc. (Stedman's Medical Dictionary)

vesicle – A small circumscribed elevation of the skin containing fluid. (Stedman's Medical Dictionary)

viable – Capable of living. (Stedman's Medical Dictionary)

viremic – The presence, as in smallpox, of a virus in the bloodstream. (Stedman's Medical Dictionary)

virulence – The capacity of a microorganism to produce disease. (The Biological & Chemical Warfare Threat) The virulence of an agent reflects the relative severity of disease produced by that agent. Different microorganisms and different strains of the same microorganism may cause diseases of different severity. (FM 8-9)

virus – Any of a large group of submicroscopic agents infecting plants, animals, and bacteria and unable to reproduce outside the tissues of the host. (USACHPPM TG 204)

visceral – Relating to an organ of the digestive, respiratory, urogenital, and endocrine systems as well as the spleen, the heart, and great vessels; hollow and multilayered walled organs. (Stedman's Medical Dictionary)

viscosity – In general, the resistance to flow or alteration of shape, by any substance as a result of molecular cohesion; most frequently applied to liquids as the resistance of a fluid to flow because of a shearing force. (Stedman's Medical Dictionary)

volatile – Capable of vaporizing or evaporating readily. (Hazardous Chemicals in Human and Environmental Health, WHO)

volatility – The tendency of a solid or liquid material to pass into the vapor state at a given temperature. (Hawley's Condensed Chemical Dictionary, 13th ed.)

weapons of mass destruction – Weapons that are capable of a high order of destruction and/or of being used in such a manner as to destroy large numbers of people. Weapons of mass destruction can be high explosives or nuclear, biological, chemical, and radiological weapons, but exclude the means of transporting or propelling the weapon where such means is a separable and divisible part of the weapon. Also called WMD. (JP 1-02)

weapon(s) system – A combination of one or more weapons with all related equipment, materials, services, personnel, and means of delivery and deployment (if applicable) required for self-sufficiency. (JP 1-02)

zoonosis – An infection or infestation shared in nature by humans and other animals that are the normal or usual host; a disease of humans acquired from an animal source. (USACHPPM TG 204)

INDEX

D

E

F

G

H

Y

FM 3-11.9
MCRP 3-37.1B
NTRP 3-11.32
AFTTP(I) 3-2.55
10 JANUARY 2005

By Order of the Secretary of the Army:

PETER J. SCHOOMAKER
General, United States Army
Chief of Staff

Official:

Sandra R. Riley

SANDRA R. RILEY
Administrative Assistant to the
Secretary of the Army
0434903

DISTRIBUTION:

Active Army, Army National Guard, and US Army Reserve: To be distributed in accordance with initial distribution number 110738, requirements for FM 3-11.9.

By Order of the Secretary of the Air Force:

BENTLEY B. RAYBURN
Major General, USAF
Commander
Headquarters Air Force Doctrine Center

Air Force Distribution: F

Marine Corps distribution: PCN 144 000154 00

Printed in the United States
207187BV00001B/1/A